The Long Evolution of Brains and Minds

Gerhard Roth

The Long Evolution of Brains and Minds

Alliant International University
Los Angeles Campus Library
1000 South Fremont Ave., Unit 5
Alhambra, CA 91803

 Springer

Gerhard Roth
Behavioral Physiology
Brain Research Institute
University of Bremen
Bremen
Germany

ISBN 978-94-007-6258-9   ISBN 978-94-007-6259-6   (eBook)
DOI 10.1007/978-94-007-6259-6
Springer Dordrecht Heidelberg New York London

Library of Congress Control Number: 2013930139

© Springer Science+Business Media Dordrecht 2013
This work is subject to copyright. All rights are reserved by the Publisher, whether the whole or part of the material is concerned, specifically the rights of translation, reprinting, reuse of illustrations, recitation, broadcasting, reproduction on microfilms or in any other physical way, and transmission or information storage and retrieval, electronic adaptation, computer software, or by similar or dissimilar methodology now known or hereafter developed. Exempted from this legal reservation are brief excerpts in connection with reviews or scholarly analysis or material supplied specifically for the purpose of being entered and executed on a computer system, for exclusive use by the purchaser of the work. Duplication of this publication or parts thereof is permitted only under the provisions of the Copyright Law of the Publisher's location, in its current version, and permission for use must always be obtained from Springer. Permissions for use may be obtained through RightsLink at the Copyright Clearance Center. Violations are liable to prosecution under the respective Copyright Law.
The use of general descriptive names, registered names, trademarks, service marks, etc. in this publication does not imply, even in the absence of a specific statement, that such names are exempt from the relevant protective laws and regulations and therefore free for general use.
While the advice and information in this book are believed to be true and accurate at the date of publication, neither the authors nor the editors nor the publisher can accept any legal responsibility for any errors or omissions that may be made. The publisher makes no warranty, express or implied, with respect to the material contained herein.

Printed on acid-free paper

Springer is part of Springer Science+Business Media (www.springer.com)

In the great majority of animals there are traces of psychical qualities or attitudes, which qualities are more markedly differentiated in the case of human beings. For just as we pointed out resemblances in the physical organs, so in a number of animals we observe gentleness or fierceness, mildness or cross temper, courage, or timidity, fear or confidence, high spirit or low cunning, and, with regard to intelligence, something equivalent to sagacity. Some of these qualities in man, as compared with the corresponding qualities in animals, differ only quantitatively: that is to say, a man has more or less of this quality, and an animal has more or less of some other; other qualities in man are represented by analogous and not identical qualities: for instance, just as in man we find knowledge, wisdom, and sagacity, so in certain animals there exists some other natural potentiality akin to these.—Aristotle, History of Animals, Book VIII.

If no organic being excepting man had possessed any mental power, or if this powers had been of a wholly different nature from those of the lower animals, then we should never have been able to convince ourselves that our high faculties had been gradually developed. But it can be shown that there is no fundamental difference of this kind. We must also admit that there is a much wider interval in mental power between one of the lowest fishes, as a lamprey or lancelet, and one of the higher apes, than between an ape and man: yet, this interval is filled up by numberless gradations.—Darwin, The Descent of Man, Chapter III.

To Bernhard Rensch and David B. Wake in deep gratitude

Preface

How did nervous systems and brains, on the one hand, and cognitive functions and intelligence, in short "mind" on the other evolve, and how are these two processes related? This is the central topic of this book—including the question to what degree humans and their minds play a special or even unique role here.

My interest in these kinds of questions dates back to my being a student of philosophy at the University of Münster, Germany and more exactly to a day in 1965, when two friends of mine took me to a public lecture in a series about philosophical problems of modern science. As lecturers there were physicists, mathematicians, chemists, and biologists, but no philosophers. One of them was the famous zoologist and evolutionary biologist Bernhard Rensch (1900–1990). He gave a lecture about the relationship between cognitive or "mental" functions and biological evolution, including that of nervous systems and brains. I was overwhelmed by what Rensch said, especially because he demonstrated that many "deep" philosophical questions such as those about the reliability of perception and the possibility of secure knowledge or the nature and origin of mind could be treated and at least partially answered in the framework of natural science, especially biology combined with psychology.

At that time, I was deeply frustrated by my study of philosophy (besides German literature and musicology) at the University of Münster, although the Institute of Philosophy offered a rather good education in classical philosophy. However, teaching was dominated by pure historicism: it was important to know exactly which of the many philosophers said what at what time—and perhaps why. Whether these theories about perception, mind, reasoning etc., were "true" in an empirical sense was of no interest. At that time, most philosophers in Münster and elsewhere had deep contempt for natural sciences, and they laughed at Rensch and other naturalists and even psychologists (at that time, the University of Münster was one of the leading centers of Gestalt Psychology) who dealt with such questions because they considered these questions to be exclusively philosophical.

A few days after the lecture, I mustered up all my courage and visited Rensch. He patiently listened to me, especially with respect to my problems with the study of philosophy, for which he had great understanding (he himself had studied philosophy besides biology). At the end of our conversation, Rensch advised me to finish my degree in philosophy first and then begin another degree in natural sciences, e.g., biology. This was among the best advice that I have ever received.

With a stipend from the German National Academic Foundation ("Studienstiftung des deutschen Volkes") I went to Rome to write my doctoral dissertation about the Italian philosopher and Marxist Antonio Gramsci, and I received my degree in philosophy in 1969. Immediately afterward, I began to study biology, again at Münster, which was made possible by another stipend from the "Studienstiftung." At the same time, I became a lecturer in the philosophy of science at the nearby University of Paderborn, and given my young age and huge ignorance, this was very good training for me in academic teaching. After two years, I began a second doctoral dissertation, this time in behavioral physiology and neurobiology, which led me, besides the University of Pisa in Italy, to the University of California at Berkeley, following an invitation from another important evolutionary biologist, David B. Wake. Under his guidance and together with him, I began studying the anatomy and physiology of feeding behavior in salamanders and frogs, which then and in the later years of collaboration led him, me and a number of graduate students (many of whom are now well-known scientists) to various places in the United States as well as to Mexico, Costa Rica, Panama, Italy, and Sardinia. I finished my dissertation in late 1974, and early in 1976, with substantial help from Rensch and Wake (which I found out about afterwards), I became professor of behavioral physiology at the newly founded University of Bremen.

I owe Bernhard Rensch and David Wake a huge debt of gratitude for introducing me to evolutionary thinking. In addition, they were—and in the case of David Wake are—wonderful teachers and highly respected and honorable personalities. Besides many conversations with Rensch at his Institute, the evening at his house together with David Wake and his wife Marvalee Wake (a renowned zoologist, too) was one of the highlights of my academic life. The dedication of this book to Bernhard Rensch and David Wake is meant to express my deep admiration for them.

The groundwork for this book was laid when I organized, together with my colleague Mario Wullimann, an International Conference on "Brain Evolution and Cognition" in Bremen in 2000. At the same time, as rector being of the Hanse Institute for Advanced Study, I was able to invite a number of leading neurobiologists such as Harry Jerison, Rudolf Nieuwenhuys, Almut Schüz, and Eric Kandel on the basis for long- or short-term fellowships to the Hanse Institute. Together with my colleague in neurobiology and wife Ursula Dicke, I wrote several articles about the relationship between the evolution of nervous systems and brains and the evolution of cognitive functions and intelligence. In 2010 I published the book "Wie einzigartig ist der Mensch? Die lange Evolution der Gehirne und des Geistes" (*"How unique are humans? The long evolution of brains and the mind"*) at Spektrum-Springer Publishing Company. In 2011 Springer asked me to prepare an English translation of that book, to which I readily agreed. While I was translating, I took the opportunity to incorporate the literature that had appeared in the meantime or had simply overlooked before. Some chapters of this book are more or less completely rewritten, and others have been modified substantially, while the general messages of the book have remained the same.

I have to thank many people who helped me in the writing of this book, either by discussions and/or critical reading of the chapters. My first thank-you goes to my colleague and wife, Ursula Dicke, from Bremen University, for close collaboration in many joint articles, providing me with literature, critical discussion, and reading the chapters of this book as well as substantial help with preparing the illustrations. I am also grateful to my brother Jörn Roth (Münster), who carefully read the German version of this book, as well as a number of colleagues: Friedrich Barth (University of Vienna), John-Dylan Haynes (Charité and Humboldt University Berlin), Onur Güntürkün (Ruhr-University Bochum), Thomas Hoffmeister (University of Bremen), Michel Hofman (Free University Amsterdam), Ferdinand Hucho (Free University Berlin), Michael Koch (University of Bremen), Michael Kuba (Jerusalem), Randolf Menzel (Free University Berlin), Martin Meyer (University of Zurich), Ulrich Müller-Herold (ETH Zurich), Michael Pauen (Humboldt University Berlin), Josef Reichholf (Munich), Helmut Schwegler (University of Bremen), Volker Storch (University of Heidelberg), Jürgen Tautz (University of Würzburg), David B. Wake (University of California, Berkeley), and Mario Wullimann (Ludwig-Maximilians-University Munich). For all still remaining errors I am the only responsible person.

August 2012   Lilienthal, Germany
Brancoli, Italy

Contents

1	**Introduction: Are Mind and Brain a Unity?**	1
2	**Mind and Intelligence**	7
	2.1 Types of Learning	8
	2.2 Types of Memory	11
	2.3 Intelligence and Behavioral Flexibility	13
	2.4 Consciousness	15
	2.5 Mind-Brain Theories	17
	2.6 What Does All This Tell Us?	22
3	**What Is Evolution?**	25
	3.1 Historical Concepts of Evolution	25
	3.2 Neodarwinism and its Problems	27
	3.3 Concepts of Evolution Beyond Natural Selection	30
	3.4 The Reconstruction of Phylogeny and Evolution	33
	3.5 What Does All This Tell Us?	37
4	**The Mind Begins with Life**	39
	4.1 What Is Life?	39
	4.2 Order, Self-Production, and Self-Maintenance	41
	4.3 Life, Energy Acquisition, and Metabolism	44
	4.4 The Origin of the First Life	45
	4.5 The Further Development of Simple Life	46
	4.6 What Does All This Tell Us?	48
5	**The Language of Neurons**	49
	5.1 The Structure of a Nerve Cell	50
	5.2 Principles of Membrane Excitability	51
	5.3 Ion Channels and Neural Transmission	54
	5.3.1 The Function of Ion Channels	55
	5.3.2 The Origin of the Action Potential	57

		5.3.3	Neurotransmitters and Other Neuroactive Substances	59
	5.4		Principles of Neuronal Information Processing	64
	5.5		What Does All This Tell Us?	67
6	**Bacteria, Archaea, Protozoa: Successful Life without a Nervous System**			69
	6.1		Bacteria and Archaea	69
	6.2		Protozoa	73
	6.3		Why Did Multicellular Organisms Evolve?	75
	6.4		What Does All This Tell Us?	76
7	**The "Invertebrates" and Their Nervous Systems**			79
	7.1		Non-bilaterians	80
		7.1.1	Sponges	80
		7.1.2	Coelenterates	81
	7.2		Bilaterians	83
		7.2.1	Acoelomorpha	84
		7.2.2	Protostomia	84
	7.3		What Does All This Tell Us?	105
8	**Invertebrate Cognition and Intelligence**			107
	8.1		Learning, Cognitive Abilities, and Intelligence in Insects	107
	8.2		Learning, Cognitive Abilities, and Intelligence in Cephalopods	113
	8.3		What Does All This Tell Us?	115
9	**The Deuterostomia**			117
	9.1		The Origin of Deuterostomes and Their Nervous Systems	117
	9.2		Echinoderms	120
	9.3		Hemichordates	120
	9.4		Chordates-Craniates-Vertebrates	121
		9.4.1	Myxinoids	122
		9.4.2	Vertebrates	123
	9.5		What Does All This Tell Us?	130
10	**The Brains of Vertebrates**			131
	10.1		The Basic Organization of the Vertebrate Brain	131
	10.2		Medulla Spinalis and Oblongata	136
	10.3		Cerebellum	138
	10.4		Mesencephalon	140

	10.5	Diencephalon	143
	10.6	Telencephalon	147
		10.6.1 Functional Anatomy of the Isocortex	155
		10.6.2 Are the Mammalian Cortex and the Mesonidopallium of Birds Homologous?	160
	10.7	What Does All This Tell Us?	162
11	**Sensory Systems: The Coupling between Brain and Environment**		165
	11.1	The General Function of Sense Organs	165
	11.2	Olfaction	168
	11.3	The Mechanical Senses and Electroreception	170
		11.3.1 The Sense of Touch, Vibration, and Medium Currents	171
		11.3.2 The Mechanoreceptive and Electroreceptive Lateral Line System of Fish and Amphibians	173
		11.3.3 The Auditory System	177
	11.4	The Visual System	182
		11.4.1 The Compound Eye of Insects	183
		11.4.2 The Vertebrate Eye and Retina	186
		11.4.3 Parallel Processing in the Visual System of Vertebrates	188
	11.5	What Does All This Tell Us?	191
12	**How Intelligent Are Vertebrates?**		193
	12.1	Cognition in Teleost Fishes	194
	12.2	Learning and Cognitive Abilities in Amphibians	196
	12.3	Cognitive Abilities and Intelligence in Mammals and Birds	198
		12.3.1 Tool Use and Tool Fabrication	199
		12.3.2 Quantity Representation	201
		12.3.3 Object Permanence	202
		12.3.4 Reasoning and Working Memory	202
		12.3.5 Social Intelligence	204
13	**Do Animals Have Consciousness?**		209
	13.1	Mirror Self-Recognition	210
	13.2	Metacognition	211
	13.3	Theory of Mind: Understanding the Others	212
	13.4	Conscious Attention	215
	13.5	How Intelligent Are Dolphins and Elephants?	219
	13.6	What Does All This Tell Us?	220

14	**Comparing Vertebrate Brains**.		223
	14.1	Brain Size and Body Size	223
	14.2	The Significance of Relative Brain Size and of "Encephalization"	227
	14.3	The Fate of the Cortex as the "Seat" of Intelligence and Mind.	233
		14.3.1 Information Processing Properties of the Cortex.	233
		14.3.2 Modularity of the Cortex.	238
		14.3.3 Specialties of the Cytoarchitecture of the Mammalian Cortex	239
	14.4	Bird Brains and Mesonidopallium	240
	14.5	What Does All This Tell Us?.	241

15	**Are Humans Unique?**.		243
	15.1	How Did *Homo sapiens* Evolve?	244
	15.2	Leaving the Jungle and Its Consequences	247
	15.3	Enlargement of the Brain and Its Consequences.	251
	15.4	Language and the Brain	253
		15.4.1 Animal Language.	253
		15.4.2 The Evolution of Human Language	256
		15.4.3 The Tempo of the Evolution of Human Language . . .	260
	15.5	Do Humans Exhibit a Special Social Behavior?.	262
	15.6	What Does All This Tell Us?.	263

16	**Determinants of the Evolution of Brains and Minds**		265
	16.1	Patterns of the Evolution of Nervous Systems and Brains	265
	16.2	The Evolution of Cognitive-Mental Functions	269
	16.3	How Do Differences in Intelligence Relate to Differences in Brain Structures and Functions?.	270
	16.4	Which Are the Ultimate Factors for Evolution of Brains and Minds?	271
		16.4.1 Ecological Intelligence	272
		16.4.2 Social Intelligence	274
		16.4.3 General Intelligence	275
	16.5	Basic Mechanisms of Evolution of Brains and Cognitive Functions	277
	16.6	What Does All This Tell Us?.	280

17	**Brains and Minds**.		283
	17.1	The Problems of Dualism	283
	17.2	Problems of Strong Emergentism	285
	17.3	Problems of Reductionism.	287
	17.4	The Anatomy and Physiology of the Mind	288

17.5	Brains and Minds in Birds, *Octopus* and the Honeybee	290
17.6	Is Mind Multiply Realized and Artificially Realizable?	294
17.7	What Is the True Nature of the Mind?	296

Literature 299

Index 317

Chapter 1
Introduction: Are Mind and Brain a Unity?

Keywords Darwin · Descent of Man · Mind-brain relationship · Dualism · Naturalism · Gradualism · Human and animal intelligence

> Mit Philosophie und speziell Erkenntnistheorie verknüpft ist die Biologie ohnehin durch die Tatsache, daß bestimmten Vorgängen in Nervensystemen und Sinnesorganen Bewußtseinserscheinungen parallel laufen bzw. diesen entsprechen. Ohne Übertreibung dürfen wir daher sagen, daß jede Weltanschauung, die biologisches Wissen nicht ausreichend berücksichtigt, der vorhandenen Wirklichkeit nicht adäquat sein kann. (B. Rensch, Biophilosophie, 1968)

> Biology is intimately connected with philosophy and particularly epistemology through the fact that certain processes in the nervous systems and sense organs are accompanied by conscious experience or are correlated with them. Without overstatement we are allowed to say that every world view which does not incorporate biological knowledge, cannot be adequate to reality. (B. Rensch, Biophilosophie, 1968)

In his epoch-making work, "*The Origin of Species*" from 1859, with a great abundance of empirical data, Charles Darwin put forward the concept of a common origin of all living beings and of natural selection as one major mechanism of evolutionary changes. However, the most delicate question of that time, whether or not this would hold for humans, too, was touched upon him only with the famous brief comment "Light will be shed on the origin of man, and his history." Twelve years later, in 1871, he addressed exactly this question in detail in his second masterpiece, "*The Descent of Man*." There he stated that humans are—of course—modified descendants of ape-like ancestors. This statement horrified many of his contemporaries, and it is reported that a British woman (some believe it was Queen Victoria herself) remarked: "Let us pray that this statement is wrong, and if it is correct, then let us pray that it won't be known."

In reading the first part of "*The Descent of Man*," the reader is not only impressed by the radicalness of Darwin, in which he states that not only with respect to the anatomical, but also to the cognitive and mental abilities are there no fundamental or *qualitative*, but only *quantitative* differences between humans and non-human animals. Likewise astonishing is the wealth of arguments, which Darwin puts forward in favor of such a view—arguments that address diverse functions like imitation, attention, thinking, decision making, tool use, memory, imagination, association of ideas, self-reflection, and reason, but also jealousy, ambition, gratefulness, magnanimity, deceit, revenge, humor, language, love, altruism, obedience, guilt, moral, ethics, and religiosity. These arguments were taken from observations of animal behavior in the wild, in zoological gardens, or in the private households by experts including Darwin himself or by laymen. Behavioral studies under controlled conditions were largely unknown at that time.

This has changed dramatically during the last 50 years, as we will see in this book. On the basis of a great wealth of new data and concepts, we have to ask ourselves if Darwin was right, i.e., whether or not human beings are distinguished by their cognitive abilities only gradually from non-human animals, or whether there are true "unique" human abilities.

In most religions, philosophies, and cultures of the world, mind, in the sense of reason and consciousness, is believed to exist only in humans. According to that view, even the smartest animals possess no mind, no consciousness, no abstract thinking, no self-awareness. Since there can be no doubt that at least in our biological nature we are closely related to (other) animals, more precisely to primates, the concept of such a "fundamental gap" between the cognitive abilities of humans and animals is best explained either by the assumption that mind, reason, consciousness, and thinking cannot be of a natural kind, but something that "transcends" the kingdom of nature. This is the *dualistic view* of the mind–body or mind–brain relationship accepting an *ontological* difference between nature and mind. An alternative view is that there was an evolutionary "leap" during the course of human evolution between our ancestors (australopithecines or later) and the great apes. Finally, some theologians and philosophers believe that this leap happens in every ontogeny of humans, for example, during the progenitive act.

Proponents of a *naturalistic* concept of the mind–brain relationship accept that at least some animals possess some forms of mind, and that the mental abilities of humans do not transcend or violate the known natural laws. Thus, in their eyes there is *no* ontological difference between mind and body/brain. However, differences exist among experts regarding the question of a uniqueness of human mental functions. Many, if not most, experts hold that there are in fact at least some qualitative differences between humans and non-human animals, including our nearest taxonomic neighbors, the chimpanzees. Only humans, so it is said by many anthropologists, behaviorists, and psychologists, possess self-reflection, a syntactical-grammatical language, a "theory of mind," religion, morality, science, and art. For such abilities, *no* preliminary stages for these abilities are found among non-human animals.

If one, as a naturalist, adheres to such a view, then he or she has to accept that such unique mental abilities must have evolved during the course of hominine evolution, starting either with the transition from our chimpanzee-like ancestors to the first australopithecies, or (supposedly) from *Homo heidelbergensis* to *Homo sapiens*. As a consequence, there are numerous attempts by psychologists, anthropologists, and philosophers to find evidence for unique genetic or cerebral traits that made such unique abilities possible.

Other experts follow Darwin in the view that there is a *strict continuity* between our non-human ancestors and us with respect to *all* mental-cognitive abilities, including consciousness and self-reflection and possibly syntactical-grammatical language. Such a view is called *gradualism*. Of course, proponents of such a gradualistic view accept that during the course of evolution novel forms and functions ("key innovations") originated like a cell nucleus, a bilateral body organization, the formation of a supraesophageal ganglion or brain, the formation of a backbone, etc., but these novelties did not "fall from heaven," rather originated from simpler versions.

In this book I will investigate, on the basis of the present knowledge of evolutionary and behavioral biology, neuroscience, and anthropology, to what extent it is possible to reconstruct the evolution of the nervous systems and brains as well as the evolution of mental-cognitive abilities, in short "intelligence," and to investigate to what extent we can correlate the one with the other, and whether or not there are truly unique human abilities. At the very end we will confront ourselves with the eminent question of whether we can arrive at a *naturalistic* concept of mind and consciousness. Is it possible, on the basis of present knowledge, to explain mind and intelligence within the framework of natural science, or do mind and intelligence, as found in humans, transcend nature?

As a consequence, much/most of this book will consist in the attempt to reconstruct the evolution of nervous systems and brains and identify the possible principles of this process. Exactly which neural features make animals and humans intelligent and creative? Numerous features have been proposed in the past, such as absolute or relative brain size (uncorrected or corrected for body size), or the size of certain "intelligence centers" inside the brains. Is the number of nerve cells inside the brain in total or in such "intelligence centers" decisive for the degree of intelligence, of mind and eventually consciousness, or is it a particular pattern of neuronal connectivity? These questions will be dealt with in detail. But which are the driving forces behind these processes? Here, there are many different answers. For some experts, the driving force is the conditions for biological survival: the more complex these conditions, the more effective need to be sense organs, nervous systems, and brains, and the stronger the trend to an increase in learning abilities, behavioral flexibility, and innovation power of animals. This is the *ecological intelligence* hypothesis. Other experts believe that the true driving force is the challenge deriving from the social life of an animal: the more complex the social conditions, the more sophisticated the abilities such as social learning, imitation, empathy, knowledge transfer, consciousness, and the development of a theory of mind and meta-cognition. This too, needs progressive changes inside the brains. This is the *social intelligence* hypothesis. Some authors distinguish *physical intelligence* as a third form of cognitive functions mostly related to tool use, tool fabrication, and understanding of the principles of how things work. However, others believe that the decisive factor in the evolution of brains and minds consisted of an increase in general intelligence, i.e., the speed and efficacy of information processing in cognitive brain centers. This is the *general intelligence* or *information processing* hypothesis. We will have to see which of these hypotheses is the most convincing one.

In this book, I will proceed as follows. In Chaps. 2 and 3, I will deal with a more precise definition of the key notions of "mind/intelligence" and "evolution." The fourth chapter is devoted to the definition of life and its origin. My general theory here is that the origin of cognitive abilities follows necessarily from the characteristic principles of living beings, i.e., self-production and self-maintenance. The fifth chapter deals with the "language of neurons" (or "language of the brain"), i.e., the principles of neuronal information processing. We will see that this "language" was formed very early in evolution, i.e., at the origin of the first unicellular organisms and, thus, long before nervous systems and brains came into existence.

In the sixth chapter we will begin our journey through the animal kingdom, beginning with bacteria and archaea as the simplest organisms. These prokaryotic organisms and later the eukaryotic protozoans already possessed and possess the equipment necessary for survival and successful reproduction, i.e., a sensorium for the perception of relevant events in the environment, a motorium for movement and behavior, and in-between mechanisms for information processing. Nothing completely new has happened in evolution since then, and this includes the equipment with receptors, ion channels and neurotransmitters, and neuromodulators. In the seventh chapter we will follow the process of the evolution of nervous systems and possibly brains, from the sponges to the non-bilateral "coelenterates" (i.e., cnidarians and ctenophorans) on the one hand, and to all bilateral animals, including the invertebrates, on the other, i.e., the acoelans, lophotrochozoans (including annelids and mollusks) and ecdysozoans (including nematods and arthropods) and the vertebrates, including mammals, primates, and humans. The first line of bilaterian invertebrate evolution leads to the cephalopod *Octopus* as the alleged most intelligent invertebrate animal, the second to the very large and diverse group of insects, in which honeybees and their brains excel in learning, memory, and cognitive abilities. In Chap. 8 we will ask how intelligent these invertebrate animals are.

From Chap. 9 on, we will deal predominantly with the vertebrates and compare the brains of lampreys, cartilaginous and bony fishes, amphibians, sauropsids (i.e., "reptiles" and birds), and mammals. We will see that the basic organization of the vertebrate brain remained unchanged for 500 million years, and that evident differences were mostly in absolute and relative sizes of the brains and of parts of it, with the remarkable exception of the covering of the telencephalon (i.e., pallium or cortex). The sense organs of invertebrates and vertebrates and their evolution are addressed in the 11th chapter. In Chaps. 12 and 13 we will ask, in parallel to the eigth chapter, how intelligent vertebrates are, and which groups of vertebrates excel in mental-cognitive abilities. In Chap. 14 we will investigate to what degree these abilities can be correlated with properties of the respective brains. Chapter 15 is devoted to the central question of whether or not, regarding mental-cognitive functions, humans are truly unique compared to all other animals, and what could be the neural basis of such unique properties, if they exist. Chapter 16, after a summary of the data presented so far, investigates the impact of ecological, social, physical, and general intelligence, and we will ask to what degree they can be considered the driving forces of the coevolution of brains and minds. In Chaps. 17 and 18 I will address the question about the degree to which we can formulate a naturalistic concept of mind and consciousness. This will include the central question as to which factors and processes in intelligent animals and in humans could constitute the neural basis of "higher" mental functions, including thinking, consciousness, and self-awareness, and whether this invariably follows the same building principles or was realized in very different ways. This eventually leads us to ask the question, whether the knowledge of such principles—once they are known—would enable us to artificially create mind and consciousness.

Chapter 2
Mind and Intelligence

Keywords Mind · Intelligence · Behavioral flexibility · Learning · Memory · Consciousness · Philosophy of mind · Dualism · Monism · Naturalism · Identism · Reductionism · Physicalism

The question of the nature, function, and origin of the mind has always been a central topic in Western philosophy. In the traditional view, mind—either in the form of a soul, reason, or consciousness—is the property that most clearly distinguishes human beings from all other creatures on earth. Most modern philosophers continue to associate "mind" and "mental" with conscious perception, reasoning, decision making, remembering, planning, etc. However, in this book I will use a much more comprehensive concept of mind, viz. cognitive abilities, which I briefly call "intelligence." This latter notion is meant to denominate, above all, the *ability of an organism to solve problems occurring in its natural and social environment*. This includes forms of associative learning and memory formation, behavioral flexibility, and innovation rate as well as abilities requiring abstract thinking, concept formation, and insight. All this may, but need not be, accompanied by explicit consciousness, and the involvement of consciousness has to be demonstrated or made likely independently.

Such cognitive abilities are found not only among humans or so-called "higher" animals like mammals or primates. Some of these abilities are already present in very simple organisms. Indeed, there is no organism on earth that responds to events in its environment in a purely reflex-like or instinctive fashion, and even unicellular organisms possess the capacity for learning, memory and related multisensory information processing, as will be shown in Chap. 6. The cognitive abilities of complex multicellular organisms, including those of humans, derive from such basic "mental" equipment. Because learning is the basis for all complex cognitive functions, we will first, at least briefly, address the different types of learning.

2.1 Types of Learning

Learning is a universally distributed capacity of an organism for medium- and long-range adaptation to its living conditions as opposed to a momentary physiological or behavioral response (for an overview see Pearce 1997; Terry 2006; Korte 2013; Menzel 2013). Generally, *associative* and *non-associative* learning and memory formation are distinguished. Nonassociative learning includes habituation and sensitization, while associative learning consists of classical (Pavlovian) conditioning and operant or instrumental conditioning. Most authors accept the existence of other and more complex forms of learning, such as imitation and insight learning, while a few of them still deny the existence of types of learning beyond classical and operant conditioning.

Habituation and *sensitization* are the simplest forms of experience-dependent behavioral adaptation. *Habituation* is the progressive *decrease* in intensity or frequency of a given behavioral or physiological response toward a repeated strong or conspicuous stimulus because of the absence of relevant negative or positive consequences. For example, a loud noise or a large dark object turns out to be not as harmful or important as it initially seemed to be. *Sensitization*, on the other hand, is the progressive *increase* of an initially weak behavioral or physiological response to a repeated or continuing weak or inconspicuous stimulus because of its negative or positive consequence. A shadow or a low noise turns out to be more important, advantageous, or negative than expected. Habituation and sensitization are based on an *evaluation* of events by a nervous system or its forerunners in unicellular organisms, although this evaluation may happen in a highly automated and/or unconscious fashion.

Associative learning is the acquisition of the experience that a certain event or object is associated with another, preoccurring or simultaneous event or object. *Classical conditioning* is a basic type of associative learning. Here, an organism exhibits a regularly occurring, often reflex-like behavior of biological meaning called unconditioned response (UR), for example, an autonomic or affective response such as salivation, changes in skin conductance, or a motor response like the extension of the proboscis in the honeybee, toward an unconditioned stimulus (US), e.g., a threatening event or food. Thus, this kind of behavior is *part of the standard repertoire of behavior* of the animal under investigation. If the unconditioned stimulus is paired several times with a hitherto neutral stimulus, for example, a sound, an odor, or a light signal, which at the beginning did *not* elicit the UR, this neutral stimulus acquires the ability to release the unconditioned response in the same or in a modified way—at least for a while. In that way, the initially neutral stimulus turns into a *conditioned stimulus* (CS) and the unconditioned response into a *conditioned response* (CR).

In most cases, the success of such classical conditioning depends on a precise temporal relationship between stimulus and response in the sense that the conditioned stimulus, e.g., a sound, must occur simultaneously with or precede for at least a few seconds the unconditioned stimulus, e.g., the presentation of food, while a CS occurring after the US has no effect or even an inhibitory one. But there are exceptions to that rule in the sense that in some paradigms, classical conditioning may be successful, if the pairing between US and CS is just statistically more likely than no pairing. Most contemporary authors assume that at least in some organisms and learning paradigms, the conditioned stimulus functions as a "predictor" of the unconditioned stimulus because of their statistically higher temporal and/or spatial co-occurrence. In this way, organisms, either consciously-explicitly or unconsciously-implicitly, learn about regular temporal or spatial relationships among events in their environment, which is of eminent significance for their survival. The specific environment becomes "ordered," and expectations are formed which then guide the future behavior.

A more complex type of classical conditioning is *context conditioning*. Here, an organism learns that some stimuli or events have positive or negative consequences only under a specific condition or in a specific *context*. Therefore, the responses of the organism may vary with that context. For example, we may feel great fear of something specific in a given context because of bad experience that had happened there and then, but not in another context, where the same event did not have negative consequences. This holds for memory recall as well, because we usually remember certain things much better, together with certain emotions, in certain contexts (e.g., places, rooms) than in others, mostly in those, where we had the initial experience (cf. Schacter 1996). Context conditioning dominates much of the daily life of animals and humans.

Operant or *instrumental conditioning* is the other basic type of associative learning. It includes changes in stimulus–response relationships. An already existing type of behavior is modified in intensity or frequency depending on the positive or negative consequences for the state of the organism. Operant conditioning in the form of *positive* reinforcement or reward learning usually proceeds as follows. A laboratory animal such as a rat or a pigeon made hungry by food deprivation is brought into a test box (often called "Skinner box" named after the eminent behaviorist, Burrhus F. Skinner). It exhibits a number of spontaneously occurring behavior, e.g., searching for food, until it inadvertently exerts a certain type of behavior, such as pecking at an illuminated disk or pressing a lever. After doing so, the animal is immediately rewarded by the delivery of food. This situation occurs several times, until in the animal an increasingly robust *association* between the hitherto randomly occurring behavior and the reward is formed. As a consequence, the hungry animal will peck at the illuminated disk or press the lever with increasingly shorter delays as soon as it is put into the box. It is assumed that it was the positive consequence that caused the change of behavior—"reinforced" it. We, therefore, also speak of "reinforcement learning."

The situation can be made more complicated in the way that the animal has to learn to peck at the disk only when the latter has a certain color or to press the lever only after a certain sound was played, or even under much more complicated conditions. In that case, experimentalists usually start by rewarding simple types of behavior and then switch to rewarding only more complex ones, which is called "shaping." Only the rewarded behavior will increase in intensity and/or frequency, and the nonrewarded behavior will be reduced in intensity and/or frequency or eventually be omitted by the animal.

In contrast to classical conditioning, where an already existing physiological or reflex-like response is elicited by a previously neutral stimulus, afterward called "conditioned stimulus," operant conditioning is not based on a preexisting physiological reaction, but on a certain type of flexible or voluntary behavior that was not shown before at that intensity or frequency or not in that context. Accordingly, in operant conditioning, the animal exhibits a *new or modified type* of behavior or an already existing type of behavior (at least in rudimentary form) in a *new context*. In both cases, however, an *association* occurs, in classical conditioning between unconditioned response and conditioned stimulus, and in operant conditioning between a behavioral response and its consequences.

In behavioral experiments as well as in daily life, operant conditioning comes in various subtypes: (1) *Punishment*: here, the intensity or frequency of an unwanted or unfavorable type of behavior is reduced by the occurrence of a stimulus experienced by the individual as harmful or annoying, e.g., a mild electroshock or a loud noise. (2) *Omission of reward*: a stimulus previously experienced as positive ("reward") will be withdrawn, whenever a certain unwanted response occurs, and the individual must learn to suppress that response. Often it must learn to carry out another type of behavior in order to regain the reward. (3) *Avoidance learning* or "*negative conditioning:*" an individual must learn to exhibit a certain type of behavior in order to *avoid* or *terminate* a negative (punishing) stimulus or situation. Avoidance or termination of a negative stimulus is then experienced as a reward. (4) *Reward learning* or "*positive conditioning:*" an individual must learn to exhibit a certain type of behavior in order to receive a reward. There are different reward strategies, such as regular, intermittent, or irregular rewarding, which have very different effects on the frequency and stability (i.e., resistance to extinction) of a given behavior.

For a long time it was debated among students of behavior whether there are types of learning beyond the types just mentioned. Besides habituation and sensitization, the hard-core followers of Ivan Pavlow (the "reflexologists") believed that there was only classical conditioning, while for hard-core followers of Watson and Skinner (the "behaviorists"), there was only operant conditioning in addition to classical conditioning—nothing else. Today, however, most experts would agree that there are more types of learning, such as imitation and learning by insight.

Imitation or "learning by observing" was long considered a primitive type of learning (often called "aping"), as opposed to "higher" forms of learning, such as insight. However, in recent years imitation turned out to be a rather complex form of learning. The appearance of novel types of behavior or of novel combinations of preexisting types of behavior is characteristic of imitation. However, among students of behavior there is no agreement about whether and in which ways animals exhibit "true" imitation. Some behavioral processes previously considered imitation are now interpreted by some authors as stimulus reinforcement, response reinforcement, or emulation, as we will learn in Chap. 12. "True" imitation is now believed to occur if the observer is not only brought to deal with a certain event or object, but if he or she solves the problem in more or less the same way as the observed individual.

While imitation is characterized by the fact that individuals more or less "slavishly" follow the sequence of the observed behavior, even if the context conditions have changed, *insight* into the principles of what is going on allows modifications of the sequence of actions. In the form of pre-meditation and mental simulation of the planned actions, it is important for a large variety of complex actions, e.g., tool fabrication.

2.2 Types of Memory

The formation of memory is the storage of results of learning processes in the nervous system or brain by creating new connections or modifying already existing ones among nerve cells—a process that will ultimately influence the execution of behavior (Schacter 1996; Squire and Kandel 1998; Korte 2013). Generally, at least three phases of memory formation are distinguished: a short-term (STM), an intermediate-medium-range memory (MTM), and a long-term memory (LTM). A special kind or aspect of short-term memory (STM) is working memory (WM); some experts assume that they are more or less identical, while others see some differences—a controversy, which is irrelevant in the present context.

Short-term or *working memory* (STM/WM) in adult humans has a span up to 30 second and appears to be based on ongoing physiological, as opposed to structural, changes in the strength of synaptic coupling between neurons involved (cf. Chap. 5). In young children and in animals, the span is much shorter. Working memory serves to hold and handle information in the mind needed for the execution of cognitive tasks such as reasoning, comprehension, learning, and carrying out sequences of actions. According to concepts developed by Baddeley (1986, 1992), in primates including humans it is composed of a number of subsystems, such as the phonological loop, the visuo-spatial sketch pad, the episodic buffer, and a central execute that distributes the tasks and the required cognitive resources. The capacity of STM/WM is notoriously limited, although in humans this limitation is slightly different for digits (around 7), letters (around 6), and words (around 5), while the complexity of single items can be reduced by "data compression," e.g., meaningful grouping. Additionally, contents of the STM/WM are very disturbance- or interference-sensitive, especially when momentarily processed information interferes with additional information similar in modality and content. This may happen, for example, when I try to keep a telephone or PIN code number in mind, and my wife tells me what time we have to be at the opera.

Although concepts of STM and WM are mostly defined along human cognition, it is clear that all animals living in complex natural and social environments must possess STM/WM in order to "keep track" of ongoing events, quickly solve complex problems, and generate appropriate responses in their natural or social environment.

Based on psychological as well as physiological and anatomical evidence, many authors see the capacity of WM as the basis of human and animal intelligence, as defined below. Also, there is a tight connection between WM, attention, and consciousness, at least in humans. It appears to be the basis of what psychologists and philosophers have described as "stream of consciousness."

The *intermediate memory* is involved in the transformation of purely physiological changes of neuronal connectivity into *structural changes* as the basis of long-term memory. It involves intracellular signaling cascades influencing, among others, the intracellular calcium level (cf. Chap. 5). This process is called *memory consolidation*. In mammals, including humans, it has a span of 30 s to 30 min. Its precise mechanisms are unknown, but it is likewise interference-sensitive, more to emotional arousal than to sensory and cognitive information load. In mammals, the hippocampus appears to be critically involved in memory consolidation.

In *long-term memory* (LTM), "traces" are stored for a period from 30 min up to decades. Its storage capacity appears to be virtually unlimited under optimal storage conditions. However, what happens there in detail is not fully known. Clearly, formation of long-term memory requires the activation of genes and protein synthesis leading to structural changes at synapses and probably at other parts of the neurons, e.g., dendrites, which influence the probability of generation and transmission of nerve impulses. Because of the structural nature of storage, LTM is much less interference-sensitive than the other two types of memory, although even LTM is a dynamic process, which means that the memory traces are constantly "re-written" and re-stored or undergo data compression. In many animals and many learning processes, the formation of long-term memory traces depends on gene expression leading to protein synthesis necessary for the mentioned structural changes. This process can, therefore, be suppressed by antibiotic drugs (cf. Korte 2013). However, in other animals and learning processes, the application of antibiotics does not lead to an impairment of LTM; therefore, mechanisms other than protein synthesis appear to be involved.

In humans, two major types of LTM are commonly distinguished, i.e., *declarative* or *explicit memory* and *procedural* or *implicit memory*. Some authors, myself included, believe that *emotional memory* is a third type of long-term memory. Declarative memory is called "explicit," because humans are able to express its contents verbally and in some detail. It is subdivided into *episodic memory*, *semantic* or *knowledge memory* and *familiarity memory*. The first one stores specific events that happened to an individual in time and space. Its core is autobiographic memory, i.e., information of some detail about what happened to me or to people I knew at some time and location, e.g., where I spent my holidays last summer and what happened there. Semantic memory, in contrast, stores factual information, i.e., what is the capital of France or when, where, and by whom Julius Caesar was murdered—*facts* that are (relatively) independent of the conditions under which we came to know them. Many authors believe that episodic-autobiographic memory is the *primary* memory, and that semantic memory in most cases is a derivative of the former, i.e., we first learn in a specific situation and from a specific person, e.g., teacher Jones, that Paris is the capital of France and that Julius Caesar was murdered by Brutus in Rome in 44 B. C., and only later—after having heard or read about event many more times—it becomes a *context-independent fact*. Familiarity memory is the impression that some person, object, place, or event is familiar or unfamiliar to us—independent of the precise content.

As we have heard, the definition of human declarative memory in a narrow sense is bound to verbal report. Therefore, in nonhuman animals the proof of existence can only be indirect, e.g., by testing whether an animal remembers something in greater detail, as revealed by behavioral responses. While declarative memory deals with "knowing what," *procedural-implicit* or non-declarative memory deals with "knowing how," i.e., skills, whether cognitive, such as fast detection of a deviation from expectation in a sequence of events, or motor actions, such as tying shoes, piano playing, bicycling, or driving a car. These skills are generally performed in an automatic way, often below the level of consciousness or with only accompanying consciousness. Also, when these memories are needed, they are automatically and effortless retrieved and utilized for the task to be executed. This process is also the basis of the formation of habits.

Categorical learning is considered another type of procedural-implicit memory. Categories help humans as well as animals to classify objects, events, or ideas according to certain common features. Nonassociative learning and classical conditioning, as presented above, are also subsumed under *implicit* learning or memory. All nonhuman animals possess procedural-implicit memory. Since it is not tied in to verbal report, it can be tested relatively easily by behavioral studies.

Emotional memory has characteristics of both declarative-explicit and procedural-implicit memory. It is mostly the result of emotional conditioning. An organism perceives a certain event (stimulus or situation) or exerts a certain action, which may have positive or negative consequences. These consequences are tied by the brain to corresponding emotional or affective states (pleasure, pain, happiness, fear, etc.), and these states are then stored together with the perceived stimulus or situation. Whenever the stimulus or situation reappears in the same or a similar way, or when the action is executed in the same or a similar way, then the corresponding emotional-affective state reappears. This situations creates *motivation* that drives animals and humans to approach or to repeat situations, events, or actions with a positive emotional connotation and avoid or terminate situations, events or actions with a negative emotional connotation. Motivation, thus, is based on the prediction or expectation of positive or negative consequences of future events or actions and requires an internal evaluation system—in vertebrates the limbic system.

2.3 Intelligence and Behavioral Flexibility

In humans, "intelligence" is usually defined as the ability for abstract thinking, understanding, communication, reasoning, problem solving, learning and memory formation, and action planning. Often, two basic components are distinguished, viz., *fluid* intelligence, which is the ability to solve problems using novel information or procedures, and *crystallized* intelligence, which concerns the use of previously learned experiences or procedures in problem solving (Cattell 1963). In humans, intelligence is usually measured by intelligence quotient (IQ) tests such as the Stanford–Binet, Raven's Progressive Matrices, or the Wechsler Adult Intelligence Scale. These tests measure abilities like quantitative reasoning, reading and writing ability, short-term memory, visual or auditory processing, processing speed, decision making, and reaction time.

A central question for this book is to what extent we can apply such a definition of intelligence to nonhuman behavior. Certainly, we cannot conduct IQ tests on animals as we do on humans, especially with respect to tasks requiring language. In order to circumvent the language problem, students of behavior have developed a variety of methods. *Comparative psychologists* investigate cognitive abilities of animals—in most cases, of groups of animals especially suited for such studies, like sea slugs, cephalopods, fruit flies, or honeybees, among invertebrates, and fish, birds, or mammals, including primates, among vertebrates. Such experiments deal with learning and memory formation, categorization, counting and numerosity, but also with problem solving using insight and understanding the principles of what is being done (e.g., during tool fabrication)—and all this under precisely controlled laboratory conditions.

In contrast, *behavioral ecologists*, also called *cognitive ecologists*, emphasize studies under natural or quasi-natural conditions and field experiments and measure the degree of *behavioral flexibility* and *innovation rate*. This addresses the ability to adapt an established behavior to a new context or to successfully deal with strongly fluctuating environmental or behavioral conditions. Here questions arise, such as: How do honeybees behave if their beehive is displaced? What is a crow doing, if tasteful food is located in a bottle with a narrow neck inaccessible to its beak? Social-communicative abilities are likewise studied in the context of questions like: Are monkeys and apes capable of deceiving conspecifics to their own advantage? Do elephants exhibit empathy toward each other? Do teleosts cooperate in prey-catching? Likewise, the ability to find novel solutions to a problem are used for intelligence testing, for example, if individuals find new ways to get better or faster access to food.

In general, what is being studied are functions in the domain of feeding, spatial orientation, maternal or paternal care, dealing with social complexity, learning of language and songs, empathy, theory of mind, knowledge representation, as well as categorization, abstract thinking, and action planning in the sense of "mental handling."

Eventually, on the basis of the results of such studies and despite considerable skepticism, one might be able to "rank" animals regarding their intelligence in the defined sense and compare them with the intelligence of humans. This will be used for answering the central question as to what extent differences in intelligence among animals, including humans, can be correlated with the properties of their brains.

2.4 Consciousness

Addressing the question of a possible evolution of consciousness we enter a field that has been hotly debated since antiquity, especially because classically, consciousness and mind, together with reason, were regarded by most philosophers as properties that distinguish humans most sharply from nonhuman animals.

In humans, consciousness includes very different phenomena, with the only thing in common the fact that we have *subjective awareness* of them. There are general states of consciousness like *wakefulness* or *vigilance* without clear content. Other general states of consciousness are fatigue, dizziness, anxiety, hunger, comfort, and the awareness of temporal duration and of spatial layout (Metzinger 1995; Koch 2004; van Gulick 2004). These conditions form a background for more specific types of consciousness; These include *conscious perception* of events happening in the world around me and inside my body, which differ in modality, submodality (quality), quantity, intensity, location in space and time, content and meaning. *Mental activities*, such as thinking, remembering, imagining, and planning, are another class of specific states of consciousness and are usually experienced differently from perceptions, as is the case with conscious *emotions*. "Background" types of consciousness include *body-identity awareness*, i.e., the belief that I belong to the body that appears to surround me, *autobiographic consciousness*, i.e., the conviction that I am the one who already existed yesterday and even earlier, *reality awareness* of what appears to happen in the world around me really happens and is no dream or illusion, awareness of *voluntary control of movements and actions*, of *being the author of my thoughts and deeds*, and finally, *self-awareness*, i.e., the ability of self-recognition and self-reflection. *Attention* is a state of (or closely linked to) increased and focused consciousness and can be driven externally or internally. In the latter case, it goes along with improved perceptual abilities (e.g., increased visual acuity or lowered auditory threshold).

Many biologists, too, were and still are undecided whether and to what degree animals possess at least some kinds of consciousness comparable to those found in humans (e.g., MacPhail 1982), and until recently it seemed impossible to reliably answer that question. The problem of "third-person consciousness" is of fundamental nature and does not only refer to animals, but to our human conspecifics as well. Only I myself know by direct experience that I am/have "first-person" consciousness or do things consciously. Whether this holds for my conspecifics, too, remains uncertain in principle. I conclude from the observation of the behavior of a conspecific, if he or she has consciousness like me while doing certain things. Here we make use of our daily life experience and of scientific plausibility (cf. Koch 2004; Seth et al. 2008). Evidence for consciousness comes from the following facts:

Firstly, humans can do many things *without* consciousness or with only *accompanying* consciousness, e.g., reflexes and highly automated actions. Also, stimuli that are too weak or too short to be perceived consciously (here we speak of "subliminal" stimuli) or are "masked" by other stimuli, can be reliably demonstrated to influence our behavior, especially when repeated. Secondly, we can perceive things consciously, but forget them a few seconds later, because our brains considered them to be unimportant or irrelevant for further processing. Recent research has found out that only those perceptions that are both sufficiently *new* and *important* are sent by the unconsciously working stages of our brain to the associative cortex, where they eventually become conscious through interaction with declarative memory. Psychologists demonstrate that consciousness is strictly needed for dealing with *complex information* in a *variable* and *detailed fashion*, e.g., in the context of action planning, understanding the content of complicated verbal information, or recalling details of past events. However, as soon as complex information and actions are automated, they become less dependent on detailed consciousness, and after long training we can manage relatively complex situations with a minimum of consciousness and awareness.

Because of such a tight link between the processing of new and important information and consciousness, mostly occurring in the working memory (see above), the presence and capacity of consciousness can be tested in a variety of animals. For this, it is necessary to confront such diverse animals as honeybees, *octopuses*, crows, elephants, or macaque monkeys with complex tasks that humans can solve *only* while being fully aware or conscious. Such experiments may consist of training monkeys to attentively follow complicated changes of an object ("morphing") on a screen and then carry out certain behavioral responses, i.e., pressing a button whenever a certain shape reappears. Chimpanzees can be taught to "mentally" find a way out of a maze by following possible tracks with the gaze. Birds can be trained to select an object in order to use it as a tool (cf. Chap. 12). It is, of course, important to exclude alternative ways of interpretation, such as unconscious conditioning or pure chance. This approach is based on the reasonable argument that it is rather unlikely that humans can carry out such complicated tasks only when conscious, while animals can do it without consciousness.

In all those cases, where the animals tested have brains similar to ours, i.e., predominantly primates, there is the possibility to study which parts of the brain are particularly activated or inhibited when dealing with such complicated tasks. If we find that essentially the same regions, e.g., dorsolateral and medial prefrontal and posterior parietal cortical areas, are involved in the same way as in humans, then it is safe to assume that these animals exert the task consciously. Thus, the test for the presence of conscious experience in animals is easier, the more closely related they are to humans, and the more difficult, the more distantly they are related to us, because of the higher or lower similarity of their brains to ours. If the mushroom bodies of honeybees (cf. Chap. 7) can be demonstrated to be highly active during complex learning tasks, this only means that the mushroom bodies are involved in this task, but we cannot conclude from this activity that honeybees have conscious experience of this learning act, because the mushroom bodies are structurally very different from our cerebral cortex. This is different from the situation where we study a macaque monkey in a difficult working memory task and record the ongoing brain activity with appropriate methods such as EEG or functional MRI. Here, we will discover that the monkey does not only show the same behavioral responses typical of focused attention, but we observe high-level activity in dorsal frontal and posterior parietal parts of their cortex in the same way that we observe this in humans during the same task. Thus, it is safe to attribute consciousness to the monkey during this task. However, whether the animal has the same or at least a very similar subjective experience to ours is perhaps an unsolvable problem—and occurs among human individuals as well (cf. Chap. 17).

2.5 Mind-Brain Theories

Such studies and their results will—hopefully—enable us to deal, both from a philosophical and neurobiological perspective, with the so-called mind-brain problem, i.e., the relationship between "mental" and "material-neural" states and processes. This is a fundamental problem in the history of philosophical thought since antiquity, today called "philosophy of mind," and is of great importance for neuroscience as far as mental states and their neurobiological basis are studied (for an overview see Guttenplan 1994; McLaughlin and Beckermann 2011).

Regarding soul-body or—in more modern terms—mind-brain theories, there are two diametrically opposed positions. One is called *dualism* and the other *monism*, each position coming in many variants. For dualism, mind, soul, or consciousness are mental states that are radically, or "ontologically," different from "material," "physical-biological," or "natural" states and do not obey natural laws; rather, as "immaterial entities" they transcend them. However, dualism of any kind struggles with the problem of whether, and—if so how—something immaterial can influence material events, e.g., brain processes, and vice versa, without violating physical laws and some fundamental principles like that of the conservation of energy or that of causal closure. In philosophy, this is called the problem of "mental causation" (cf. Davidson 1970). While epiphenomenalists avoid these problems by making the—utterly implausible—assumption that mental states are causally ineffective byproducts of brain states, interactionist dualists grant mental states causal relevance at the cost of having to deal with some really difficult problems.

The eminent French philosopher René Descartes (1596–1650) was the first to develop the modern concept of an "interactive dualism," which means that the "mental" and the "material-natural" world, including our bodies, do indeed interact despite their different ontological statuses. When we intend to move our arm, our mental-immaterial states act upon our brain in such a way that the appropriate muscles are activated. When we see something frightening or cut our finger, then material-bodily things happen, which then are turned into immaterial sensations of perception (vision) and emotion (fear or pain).

Descartes believed that mind and body interacted in the pineal organ (epiphysis) as a kind of interface, where the transformations from mental to material states, and vice versa, happen. This concept of the pineal organ was put forward in his unfinished "La Description du Corps Humain" ("Description of the Human Body") from 1647. Descartes misunderstood the structure and function of this organ inside the brain and likewise misunderstood the ancient brain-ventricle model, where a "vermis," roughly in a position of the pineal gland, regulated the flow of information from the first ventricle to the second one (as the alleged seat of mind and soul). Thoughts and acts of will would put the pineal organ into vibration, and these activate different nerves, which were (erroneously) conceived as being filled with a nerve fluid or little corpuscles, which eventually inflated the muscles. For Descartes, this interaction between the mental and material-physical including brain world happens beyond physical causality, and consequently does not underlie natural laws. Rather, it is a "mental causality" that influences brain activity. Descartes left the problem—how such mental causation could function without violating physical laws—unanswered.

The great German philosopher Wilhelm Leibniz (1646–1716) recognized this serious problem and tried to avoid it by radically denying that there was any factual interaction between the two worlds. What seemed to be an interaction was the product of "pre-established harmony" between them, which God had arranged from the very beginning.

Immanuel Kant (1724–1804), another eminent philosopher, offered another solution to this problem and of the mind–body problem in general by stating that there are two worlds, i.e., a supernatural or "intelligible" world not bound to physical laws, and a natural world obeying physical laws. For Kant, humans are "citizens of both worlds:" when we think and act *morally*, we are citizens of the "intelligible" world, while in all other aspects of our lives, including all psychological states (!), we are citizens of the natural world characterized by causality and determinism in the Newtonian sense (Kant was a great admirer of Newton). However, the relationship between these two worlds remains undefined. Thus, there remains the problem of how within a dualist concept immaterial mental acts can cause physical events, because if morality should have any impact on our real lives, it must be turned into bodily actions in the physical world in order to allow humans to *behave* morally.

A modern variant of interactive dualism was proposed by the neurophysiologist John Eccles (1903–1997), who in 1963 won the Nobel Prize in physiology-medicine and was what one would call an amateur philosopher. For Eccles as a dualist, the mental and the material belonged to two fundamentally different worlds, like for Kant. Still, he followed Descartes by believing that mental and material-neurophysiological events can influence each other. Particularly, our will (for example to move our arms) *needs* (strangely enough) the neurophysiological processes to get "materialized," i.e., to actually move the arm, and when I cut my finger, this will induce the sensation of pain in my mind or soul. For Eccles, this interaction takes place inside the "liaison brain," which is a modern version of Descartes' pineal organ, but now predominantly located in the dorsal frontal lobe of the cortex, the "supplementary motor area" (Eccles 1994).

Dualism corresponds well with our everyday psychology in the sense that we experience our mental states like thinking, imagining, remembering, etc., as something "fundamentally different" from objects and processes in the material world, but they likewise differ in a strange way from our bodily states although, especially as emotions, they seem to represent somewhat of a mixture of mental and material states. While the latter can be investigated using scientific methods and apparently underlie natural laws, such an approach seems to be inapplicable to the former, because mental states seem to have no exact location in the world, their time properties are enigmatic, and they appear to have no spatial extension or weight. There seem to exist no universally accepted "laws" of mind or emotions in the sense of physics, although many psychologists would insist that our mental activities are by no means "lawless." But if there are laws of thoughts and memories, they seem to differ from physical laws, for example, those described by Gestalt psychology (cf. Metzger 1975).

A special blend of dualism and naturalism-monism (see below) is "emergentism." The concept of "emergence" has been discussed by philosophers of science at length, including the distinction between "strong" and "weak" emergence (cf. McLaughlin 1997; Beckermann et al. 1992). "Strong" emergence of properties, as defined by the British philosopher C. D. Broad in his influential work *"The Mind and its Place in Nature"* from 1925, means that the properties of a system can in no way be explained by the properties of the components. Something "completely new" arises by the interaction of the components. The Austrian–British philosopher Karl Raymund Popper (1902–1994) adopted such a "strong emergentism" in the framework of a "creative universe" and an evolution that is intrinsically goal-directed (i.e., not extrinsically by God). Popper, in contrast to his friend Eccles, accepted that mind is of *natural* origin. During the evolution of the human brain, however, some "completely new" properties appeared that constitute the mind and thus transcend the natural-material world. Accordingly, these properties cannot be reduced to and not explained by the laws of physics (Popper and Eccles 1984). In this way, the mind "frees itself" from the material conditions inside the brain. The Austrian-German biologist-philosopher and Nobel Laureate Konrad Lorenz (1903–1989) used the term "fulguration" for such leaps of the mind during human evolution (Lorenz 1973). Just recently, such a strong emergentism regarding life and mind was proposed by the American anthropologist Terrence Deacon (1997, 2011).

In contrast, *monistic*, *naturalist*, or *physicalist* mind-brain concepts start with the general assumption that there is *no* ontological difference between the mental and the material/natural/physical world. Thus, the world can, at least in principle, be described using solely terms and methods of natural sciences. Accordingly, mental events do not violate, but "obey" natural laws. There is just one world composed of different entities interacting with each other according to known principles. At the same time, substantial differences exist between different monist positions.

The most radical position is "eliminative materialism" of the philosophers Patricia and Paul Churchland (cf. Patricia Churchland 1986; Paul Churchland 1995). These two authors deny that mind and consciousness are real entities, rather they are mere phrases taken from everyday psychology. Therefore, such phrases should be "eliminated" and replaced by more exact descriptions of the neuronal processes ("love is nothing but a neuronal event like…"). Other "identists," like the American philosopher Daniel Dennett (1991) do not deny the existence of mental or conscious states, but reject the idea that they are ontologically different from brain states. They arise as something like labels in the brain in the context of information processing, to the extent that they are nothing but brain states themselves, and their phenomenal uniqueness is an illusion.

Like dualism, "strong reductionism" is confronted with a variety of serious problems. It cannot be denied that mental-conscious states on the one hand and neuronal states on the other are *experienced* by us as being fundamentally different. When I, as a neuroscientist, study brain processes assumed to underlie mental states, I have no direct access to and awareness of these mental states. What I do is to refer to verbal reports in case of experiments with human subjects and to behavioral responses in the cases of human and animal subjects in order to investigate the relationship between mental and neuronal states. This holds even for a self-experiment, when I record from my own brain (e.g., by means of electroencephalography or functional magnetic resonance imaging) and being aware of my own mental states (perceptions, imaginations, act of will, etc.). While experiencing these mental states and looking at the recording data, I do not see their identity but at best their strict correlation. The two domains of "first-person" and "third-person" experience *do not overlap phenomenally*. This is different in the realm of third-person perspective. Here I may perceive the identity of the morning star *Phosphorus* and evening star *Hesperus* or *Venus* or that of *Mark Twain* and *Samuel Langhorne Clemens*—to use two famous philosophical examples for identity.

A much-discussed concept in the realm of monism-naturalism is the "supervenience theory" (Davidson 1970; Kim 1993; McLaughlin and Bennett 2005). The somehow awkward term "supervenience" essentially means that certain properties at a higher, e.g., mental, level are strictly determined by corresponding properties at a lower, e.g., physical-neural level, such that if two people have the same physical-neural properties, they also have the same mental properties, although not vice versa: Physical properties may differ even when mental properties stay the same, because the mental properties could be "instantiated" by different physical-neural problems. In this case, mental properties would "supervene" on physical properties.

The concept of supervenience with respect to the mind-body problem has been the focus of the influential book *The Conscious Mind. In Search of a Fundamental Theory* by the Australian-US-American philosopher David Chalmers (Chalmers 1996). His central tenet is that even if we adopt a naturalistic point of view and believe that there are no metaphysical entities such as an immortal soul, the properties of mind, as we experience them *phenomenally*, cannot be explained by (or reduced to) laws and properties of physics. The reason is that mind and consciousness do not follow *logically*, i.e., *with absolute necessity*, from them. It may well be that in our earthly world, mental-phenomenal experience (the sensation of colors, sounds etc., called "qualia" by philosophers) may be strictly linked to brain states and processes, but such a strict correlation would not *with necessity* explain how that happens, what the qualia really are, and why they exist at all. Such an

2.5 Mind-Brain Theories

explanation is—in the eyes of Chalmers—in principle possible in the physical world only when the properties of a compound, e.g., water, follow necessarily from the properties of their components, i.e., oxygen and hydrogen: the latter "supervene" on the former.

For Chalmers this is impossible in the case of mind due to the unique nature of the mental phenomena (qualia) being fundamentally different from the nature of physical-neural events. We would not even know to which physical-neural components mind could be reduced. Thus, at least at the phenomenal and explanatory level, the material and the mental world are completely different—a concept called *phenomenal* or *property* dualism, although for Chalmers this does not necessarily imply an *ontological* dualism, because mind and matter could just be two different aspects of a hidden, unrecognizable world.

In order to illustrate his point of view, he develops the scenario of a "zombie world," where mindless and consciousless persons exist who in any physical aspect are identical to (i.e., indistinguishable from) a person having mind and consciousness. Although Chalmers admits that there is no empirical evidence for such a "zombie world," he makes the strong statements that this world is logically possible, because we can conceive it without running into any inconsistency. This conclusion from conceivability to metaphysical possibility is highly controversial, but this brings Chalmers in some way back to a kind of property dualism, as he himself admits, or "epiphenomenalism." Indeed, if a Zombie is indistinguishable in any aspect including behavior from a mindful person, then—at least in the physical world—mind does not have a causal effect. Interestingly, Chalmers sees in panpsychism (see below) a possibility to overcome the dangers of both dualism and epiphenomenalism.

There are types of monism that try to avoid these difficulties, for example, "non-reductionist physicalism" (cf. article "physicalism" in Stanford Encyclopedia of Philosophy). "Physicalism" here means that mental states and phenomena can be regarded as *physical states* insofar as they interact with known physical states without violating basic physical laws or principles like the law of conservation of energy. To conceive mental states as "physical" does *not* require that their properties be fully explicable in terms of *known* physics or reducible to other physical states (to solid body physics or electromagnetism, for example). Rather, it is sufficient that that they can be proven to *interact* with other physical states while *not violating* known physical laws. We can accept "special laws" being valid only for mental states, and we can easily live with the fact that there are mental states that we do not yet understand.

Present-day physics, therefore, would have to be enlarged by a new domain, namely "mental physics." Similar enlargements have taken place several times in the history of physics (e.g., in the case of electromagnetism or quantum physics, each of them having their specific laws). The situation would be completely different if certain experiments demonstrated phenomena that undoubtedly *contradict* known physical laws or principles, e.g., the conservation of energy. This, however, has not been the case so far.

Another attractive way to escape both dualism and reductionism is "panpsychism" as taught by eminent philosophers and scientists such as Giordano Bruno, Benedict Spinoza, Gottfried Wilhelm Leibniz, William James, Bertrand Russel, Ernst Haeckel, Albert Einstein and my teacher, Bernhard Rensch. Panpsychism holds that matter and mind coincide from the very beginning and even at the lowest level of complexity, i.e., of elementary particles, and that both matter and (proto-)mind increase with an increase in systems complexity. Rensch characterizes this view in his book "*Biophilosophie*" (1968b) when he writes, "We ought to describe to all 'matter' a proto-psychic nature. The protophenomena of elementary particles, while forming atoms and molecules, constitute new relationships, which leads to new system properties. But only at the complicated structural level of nerve cells sensations can originate, and only when inside a larger central nervous system these sensations form a continuous process of consciousness and generate memories and imaginations, self-consciousness and ego-awareness can arise, which eventually at its highest phylogenetic-developmental level and in combination with logical reasoning, can reach objective knowledge of the world" (p. 236, translation by G.R.).

Panpsychism avoids the above-mentioned problems of dualism as well as of identism/reductionism by assuming that mental states necessarily arise together with, or are just a different aspect of material/physical states. Both evolve to higher forms in parallel by increasing in complexity. However, Rensch and other panpsychists do not explain *which* properties of elementary particles, atoms, molecules, and on eventually nerve cells lead to the origin of mind, and it appears as a mere play of words to call certain properties of molecules ion channels, or membranes "protopsychic." Furthermore, neuronal complexity *per se*, e.g., in the cerebellum, does not automatically lead to consciousness, and the conditions for the appearance of consciousness seem to be highly specific.

2.6 What Does All This Tell Us?

When I write in this book about the evolution of mind(s) in relation to the evolution of nervous systems and brains, then I do not use the term "mind" in the narrow sense of conscious experience, but rather in the wide sense of cognitive functions, from simple forms of learning to insight, problem solving, knowledge attribution, symbolic representation, and thinking-reasoning. These various cognitive functions represent central aspects of "intelligence," i.e., behavioral flexibility and innovation capacity. Such abilities can be studied not only in humans, but also in animals under laboratory conditions as well as in the wild. In the recent past, there has been enormous progress in such studies.

Mind as defined this way may, but need not, include conscious experience. For a long time it seemed impossible to prove or disprove the existence of consciousness in nonhuman animals, but new methods and empirical data are available that make the existence of at least certain states of consciousness very likely in a number of animals. The starting point is the fact that we humans can

2.6 What Does All This Tell Us?

accomplish certain cognitive tasks such as the recognition of complex processes, solving novel problems, mirror self-recognition, middle- and long-term action planning, or the understanding of and acting according to detailed instructions only when we are aware of these things. At the same time, specific parts of our brains are active. If we discover that animals can accomplish comparable cognitive tasks and that more or less the same parts of their brains are active, then it appears justified to ascribe consciousness to them—in the same way as we do to our conspecifics. This, of course, is easier the more closely these animals are related to us.

In the last part of this chapter I discussed—in a highly abbreviated fashion—the main positions regarding the mind-body or mind-brain relationship, which is now discussed under the heading of "philosophy of mind," i.e., dualist and monist-identist concepts, physicalism as well as panpsychism. At the end of this book we will ask ourselves to what degree the empirical data and neurobiological concepts will help us answer or at least further clarify the "eternal" problem of the relationship between mind and brain.

Chapter 3
What Is Evolution?

Keywords Darwinism · Neodarwinism · Natural selection · Non-darwinian evolution · Mass extinctions · Canalization · Reconstruction of phylogeny · Deep homologies · Homoplasies · Convergent evolution

If we are going to study the evolution of nervous systems and brain and its relationship to the evolution of minds (if there is such an evolution), we have to say something about the principles of biological evolution. This, however, is not an easy task, because there is no unanimously accepted concept of evolution, let alone a concept of evolution of nervous systems and brains, as we will see. For the "classical" Neodarwinian concept of evolutionary biology, see Futuyma (2009).

3.1 Historical Concepts of Evolution

The study of biological evolution deals with three main topics: (1) the origin of life (*biogenesis*); (2) the origin of existing species (*cladogenesis*) including mechanisms of speciation; and (3) the mechanisms of changes of organisms in form, function and behavior over time (*anagenesis*). Traditionally, and often even today, such changes and with this, biological evolution in general, are often understood as unilinear *progress* from simpler to more complex and better states. We will see that such a view is incorrect, because the evolution of nervous systems and brains reveals at least as many cases of "backward evolution," for example in the way of secondary simplification, as those of "forward evolution" in the sense of increases in complexity, and in the vast majority of cases, organisms and their nervous systems and brains remain more or less unchanged over long periods of time, often many millions of years.

The Greek philosopher Aristotle (384–322 B.C.), who can be considered the first naturalist-biologists in modern terms, and following him many theologians and philosophers until modern times assumed that there is a ranking order of living beings, a *scala naturae*, starting with the most primitive organisms known at that time (e.g., worms) and ending with humans (and sometimes continuing beyond over angels of various classes to God), and that these forms existed (or were created) independently of each other (Lovejoy 1936). Such a view prevailed until the eighteenth century. Each of these forms was considered perfect because they were created by God. Therefore, no modification or extinction was necessary and even conceivable, because this would—in the eyes of the philosophers of those days —contradict the superior creative force of God. The invariability of species was accepted even by the eminent Swedish biologist Carl Linnaeus (1707–1778), the founder of modern biological taxonomy. He grouped animals into taxa based on anatomical and morphological similarity without asking where such a similarity would come from. Siblings are similar to each other because of common descent, but such a common descent did not exist among biological taxa.

Until the nineteenth century, the prevailing view was that all organisms were created or came into being at the same time. James Ussher (1581–1656), Archbishop of Armagh, calculated that to have been in 4,004 B.C. (more exactly on October 23). But in the eighteenth and nineteenth centuries, an increasing number of "fossils," i.e., petrified remnants of animals, mostly bones, were found which often showed a striking resemblance to living forms. While at first such fossils were interpreted as "monstrosities" of existing forms, more and more of them appeared rather normal, and this meant that many thousands or even millions of years ago forms of animals resembling the existing ones must have existed but became extinct. Because some of them were found in deeper and older, and others in upper and more recent geological strata, they could not all have existed at the same time. However, the question remained whether forms of organisms developed independently of each other constituting separate lines of evolution, or developed in a family- or tree-like fashion based on common ancestry.

In the first half of the nineteenth century, it came to a dramatic dispute between the followers of the idea of an independent evolution of forms on the one hand, most prominently led by the French paleontologist George Cuvier (1769–1832), and those of the idea that all living forms have a common origin and/or that the existing forms descended from an ancestor by accumulation of modifications leading to increasingly greater dissimilarities, prominently led by the French biologist Etienne Geoffroy de Saint-Hilaire (1772–1844). For Cuvier, there was no true evolution despite the existence of fossils, because he thought that there had been regular catastrophies leading to the extinction of existing organisms and the subsequent spontaneous formation of new forms. Another important personality in this debate was Jean-Baptiste Lamarck (1744–1829), who believed in evolution as a process with improvement in forms and functions of different, independent lines of organisms, i.e., without common origin. Differences among organisms, according to Lamarck, are due to differences in environment, different needs for survival, and different use of organs that become heritable (a concept called "Lamarckism"). Interestingly, Lamarck believed that evolution was driven forward or upward by a "nervous fluid" inside the organisms.

3.1 Historical Concepts of Evolution

This dispute about common origin, speciation, and mechanisms of modification of organisms remained undecided until Charles Darwin (1809–1882), in his *"Origin of Species"* of 1859, and Alfred Russel Wallace (1823–1913) at the same time, presented a comprehensive concept about the common origin of all extinct and living beings. Both scientists proposed theories about speciation as well as about mechanisms underlying evolutionary changes known as "natural selection." In the case of Darwin, this latter concept departs from the following premises: (1) Organisms produce more offspring than can possibly survive due to scarcity of resources. (2) Traits that are both heritable (genetically controlled) and relevant for survival, such as visual acuity, length of limbs or color of body surface vary among individuals of a population. (3) In the "struggle for existence," such differences can lead to differential rates of survival and reproduction such that some individuals survive better and have more offspring than others, and therefore are "fitter" or "more adapted" than others. (4) Repetition of this selection process over many generations leads to an increase in the number of carriers of more favorable traits and a decrease in the number and eventual disappearance of carriers with less favorable traits. This is called *fixation* of a given trait within the gene pool of a population.

Darwin's, and to a more limited degree Wallace's, idea of a common origin was readily accepted, because there was overwhelming evidence from paleontology, comparative anatomy and embryology, but many experts of that time remained strongly opposed to the concept of natural selection, especially because it remained unclear by which precise mechanism heritable modifications of traits took place. The rediscovery of Mendel's laws of inheritance of traits at the beginning of the twentieth century led to the development of the concept of the "gene" as the basic unit of heredity, but only the discovery of the chromosomes and the decoding of the molecular structure of ribonucleic acid (RNA) and deoxyribonucleic acid (DNA) as carriers of genetic information in the 1950s of last century gave this concept a precise molecular and functional meaning. Today, a "gene" stands for stretches of DNA and RNA that code either for the production of proteins and with this the formation of certain structures or the regulation of certain functions or for the control of expression of other genes (control genes). In most organisms, each gene comes in two forms called "alleles" due to the double-helix structure of DNA.

3.2 Neodarwinism and Its Problems

The linkage between Darwin's theory of selection with modern genetics is called "Neodarwinism" or "Modern Synthesis," which was developed between 1930 and 1950 by interaction of eminent population geneticists and evolutionary biologists like R.A. Fisher, J.B.S. Haldane, B. Rensch, J. Huxley, Th. Dobzhansky and E. Mayr. The outcome of work of this era was the understanding that the change of gene (or allele) frequency in a given population is the backbone of evolution:

Certain genes or alleles prevail due to better "fitness," while less fit genes or alleles disappear. The basis of this process is genetic variability of a trait. This can be due to a change in the molecular structure of a gene, i.e., its sequence of base pairs, which in turn are induced by effects of radiation, viruses, horizontal gene flow, mutagenic chemicals, or errors occurring during meiotic cell division or DNA replication or—most frequently—recombination of the parental genes. It could also happen that the genes themselves remain unchanged, but the gene expression mechanisms are modified, leading to different structures and functions inside the organism.

The strong tenet of these founders of the "Modern Synthesis" was *gradualism*, i.e., the belief that evolution is based on slight changes due to steady natural selection. There are no leaps in evolution, no differences between microevolution and macroevolution, as opposed to what the German-born American geneticist Richard Goldschmidt (1878–1958) had put forward at the same time with his concepts of "hopeful monsters," i.e., *macromutations*.

The great success in the elucidation of the molecular basis of evolutionary genetics cannot hide the fact that even until today the mechanisms underlying organismal evolution are largely not understood. Best confirmed is the concept of descent of all living beings from one common ancestor or gene pool. Only in this way can the similarities among all organisms regarding the structure and physiology of the cell, mechanisms of reproduction and metabolic pathways be explained in a satisfactory way. On the other hand, there is a dispute as to whether or not natural selection, in combination with genetic variability, is the only, or at least dominant basis of evolutionary changes in form and function.

This same assumption has been criticized by other biologists who emphasize that such "Darwinian" selection can explain only what is called "microevolution," i.e., changes of traits in small steps. This microevolution can be reproduced in the laboratory using organisms with fast succession of generations, such as bacteria, fruit flies, or zebrafish and creating a strong selection pressure over many generations. Neodarwinists believe that many small changes add up to larger changes, until new species appear, especially in combination with geographic isolation. For them, *macroevolution*—the origin of higher biological taxa such as phyla and classes and with this new construction principles ("*baupläne*")—is based on the same principles and mechanisms as microevolution at the population or species level.

One of the most crucial and at the same time most problematic concepts of (Neo)Darwinism is "adaptation." Even among neodarwinists, it is disputed whether "adaptation" has to be understood as a *trait* or a *process* (or both; cf. Dobzhansky 1970). As a *trait*, adaptation means that organisms possess certain traits that appear to promote their survival and the generation of offspring in a given environment. Therefore, we may speak of an "adaptive" trait or the "adaptedness" of a trait. Such a definition is rather unproblematic, because in many cases we can show a clear correspondence between traits and living conditions, particularly in the domain of senses (cf. Chap. 11). However, the correlation between the adaptive trait and reproductive success has to be proven empirically, which in most cases has not happened so far.

As a *process*, adaptation means an evolutionary process driven by natural selection, whereby an organism becomes better fit to its habitat. It is here where the problems begin. First, for neodarwinists, there is a close link between adaptation and fitness (i.e., reproductive success). However, an increase in adaptation of traits to certain environmental conditions in the short run often decreases fitness at least *in the long run* of evolution, namely when the environmental conditions change dramatically. It is the specialists, i.e., the more adapted species that become extinct first. In this sense—as has been stated quite often—natural selection works always in an opportunistic way in the short run, along the "path of least resistance" and not in the long run.

A frequent argument against orthodox neodarwinism is that the concept of adaptation is essentially based on a circular argument: we may state that a certain species of animals is "well adapted" to its habitat, e.g., an owl to nocturnal foraging or a weak electric fish to live in muddy waters. From this it is concluded that this adaptedness evolved under a specific "selective force," that animals which do not have such adaptive traits have less reproductive success—which mostly is not demonstrated. In that way, natural selection becomes a hypothetical (i.e., unobservable) *cause* by which the adaptedness is explained. This is a classic "petitio principii": a hypothetic factor concluded from an observed phenomenon is turned into the explanatory factor for that phenomenon. This often leads to ridiculous formulations of evolutionary biologists, such as like "this astonishing morphological trait must have evolved under very strong, albeit still unknown selective pressure." If this selective pressure is unknown, it cannot serve for an explanation of adaptedness.

Adaptation to the environment, which cannot be denied as a phenomenon particularly in the morphology and function of sense organs (cf. Chap. 11), may have come up by several processes. The first is Darwinian natural selection: there is genetic and phenotypic variation, and the carriers of certain traits have greater reproductive success; in this way, genetic changes become fixed in a population. The carriers of that trait have won the competition within a population and its habitat. Another and apparently frequent process is that modifications in certain traits enable their carriers to *escape* tight competition in a given habitat by invading a new habitat, from which the competitors are excluded. This may again have occurred quite often, for example when our ancestors left the tropical rain forest for our generation in the savanna (cf. Chap. 14), or when animals developed wings for immersing the air.

One useful concept for correcting pan-adoptionism is that of "satisfycing." This term (a combination of "satisfy" with "suffice") was introduced by the psychologist Herbert A. Simon in 1956 in the context of decision making (Simon 1956). Simon tried to point out that under most circumstances of decision making, the optimal decision cannot be made because it is either unknown or unreachable—for whatever reason. Under realistic conditions and in clear opposition to the concept of "rational choice," humans accept solutions that are "good enough" for the moment, although it is likely that better solutions exist. In evolution this would mean that in most cases "adaptation" will not lead to optimal solutions for survival, but to those that are "good enough" (Nonacs and Dill 1993). This is identical with the view that evolution is "opportunistic" or takes the "way of least resistance."

Another useful concept in the context of evolutionary changes of traits is that of *exaptation*, put forward by Stephen Gould and Elisabeth Vrba (1982). This means that during evolution a trait serving a certain function undergoes a shift in function without major changes, especially when the first function becomes obsolete or less important. Processes of exaptation happen quite often in evolution. A classical example is bird feathers, which apparently evolved first for thermoregulation. In other words, a preexisting trait is co-opted for a new function. Therefore, instead of "exaptation" some evolutionary biologists prefer the term "co-option." By exaptation or co-option, biologists can better explain why seemingly imperfect traits (with low "fitness") could survive and develop into complex new traits.

3.3 Concepts of Evolution Beyond Natural Selection

A well-accepted process underlying both micro- and macroevolution is genetic (or allelic) drift. This phenomenon occurs when a low number of members of a species, in extreme cases one pregnant female, enters a hitherto unsettled biotope and founds a new population. Which genes or alleles are present in the "founder gene pool" may be completely random, but in any case they represent only a fraction of the original gene pool, to which the founder fathers or mothers belonged. Accordingly, a genetic bottleneck is formed, in which only a small selection of the original gene or allele pool can be subject to natural selection, and the result of this process may be strongly biased.

Today, the concept of genetic drift is often discussed in the context of so-called neutral evolution. According to this concept, first developed by the Japanese evolutionary biologist Motoo Kimura, most genetic changes are *fitness-neutral*, which means that they have neither a positive nor a negative effect on the survival of their carrier. Such neutral genes can be "fixed" in the gene pool of a population by genetic drift. Many evolutionary biologists assume that in addition to gene drift and "founder effect," other factors are relevant for macroevolution. One of these putative factors is "canalization" of evolutionary modifications due to developmental and/or structural-functional "constraints" (Waddington 1956; Gould 1977). The way in which a given organism is constructed and functions or develops ontogenetically does not allow further evolutionary changes with equal probability, but makes some changes and developmental-evolutionary lines more and others less likely. This often results in *evolutionary trends* lasting for millions of years, e.g., increase or decrease in body size, reduction of limbs, increase or decrease in complexity of nervous systems. The effect of canalization is best documented by the fact that all existing animal phyla, and with this basic organismal plans, originated about 530 mya in the so-called Cambrian explosion or radiation, and no new phylum has arisen since then. The same is true for the origin of subphyla, classes, and families in the sense that the evolutionary dynamics shifts increasingly to lower taxa, giving the impression that it diminishes.

This process can be illustrated by studies on functional morphology demonstrating that certain adaptive changes may *constrain* the possibility of further adaptive changes, as is the case in those vertebrates like bird or bats that use the forelimbs for the formation of wings. While this is highly advantageous for conquering a new ecological niche (the air), it rules out an elaborate use of hands, which is typical of many mammals, e.g., rodents and especially primates. In addition, certain structural or functional traits which arose independently of each other can become increasingly intertwined, so that a further simultaneous optimization of both traits becomes more difficult because an increase in efficiency of one trait may lead to a decrease in efficiency of the other. One example studied by David Wake, a number of colleagues and me is the structural and functional coupling between air-breathing by lungs and tongue feeding in salamanders, because both functions involve the hyobranchial apparatus. Only after the loss of lungs in the specious group of lungless salamanders (family *Plethodontidae*) was it possible for these animals to develop the hyobranchial apparatus into a projectile tongue—apparently several times independently (Roth and Wake 1989). Frogs decoupled these two functions in another way, i.e., by using only the muscular tongue tip for feeding and specializing the hyobranchial apparatus for breathing and calling. Much of the high conservatism of the vertebrate brain can be understood as a consequence of structural, functional and genetic coupling.

From the point of view of many biologists, a particular case of *canalization* is found in the ontogenetic development of organisms including "homeotic genes" which determine the basic organization of an animal body, including brains. For these developmental biologists and geneticists, evolution is a sequence of changes in ontogenies, and this view is called "evolutionary developmental biology," or in short "evo-devo" (Kirschner and Gerhardt 2005; Mueller and Newman 2003; Schlosser and Wagner 2004). Early stages of animal ontogeny, such as the first cell divisions of the fertilized egg, the *zygote*, giving rise to the blastula or gastrula stage, the formation of a primitive gut called "archenteron," the determination of the dorso-ventral and rostro-caudal body axes and of extremities, are considered "bottlenecks of evolution." Here changes are critical, because they can endanger general functionality of the organism.

This means that any evolutionary change of the "blueprint" or "bauplan" of an organism needs to be compatible with these bottlenecks. This results in the fact that—with the exceptions of a few early processes—earlier developmental stages tend to be more conservative than later ones. This is best demonstrated by the impressive similarity between phylogeny, the history of organisms, and ontogeny, the development of individuals, as discovered and described first by Karl Ernst von Baer (1792–1876) and later (1874) called "biogenetic law" or "theory of recapitulation" by Ernst Haeckel. For Haeckel (1834–1919), one of the most influential biologists of the late nineteenth and early twentieth centuries, ontogeny was an abbreviated recapitulation of phylogeny. Although this concept was strongly criticized for a long time, it experienced a revitalization by the discovery of universal "bauplan genes" across the animal (and even plant) kingdom, constituting "deep homologies" in form and function (see below). In short, very early in evolution, certain developmental mechanisms limited further variation.

An interesting evolutionary process is the increase in genome size, as has happened many times in plants, animals and vertebrates in lungfish and amphibians, here most prominently in salamanders. Lungfish and some salamanders have genome sizes about 100 times larger than the average vertebrate genome size, and this process is understood as nonadaptive or an example of neutral genomic evolution. Genome size, i.e., the amount of DNA, is correlated with an increase in cell size, a decrease in cell metabolism and, probably as a direct consequence, a retardation in the division cycles of cells including neurons and the development and differentiation of cells. Accordingly, animals with large genomes have much larger and fewer cells, including neurons, and exhibit a retardation and often failure of late developmental stages (Roth et al. 1993). In the brains of lungfish and amphibians, especially salamanders and caecilians, this leads to secondary simplification of morphology, i.e., restricted cell migration and differentiation, poor or complete absence of lamination, e.g., in the mesencephalic tectum, and torus semicircularis, cerebellum or telencephalic pallium (cf. Dicke and Roth 2007).

Undoubtedly, important factors for the course of macroevolution are *mass extinctions*, in which a high percentage of existing species disappeared. Many causes for such large catastrophies are being discussed, such as the impact of meteorites, dramatic climatic changes (e.g., glaciations), volcanism, oxygen depletion, etc. Presently, six mass extinctions are recognized:

1. A series of mass extinctions during the transition from Cambrian to Ordovician.
2. Ordovician mass extinction (in fact, two mass extinctions), erasing 27 % of all families and 57 % of all genera.
3. Late Devonian mass extinction with the disappearance of about 70 % of organisms.
4. Permian-Triassic mass extinction about 248 mya, which is the greatest known mass extinction of all times, in which 57 % of families and 83 % of genera became extinct. This event is believed to have prepared the dominance of the dinosaurs during the Mesozoic.
5. Triassic-Jurassic extinction about 200 mya wiping out 23 of families and 48 % of genera.
6. End-Cretaceous (K-T) extinction about 65 mya, which marked the transition from the Cretaceous to the Tertiary. It destroyed about 17 % of families and 50 % of genera. It terminated the dominance of the dinosaurs and opened the way to the further evolution of mammals and birds.

These mass extinctions mark most of the transitions between the geological ages. The mass extinctions did not affect organisms indiscriminately; rather, at one time they hit predominantly terrestrial and at other times predominantly marine groups of animals. It is plausible to assume that the ecological specialists suffered more from dramatic changes in living conditions than the generalists, large creatures more than small ones. Mass extinctions gave a chance to those groups that had been dominated by other groups, as the case was with mammals which began thriving only after the disappearance of the dinosaurs at the sixth, K-T catastrophe. Without a doubt, mass extinctions gave evolution new, macroevolutionary directions not determined by microevolution.

3.4 The Reconstruction of Phylogeny and Evolution

It is now generally accepted that all organisms on earth share a common ancestor which existed about 3.6 bya ago and from which all organisms descended by continuous branching. The number of species that underwent extinction during that process by far exceeds the number of existing ones—some experts speak of 99 %. However, it is still unclear *why* groups of animals died out at all, if they were not destroyed by mass extinctions. It is clear that unicellular prokaryotes—bacteria and archaea (previously named archaebacteria) which lack a cell nucleus—developed first. About 2.7–1.6 bya the first eukaryotes originated, i.e., unicellular organisms with a cell nucleus and cell organelles like mitochondria and chloroplasts. Probably, this happened due to a fusion of different types of bacteria and/or archaea—a process called "endosymbiosis" (Margulis 1970).

In this book we deal with the evolution of nervous systems and brains and of their cognitive functions. By doing so it is important to assess whether similarities of certain structures and functions among groups of animals (e.g., concerning the eye) are due to common ancestry (as father and son resemble each other in many traits), or whether these similarities arose independent of each other because of similar selection pressures. In the former case we speak about *homologies* and *homologous* traits, and in the latter case about *homoplasies* and *homoplasious* (or *homoplastic*) traits. The latter can occur either as *parallel* or as *convergent* evolution, which in the former case is based on similar developmental genetic mechanisms, and the latter on different ones (cf. Wake et al. 2011).

In order to answer the question of homologous vs. convergent evolution of similar structures and functions, we need "robust," i.e., reliable information about the evolutionary relatedness of the organisms compared. In the cases of nervous systems and brains, these should be based on non-neural traits in order to avoid circular conclusions. Since nobody was present during evolution, we must *reconstruct* the evolutionary relatedness—an often painstaking procedure called "phylogenetics." The oldest basis of such a reconstruction is the paleontological analysis of fossils, i.e., preserved remnants of animals or plants, in the case of vertebrates mostly of skeletal parts, less often of completely mineralized (or "petrified") animals or those enclosed in fossil resin ("amber"). Early in the history of modern science it was observed that certain fossils were found in certain rock strata, and in the nineteenth century geologists were able to derive from that a geological timescale which allowed paleontologists to determine which fossils came first and which later. The later development of radiometry allowed geologists and paleontologists to determine the age of the strata and the included fossils.

Often, however, fossil records of a given group of animals (including human ancestors) are poor, and today phylogenetics largely makes use of molecular data, e.g., protein or DNA sequences, to reconstruct phylogenetic trees. These molecular data are not only taken from living organisms, but it is now possible—due to advanced sequencing techniques—to use well-preserved body tissue of extinct forms to extract molecular data. Using various mathematical methods, one can construct phylogenetic trees with higher or lower probability.

Today, the most used method of distinguishing homologous and convergent/parallel traits is "cladistics" (from the Greek word "klados" meaning "branch"). Cladistics originated from the work of the German entomologist Willi Hennig (1913–1976). His methods classify species into groups called "clades," which include an ancestor organism and all its descendants and only those, which then form a *monophyletic group* (Hennig 1950). Together, monophyletic groups form a tree of phylogenetic relationships or "cladogram." An example for such a cladogram is given in Fig. 3.1. Living sauropsids, i.e., reptiles and birds, share an amphibian-like "proto-sauropsid" as a common ancestor, which already differed from the amphibian-like ancestor of later mammals. Thus, sauropsids are considered to form a monophyletic group. However, within the sauropsids, the group traditionally called "reptiles" does *not* represent a monophyletic, but rather a *paraphyletic* group, because they include the "crocodiles," but exclude the birds, although crocodiles are more closely related to the birds than to other "reptiles," i.e., the squamates (lizards and snakes), and the tuatara (genus *Sphenodon*, with two species living in New Zealand), and the turtles. Birds and crocodiles form the

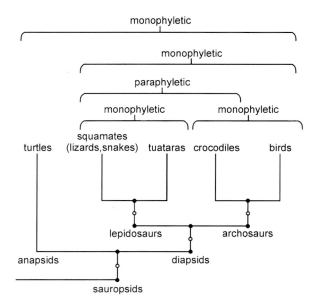

Fig. 3.1 Cladogram of sauropsids ("reptiles" and birds) as an example of the cladistic method. For further explanations see text

3.4 The Reconstruction of Phylogeny and Evolution

monophyletic group "Archosauria," the squamates and the tuatara form the monophyletic group "Lepidosauria," and the (putatively) monophyletic group "Testudines" comprises all living turtles, tortoises and terrapins. Recent analysis suggests that turtles are the sister group of the Archosaurs.

For the reconstruction of a cladogram, morphological, physiological and molecular–genetic traits and data all derived from living species are used, and by means of various mathematical-statistical methods one establishes a *dichotomous* branching scheme, i.e., a scheme, in which, at least ideally, each line can split only into two (and no more) new lines. Each bifurcation point represents the appearance of at least one new trait called *apomorphic*, while the other is retained, or two new derived traits which then are both apomorphic, which also means that the new trait or new traits originated from a common ancestral trait. Such a cladogram reflects only the phylogenetic relationship and makes no statements about the time scale and strength of evolutionary changes with respect to extant species. However, when using certain reference data (e.g., mutation rates of "neutral" genes as a sort of "molecular clock"), one can estimate the point of time of a bifurcation and express the results in "distance lengths" such that the presumed chronological sequence of phylogeny can be determined with some probability.

However, even when using all available information about evolutionary changes of traits, one does not reach just one cladogram consisting of 100 % probability only of monophyletic groups, but one ends up with several to many cladograms with different probabilities. In these cases, the "maximum parsimony (MP) principle" has often been applied, which identifies the phylogenetic tree that assumes the minimum number of evolutionary changes. This procedure is based on the reasonable, albeit not always correct, assumption that simpler solutions are more likely than more complex ones. In the most favorable case one finds just one "shortest" or most probable tree, but often there are several or very many trees of equal probability. For such cases of equally probable trees, there are other criteria that can be used to select the most useful tree. One major problem with the MP principle is the apparently large number of homoplasies (see below). Besides the MP method, other procedures exist for the construction of phylogenetic trees like Bayesian statistics.

In the context of reconstructing the evolution of nervous systems and brains, we are confronted with the fact that soft tissue, including nervous systems and brains, do not fossilize. Luckily, however, in many cases of vertebrates, their size and surface can be estimated through the size of the cranial cavity and the impressions the brain made on its surface—a method that has been extensively used by Harry Jerison (cf. Jerison 1973). With respect to an evolutionary comparison of brains, it is crucial to ask whether a given neuronal trait found in a number of species is "primitive" or "plesiomorphic," i.e., "inherited" from the ancestor, or "derived" or "apomorphic" and therefore newly evolved.

In order to decide this question, we need well-established phylogenetic trees and to insert the trait under consideration. As an example, let us ask whether the six-layered cerebral cortex (called "neocortex" or more precise "isocortex"—see Chaps. 10 and 14), as found in most, but not all mammals in a narrow sense (the

Eutheria or *Placentalia*, see Chap. 9) is a plesiomorphic trait, i.e., one found in the last common ancestor of the eutherian mammals or the product of independent evolution. If the former is the case, then the absence of that trait in some mammals, e.g., in insectivores (e.g., hedgehog) and cetaceans (e.g., whales, dolphins and porpoises), which have an essentially five-layered cortex, must be due to loss or simplification from ancestral six layers (for whatever reason). We also can assume, however, that a five-layered cortex as found in insectivores and cetaceans represents the *primitive-ancestral* condition, and a six-layered cortex originated later. In that case, there are two possibilities. The first is that those groups of mammals that exhibit a six-layered cortex have a common ancestor and consequently form a monophyletic group inside placental mammals. This, however, is not the case, because cetaceans are assumed to originate from ungulates (hoofed animals), which are placed between the mammals, while insectivores are definitely are unrelated to ungulates and cetaceans and considered to resemble the ancestral mammalian condition. The second possibility, then, is to assume that simplification of the six-layered cortex took place in insectivores and cetaceans independently.

But it could also be that a six-layered cortex originated *after* the more recent eutherian mammals split from the insectivore-like stem mammal and that a non-six-layered cortex found in insectivores simply represented the ancestral condition. This question can be answered if we make a *sister-group comparison*. A sister group is the group or taxon most closely related to a given group. In the case of eutherian-placental mammals, this is the metatherians or marsupials (i.e., pouched mammals). Since animals possess a six-layered cortex (cf. Wong and Kaas 2009), it is more parsimonious to assume that the common ancestors of eutherian and metatherian mammals already possessed a six-layered cortex and the situation found in insectivores is due to simplification and happened independently in cetaceans. It appears less parsimonious that the marsupials developed a six-layered cortex independently of the eutherian mammals. However, to make this conclusion even more convincing, we can look at the sister group of therians + eutherians, which are the Prototheria or monotremata (i.e., egg-laying mammals). They likewise exhibit a six-layered cortex, which makes it very likely that the last common ancestor of all mammals (prototheria, metatheria and eutheria) already had a six-layered cortex, rather than assume that they developed it independently.

The question, now, could be whether a six-layered cortex was even older than the mammals and was an "invention" of terrestrial vertebrates, i.e., amphibians, sauropsids, and mammals. The sister group of mammals are the sauropsids, and it is known that mammals and sauropsids descended from different amphibian-reptile-like ancestors. By comparison, we find that neither sauropsids nor amphibians have a six-layered cortex. We could even go beyond the terrestrial vertebrates and have a look at the most closely related fish-like vertebrates, the lungfish. Here we find that they, too, lack a six-layered cortex, as is the case in all other bony and cartilaginous fishes. Thus, we come to the conclusion that a six-layered cortex developed together with the evolution of prototherian, metatherian and eutherian mammals, and that a cortex with less than six layers, as found in insectivores and cetaceans, is most probably due to simplification.

3.4 The Reconstruction of Phylogeny and Evolution

This example tells us how to answer the question of whether similarities in form and function are due to common ancestry and have to be regarded as homologous, or are due to convergent evolution or homoplasy. There is, however, a caveat. First, there are many structures and functions that are astonishingly similar but occur in unrelated species, and therefore were considered to be the result of independent evolution or homoplasy, for example as the result of the same or very similar selective pressure (e.g., the formation of limb-like bodily appendages or lens eyes). However, the strict opposition between homologous and homoplastic traits has become weaker since the discovery of "deep homologies" in the regulatory or developmental genetic networks (e.g., "homeotic genes") that apparently underlie these evolutionary processes. These developmental programs may lead either to highly similar morphologies such as the eye of *Octopus* and the vertebrate eye or to "dissimilar" solutions such as the lens eye of *Octopus* and vertebrates and the compound eye of arthropods. In any way, we are led to assume that such "deep" developmental genetic programs *canalize* the development of structures and functions to a much larger extent than previously thought. However, even if traits have the same genetic developmental basis, they may be regarded as morphologically homoplasious. The frequency and importance of convergent evolution and the role of "deep homologies" based on common homeotic genes is becoming a "hot topic" in evolutionary biology (Wake et al. 2011) and of high importance for an understanding of the evolution of brains.

3.5 What Does All This Tell Us?

In this book we investigate the relationship between the evolution of nervous systems and brains on the one hand and the evolution of intelligent-cognitive functions on the other. In order to do so, I have tried to briefly summarize the present state of evolutionary concepts. While the common origin of all living beings is widely accepted, since Darwin there is a dispute about the mechanisms underlying the observed evolutionary changes. Many "neodarwinists" propose a gradualist concept saying that small changes ("microevolution") eventually lead to large changes ("macroevolution") and that these processes are primarily based on natural or "Darwinian" selection. Other evolutionary biologists insist that the principle of natural selection or "survival of the fittest," together with genetic drift, is just one of many other factors and mostly found in microevolution, while macroevolutionary processes are mainly dominated by other factors like mass extinction, neutral evolution and "canalization," especially when affecting the ontogenies of organisms. At the end of this book, we will see to what degree these different factors may have determined the evolution of nervous systems and brains and consequently of cognitive functions.

The reconstruction of the evolution of nervous systems and brains presupposes robust phylogenies based on non-neural characteristics, which can then be used to decide whether certain neuronal or behavioral characteristics have to be regarded as homologous, i.e., due to common ancestry, or the result of independent evolution. As we have learned, there are relatively reliable methods for such estimates. However, there is always the possibility that behind seemingly independent or homoplasious evolution of similar traits there are developmental or regulatory genes constituting "deep homologies." Also, we have to accept the possibility that secondary simplification of initially complex structures may be at least as common as increases in complexity of initially simple structures.

Chapter 4
The Mind Begins with Life

Keywords Vitalism · Living systems · Self-production · Self-maintenance · Self-organization · Origin of life

In a famous speech in Berlin in 1880, the founder of modern neurophysiology, Emil du Bois-Reymond (1818–1896), gave a list of seven "world riddles" containing—besides the problem of how consciousness can originate in the brain—the "origin of life". About 130 years later, we still do not know exactly how and where life originated, rather as the majority of scientists does not regard this question as a "world riddle" anymore, but as a problem that can and will be solved step by step through patient research; at least it has lost its mystical appeal, and this may happen with respect to consciousness as well.

4.1 What Is Life?

Since antiquity, much has been thought and written about the question what life is, how it originated, and how it can be distinguished from non-living entities. Until modern times the most widely accepted view was that organisms come to life by a specific principle or force called *pneuma, anima,* or *spiritus.* All these words mean "breath" or "breath of life" and later acquired the meaning of "soul" or "mind." In Aristotle, we find a subdivision of organisms into three "kingdoms" in ascending order, i.e., the kingdom of plants, of animals, and of man. *Plants* represent the lowest level of life and possess only the ability for nutrition and reproduction. The underlying principle was later called *"anima vegetativa"* or "plant soul." In *animals* we find the additional ability for self-motion and perception, which later was called *"anima animalis"* or "animal soul." Finally, humans have an *"anima rationalis"* or "rational soul" in addition. Thus, humans have something that all other organisms lack, i.e., reason and insight (*ratio* or *intellectus*).

Related to this concept of pneuma or anima is the *vitalism* of modern times. Its adherents believed and still believe that life is based on a specific force called "*vis vitalis*" or "*élan vital*" (the latter by the French philosopher Bergson). In contrast to the ancient pneuma concept, this force was believed to work beyond known physicochemical principles and not explained by natural laws. Even at the beginning of the twentieth century, eminent embryologists like the German Hans Driesch (1867–1941) proposed such ideas which—in direct recourse to Aristotle— was called "entelechy" as a form of life force. Closely connected to the vitalism of the eighteenth and nineteenth centuries was the idea that the chemical processes found in organisms, which accordingly were called "organic chemistry," are fundamentally different from the chemistry of "dead" matter, called "inorganic chemistry." However, in 1828 the German chemist Friedrich Wöhler produced the organic chemical *urea*, a constituent of urine, from the inorganic ammonium cyanate (NH_4CNO). This process is now called the "Wöhler synthesis." This has often been interpreted as a turning point in "natural philosophy." Today the term "organic chemistry" denotes the chemistry of carbon compounds. But even after Wöhler, many biologists and physicians, such as the famous Louis Pasteur, stuck to the view that living beings were characterized by unique abilities or forces transcending the realm of scientific explanation.

Today, at least among scientists, the generally accepted view is that life is made possible by a highly specific arrangement or "organization" of non-living components (Alberts et al. 2002). The major components are hydrogen, oxygen, carbon, nitrogen, sulfur, and phosphor as well as sodium, potassium, chloride, iron, iodine, calcium, and magnesium. These substances, especially hydrogen and carbon, connect themselves to extended structures, i.e., nucleic acids, proteins, fatty acids, and carbohydrates, which usually reveal very special functions and interact in a specific manner.

The question of whether this chemical composition of existing life is a necessary one (i.e., without alternatives) or whether life could exist on the basis of other components as well, is still undecided. Alternatives to hydrogen and oxygen that have been discussed could be silicon or aluminum, which are sufficiently abundant on earth and have similar structure-forming abilities, but for their other chemical properties are not suitable for systems comparable to earthly life. Because of a lack of knowledge about an *alternative chemistry* of life, it is impossible to define life on the basis of its chemical components in a satisfying way. Life in its present form on earth apparently has originated only once, but it is possible—or even likely—that it eliminated other coexisting forms of life. It is likewise possible that life could exist on another planet in form a completely different from terrestrial life. This problem leads us to an alternative approach, namely, to define life not materially but *formally*, i.e., as a specific pattern of interaction of specific non-living chemical components. In a fundamental way, organisms can be defined as *self-producing* and *self-maintaining systems*. This means that *self-production and self-maintenance* are the formal defining properties of living beings (an der Heiden et al. 1984, 1985a, b).

Self-production means the origin of a certain state of order, which is realized predominantly through the internal interactions of the components of the system and is not critically induced by external agents. With very few exceptions, organisms produce their own components, and these components assemble themselves in a "self-organizing" fashion to a structural and functional order that allows life. In addition, multicellular organisms have—at least for a certain period of time—the capacity to continuously repair and restore this order, until they eventually decompose. An exception are unicellular organisms which can repair their life-giving order completely and can live forever, unless they are eaten up or otherwise destroyed. We thus recognize three important properties of living beings: the production of their own components, the correct assembly of these components, and the continuous repair of their order and maintenance of their own existence—a phenomenon which the two Chilean biologists Maturana and Varela have called "autopoiesis" (Maturana and Varela 1980).

Self-production of a complex order state is not exclusive to organisms. Rather, in the inanimate nature there are many processes that come close to the phenomenon of self-production, and organisms have developed a special case of self-production that includes self-maintenance. In order to understand the prebiotic kinds of self-production of order states, we need to address the difficult notion of *order* in greater detail.

4.2 Order, Self-Production, and Self-Maintenance

Order can be static or dynamic. There is, of course, nothing in our universe that is absolutely static. Everything changes, but some things change faster and others more slowly, but at least at a macrophysical level all of them will eventually decay. Only one macrophysical entity exists in our universe that seems to withstand this universal decay, i.e., living beings. Life originated about 3.6 billion years ago and has survived all catastrophies on earth, and there is no hint that life will completely disappear from earth before the end of the earth itself. Even if some or many species will die out—and humans may well be among them—others, like bacteria or some worms, will survive.

In contrast to crystals, organisms are not static, but highly dynamic systems. An organism exists and maintains itself while it undergoes changes, but these changes need to be compatible with its self-maintenance. More precisely, these continuous changes are necessary, because only through continuous flow of matter and energy for repair and growth can an organism oppose the otherwise inevitable decay. The principle of this fundamental process is what the Austrian-born biologist and founder of the "General Systems Theory," Ludwig von Bertalanffy (1901–1972), called "Fliessgleichgewicht," i.e., a dynamic quasi-stationary state, in which substances and energy are continuously brought into a system and reaction products are removed from the system. Thus, there is an equilibrium between composition and decomposition, called *homeostasis*, or a continuous repair, which need not always lead back to the original state, but allows minor and, in the long run, major changes.

While organisms are the only known self-producing and self-maintaining systems, some non-living, self-producing systems show signs of self-maintenance as well, at least for a certain period of time. This is found in all dynamic physico-chemical systems exhibiting a spatio-temporal pattern. This pattern is due to the fact that complex chemical processes occur that influence each other in a *cyclic* manner such that after a certain amount of time they return to an initial state, which, however, need not be absolutely identical to the previous initial state. Chemical self-organizing systems can be described by reaction-diffusion equations including the existence of positive and negative feedback and autocatalytic processes. They all belong to the class of non-equilibrium processes such as the Belousov-Zhabotinsky reaction (BZR), the Winfree oscillator or the Rayleigh-Bénard-convection, but well-known natural phenomena such as cloud formation, growth of crystals, or the candle flame belong to this type of dynamic systems.

Typically, such self-organizing, non-living dynamic order states decay after a while, and this happens the faster and more complex they are because they represent a thermodynamically "improbable" state of high order. Living beings, too, represent such a thermodynamically "improbable" state of high order, but they are, as already mentioned, capable of maintaining this order for a long time because they do not only supply themselves with matter and energy, but with them they "import" order at the expense of their environment. This is usually interpreted in the sense that order increases internally, while it decreases outside the system. This appears to be the trick of living beings and ensures their *autonomy*, i.e., their (relative) independence from their environment, while non-living, self-organizing entities are *heteronomous* in the sense that after a short period of time their internal order decays because they cannot continuously guarantee matter and energy supply.

The so-called "Bénard cells" are generated by the upwelling of warmer liquid from the heated bottom layer, and this structure disappears as soon as the heating from below stops. This system is heteronomous, because it critically depends on that heating as an external factor. The same holds true for the BZR or the Winfree oscillator as examples for non-linear chemical oscillators. The BZR exhibits its color oscillation only as long as there is sufficient potassium bromate and malonic (propanedioic) acid, and these substances have to be substituted from the outside. The candle flame burns only as long as the candle with the wick and the wax are used up. Importantly, the Bénard cells do not control the heating on which they crucially depend, the BZR does not supply itself with potassium bromate and malonic acid (and the other substances required), and the candle flame does not produce the wick and supplies itself with wax. If they were to do so, then they would come close to our definition of living beings as self-organizing and self-maintaining, or "autopoietic" systems.

An important feature such "autopoietic" systems is that the components of the systems may undergo changes during interaction, and in that way, "radically new" system properties could "emerge" (Chap. 2). This has often been misinterpreted as a "top-down effect" of the system as a "whole" on their components (Deacon 2011). However, this is misleading, because the "system as a whole" can do

4.2 Order, Self-Production and Self-Maintenance

nothing beyond what single components do when interacting. The phenomenon that within a complex system, components may change as a consequence of their interaction, is particularly important for the nervous system and the brain in the context of plasticity.

A critic will object that even organisms are not completely autonomous. Many of them depend directly or indirectly on the energy of sunlight, and if the sun would stop shining for a while, they would die, and this would happen with those organisms that are directly dependent on sunlight, i.e., the photoautotrophic unicellular organisms and the plants, and, as a consequence, with all heterotrophic organisms, i.e., animals and fungi that feed on the photoautotrophic ones and those that feed on these heterotrophic organisms. There are organisms that exist independent of sunlight while they exploit other energy sources, mostly energy-rich chemical compounds such as hydrogen sulfide, but they, too, are completely dependent on these sources. A living being is not a *perpetuum mobile*, which in fantasy can function without external energy supply by generating for itself the necessary energy, but rather exploits energy sources in its environment. The crucial difference between living beings and the above-mentioned physical and chemical self-organizing and self-producing systems consists of the fact that the former *actively* manage the energy and matter supply needed for their existence, which in the latter must be delivered by another agent (in most cases, a human being).

In the most favorable case, there is abundant energy and matter in the immediate environment, and the only thing the organism has to do is to incorporate both. In the case of substances needed, there has to be a *concentration gradient* between outside (high) and inside (low), and the substances enter the organism along this gradient through the membranes. However, there might be substances that are not favorable or even harmful. Therefore, organisms need membranes that are *selective* for certain substances in the sense that some substances can pass the membranes and others cannot. This is the prototype of selective interaction of organisms with their environment.

Often, however, the substances needed do not always exist in the immediate environment, but are found at some distance, which means that the organism must either move toward them or possesses mechanisms that bring them nearby. But there are organisms that come rather close to a life in *Cockaigne*, the "land of milk and honey." Plants, for example, extract carbon dioxide from the air and water from the earth and from both produce glucose and metabolic energy using sunlight in the process of photosynthesis. They have a light-dependent, day-night rhythm, exhibit photo—and geotropism, i.e., they direct themselves or parts of themselves (e.g., leaves) toward the sun or into the earth (e.g., roots), guided by light or gravity. Another example are the so-called endoparasites, such as tapeworms, which "latch" themselves onto the metabolism of their host. These endoparasites all have—as we will see—a highly reduced nervous system and very simple sense organs, because food comes directly to them. But even in this easiest way of energy and matter supply, some basic mechanisms of information processing and communication are necessary.

4.3 Life, Energy Acquisition, and Metabolism

Life as a dynamic non-linear system depends on metabolism, i.e., the continuous import, processing, and export of matter and energy (Alberts et al. 2002). On the one hand, metabolism means "catabolic" metabolism, i.e., the acquisition of energy by decomposing energy-rich matter, and on the other hand, "anabolic" metabolism, i.e., the construction of components of cells such as proteins and nucleic acids needed for repair and growth. The earliest form of catabolism is *chemotropism*. Here, the sources of energy are energy-rich inorganic chemical compounds such as hydrogen sulfide, ferrous iron or ammonia. These compounds are "broken up," i.e., oxidized, and the released energy is taken up by specific mechanisms. Important is the gain of electrons and protons (H^+). In organisms, common energy acceptors and energy carriers are usually nucleoside phosphates, because they readily accept and release electrons and protons, for example, adenosine and guanosine phosphate, which come as triphosphates (ATP, GTP), diphosphate (ADP, GDP), or monophosphate (AMP, GMP). One energy-rich form is the triphosphate, and the decomposition from tri- to di- and eventually monophosphate releases energy, which then can be used to build up cellular structures. Thus, ATP and GTP, as well as ADP and GDP, are very good energy transporters, as is NAD^+ (nicotinamide adenine dinucleotide), which can be reduced to NADH.

For most organisms, sunlight is the main energy supplier. Plants use sunlight energy for the cleavage (oxidation) of water molecules (H_2O) in the process called *phototrophism*, which means "nutrition using light." This is a very clever process. The first organisms using phototropism probably were the cyanobacteria ("blue algae"), and this was an event with enormous consequences for the biological evolution. Firstly, sunlight represents a virtually unlimited energy source, and secondly, in the process of *photosynthesis*, which is the heart of phototrophism, carbon dioxide and hydrogen are converted into glucose (sugar), while oxygen (O_2) is released as a "waste product." Oxygen then, together with nitrogen, becomes the major component of our atmosphere, and plants and animals, including humans, cannot live without this atmosphere.

Both glucose and oxygen are used in a highly sophisticated way. On the one hand, glucose is needed as an energy storage and supplier. When needed, the energy stored in a glucose molecule can be released by decomposing it into pyruvate—a process called glycolysis. The free energy released is used to form ATP and NAD as high energy compounds. Glycolysis yields two molecules ATP per molecule glucose, which is a rather poor energy budget typical of all organisms living without an oxygen atmosphere. In the so-called *oxidative metabolism* (i.e., citrate cycle plus oxidative phosphorylation), one glucose molecule yields—besides water and carbon dioxide—36 molecules ATP, which is an enormous energy gain compared to glycolysis. Thus, as soon as oxygen was available through the activity of cyanobacteria, other uni- and multicellular organisms had a much better way of gaining and storing energy.

4.4 The Origin of the First Life

There is no full agreement among experts about the origin of life, although it is now accepted that life on earth originated about 3.6 billion years ago, which would be roughly 1 billion years after the origin of the earth. Also, most experts believe that life on earth developed rather slowly by synthesis of simple molecules, i.e., methane, ammonia, water, hydrogen sulfide, carbon monoxide, and carbon dioxide, into small compounds called organic monomers such as amino acids and nucleotides. These in turn have the tendency to assemble themselves into long molecule strings, such as, peptides, or the nucleic acids RNA or DNA. Exactly how all that happened is disputed. While some experts believe that the organic monomers, were formed on earth, albeit under still unknown conditions, others believe that organic monomers, or even simple forms of life, came to us from space (see below), which, however, only shifts the problem of the origin of life to other places in the universe.

Charles Darwin had already suggested that life may have originated in a "warm little pond, with all sorts of ammonia and phosphoric salts, lights, heat, electricity, etc., present, so that a protein compound was chemically formed ready to undergo still more complex changes." A similar and long dominating view was that the primordial atmosphere on earth was rich with ammonia, methane and hydrogen. In 1953, Professor Harold Urey and his student Stanley Miller carried out a famous experiment. They started with a "primordial soup" consisting of hydrogen, methane, ammonia and steam, and applied electric sparks simulating thunderstorm flashes. Eventually they found organic compounds such as formaldehyde, hydrocyanic acid, amino acids and long chains of carbohydrates. In addition, they could show polymerization, i.e., the formation of long chains of molecules from monomers, and demonstrated that complex biological structures can form spontaneously, in a self-organizing fashion, under favorable energetic conditions. In this way, short proteins can originate from amino acids, RNA and DNA strands from nucleotides and double-layered phospholipid membranes from phospholipids of appropriate length. As cell membranes, such structures were of great importance for the further evolution of organisms. However, it is debated whether high-energy conditions ever existed on earth, as assumed in the Miller-Urey experiment, or whether organic monomers were really formed giving rise to longer molecular chains and eventually to highly complex structures and functions.

An alternative hypothesis put forward by Michael J. Russell, which does not rely on high-energy and high-temperature conditions or ultraviolet radiation, proposes that life originated in the deep sea in close proximity to so-called hydrothermal vents, where hydrogen-rich streams from the underground meet the carbon dioxide-rich ocean water (Russell and Hall 1997). Still another alternative, proposed by the German chemist Günter Wächtershäuser (Wächtershäuser 1988, 2000) is based on the idea than in an "iron-sulfur world," metal or mineral compounds supplied energy for the spontaneous formation of macromolecules, which eventually achieved self-replication.

In any case, a necessary precondition for the origin of life on earth was the mutual replication of nucleic acids (RNA, DNA) and a mechanism for the synthesis of amino acids and proteins. However, there are two competing concepts: The first one says that life began with RNA or RNA-like molecules, which, due to their specific structure were able to store information *and* replicate themselves. In such an "RNA world," neither DNA nor proteins existed. Under specific conditions, e.g., inside "microspheres" (see below) or on a clay substrate, "ribozymes" developed which could catalyze themselves. A problem here is to explain the existence of the four RNA nucleotides adenosine, thymidine, cytosine and uracil, because it is still unclear how the latter two bases could have developed under abiotic conditions. Later—it is assumed—this self-replicating RNA came into contact with proteins that could function as enzymes and were able to synthesize RNA in a more efficient way than did the ribozymes. From such RNA-protein combinations, the ribosomes as production sites for proteins eventually developed. The interaction between RNA and proteins is the central element of the "hypercycle" theory of the German chemist and Nobel Laureate Manfred Eigen, together with his Austrian colleague Peter Schuster (Eigen and Schuster 1979). Later, in eukaryotes and some prokaryotes, RNA was partially replaced by DNA, which, as a double strand, is chemically more stable. As a single-strand, RNA, as messenger and transfer RNA, is still present.

The alternative concept departs from the hypothesis of "metabolism first," i.e., the idea of a prebiotic world without RNA or DNA. The basic idea is the spontaneous formation of "coacervates" or "microspheres" as developed by the Sowjet biochemist Alexander Oparin and the US-American biochemist Sidney Fox. Coacervates and microspheres are thought of as little spheres with a kind of membrane enclosing a simple metabolism. These little spheres could divide after having reached a certain size and each new part had its own metabolism. Only later—it was assumed—RNA and DNA were formed which made the transfer of genetic information much easier.

Again, another theory, which presently finds many supporters, departs from the idea that either the inorganic components of life or life itself did not originate on earth, but came to it from the far universe by meteorites or from Mars. Such a concept of an extraterrestrial origin of life, or at least of prebiotic states, is corroborated by the fact that organic components are frequent in the universe, particularly in the outer parts of our solar system, where such components are not immediately destroyed by the warmth of sunlight.

4.5 The Further Development of Simple Life

Organisms on earth are divided into three major domains: bacteria, archaea (together forming the prokaryotes) and the eukaryotes (Fig. 4.1). All available evidence tells us that these three basic forms of life have a common origin, independent of how and where life developed first. The first fossilized organisms, the prokaryotes are at least 3.5 billion years old. They lack a cell nucleus and cell organelles such as chloroplasts (necessary for photosynthesis) or mitochondria (necessary for energy metabolism), all enclosed by a membrane. In these organisms, metabolic processes such as photosynthesis or oxidative phosphorylation occur directly at the cell membrane.

The metabolism of prokaryotes is extraordinarily diverse. Besides energy gain by photosynthesis and the decomposition of organic compounds, both of which are typical of eukaryotes, prokaryotes can exploit inorganic compounds. This makes life in biotopes possible where eukaryotes are unable to survive, such as antarctic snow, hot wells or hydrothermal vents in the deep sea.

Eukaryotes probably originated by endosymbiosis, i.e., the fusion of several prokaryotic individuals (Margulis 1970). Eukaryotes existed for at least 1.7, possibly 3 billion years. The phylogenetic relationship among the three domains of life—bacteria, archaea, and eukaryotes—is unclear. It is debated whether bacteria or archaea represent the basis of life. The majority of experts now accept the view that bacteria (about 9,000 species described, but in reality many more) originated first and that archaea (about 260 described species) and unicellular eukaryotes (protozoans) developed from them, and the latter then split up into large domains (kingdoms) of other protozoans, fungi, plants, and animals (Fig. 4.1).

The key innovations of eukaryotes are membrane-covered organelles, particularly a cell nucleus, mitochondria, endoplasmic reticulum and Golgi apparatus, and—in plants—chloroplasts. Cell division (mitosis) of eukaryotes involves the replication of each pair of chromosomes (one from the father and one from the mother organism) and the subsequent separation of the two sets of paired chromosomes. In meiosis a special type of cell division for the production of gametes, i.e., sperm and egg cells, the genes in the chromosomes undergo a recombination producing a different genetic combination in each gamete. Besides gene mutations, this recombination of genes during meiosis is one important basis of evolution.

It is assumed that multicellular organisms have reproduced independently many times among bacteria (e.g., cyanobacteria) and protozoans (Rokas 2008). Choanoflagellates are believed to be the origin of the evolution of metazoans that gave rise to sponges, "coelenterates" and bilaterials, as will be described in Chapter 6.

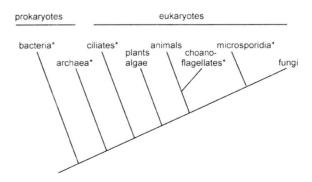

Fig. 4.1 Hypothetical "tree of life." All organisms are divided into prokaryotes, i.e., unicellular organisms without a cell nucleus, and eukaryotes, possessing a nucleus, which are either unicellular (marked with *asterisk*) or multicellular organisms (fungi, plants and animals)

4.6 What Does All This Tell Us?

Most philosophers, from antiquity to modern times, could not imagine that highly ordered and complex entities like organisms could come into existence without a creator-god or an equivalent mystical force (*élan vital*). Some philosophers, however, already had the idea that spontaneous formation of order could occur under certain favorable conditions. In Kant's "Critique of Judgment," there is the famous remark that organisms are "self-organizing" systems. Today we can name and study physical principles that underlie the origin of prebiotic and biotic structures. All that is very complex, and many details are not yet understood, but there is nothing truly enigmatic, a "Welträtsel," as Emil du Bois-Reymond had called it.

I have characterized organisms as self-producing and self-maintaining, or "autopoietic" systems. Cases of the spontaneous occurrence of orderly states or processes are already found in the abiotic nature, such as autocatalytic processes, oscillatory chemical systems, or the flame, and organisms add the ability for self-maintenance which is based on an active supply of matter and energy and continuous repair of own structures. Continuous repair functions at least for a while, and when combined with cell division, in all extant unicellular organisms including our own germ cells, this has already lasted for several billion years.

One of the most important characteristics of organisms as self-producing and self-maintaining systems is that during their interaction the system components may undergo changes which then lead to changes in the pattern of interaction and with this of properties of the system as a whole. This plasticity need not be the same for all components: many remain relatively constant, e.g., those that are important for basic metabolism, while others are highly modifiable, like muscles and above all nerve cells (Chap. 5). The plasticity of at least some of the components is one of the major sources for "emergent properties" of the system. However, there is nothing mystical with such processes, and the much-cited statement that "the system is more than its properties" addresses the almost trivial, but at the same time important fact that system components like nucleic acids and proteins *in isolation* do not reveal certain properties that come up only during interaction with other components.

To date, there is no universally accepted view of how life came into existence on earth, or whether at least some necessary chemical compounds arrived from the universe. Likewise, it is unclear whether life can exist only in the present earthly form, or whether there are physicochemical alternatives which realize the principles of self-production and self-maintenance. Besides a continuous supply of matter and energy, information gathering, processing and transformation into behavior are necessary prerequisites. No organism can exist without sensory mechanisms and a minimum of processing of sensory information used for guidance of behavior. This what I mean when I say that mind began with life.

Chapter 5
The Language of Neurons

Keywords Structure of nerve cell · Membrane excitability · Ion channels · Neuronal transmission · Synapse · Action potential · Neuronal information processing

As discussed in the previous chapter, life is based on a continuous supply of energy and matter. In order to guarantee this continuous supply and to maintain other functions necessary for survival, the organism must gather and process *information* about its environment and turn it into adequate behavior. However, while widely used, "information" is an ill-defined notion. A strict definition of "information" exists only in its technical sense of an ordered sequence of symbols which is recorded, transmitted, or stored. As already stated by the "fathers" of information theory, Shannon and Weaver (1949), this sequence of signals is either inherently void of *meaning* or can have arbitrary meaning.

The problem is that in the bio- and neurosciences as well as in psychology, the notion of "information" is understood as a *meaningful* signal, while at the same time, there is no generally accepted theory of *information as meaning*. In this book I will use an operational definition of "information" in the sense that the informational content, or *meaning* of a signal or sequence of signals from the environment (including the body) is its effect onto the nervous system leading to a certain internal state which, sooner or later, results in a certain *behavior*. The signals could likewise come from other parts of the nervous system, e.g., memory or the limbic system. More precisely, meaning is constructed by the nervous system or brain via the interaction of incoming signals with current cognitive and/or emotional states, mostly via a cognitive or emotional *evaluation process*, which then modifies the current states. The resulting meaningful states may, but need not be, accompanied by consciousness, and states of consciousness are relevant only insofar as they influence behavior. If signals from the environment or the body or from other parts of the nervous system do not induce such changes of internal states and of behavior, then they contain *no* information and they are *meaningless*. It is clear that such an operational definition by no means covers the entire spectrum of phenomena related to "information" and "meaning," but it may suffice for the present context.

5.1 The Structure of a Nerve Cell

Gathering and processing of information about the environment and transformation of the results of these processes into behavior is the scope of nervous systems and brains. Nervous systems and brains are built of two major types of cells, nerve cells and glial cells (often called "neuroglia") (Zimmermann 2013). The latter have rather diverse functions, e.g., to supply nutrients and oxygen to the nerve cells, insulate one nerve cell from the other, destroy harmful material, and remove dead neurons. Glial cells also take part in neurotransmission, although their precise role is not fully understood (Götz 2013).

Nerve cells or *neurons* are cells that process and transmit signals in the context of sensory, limbic, cognitive or motor functions, or the activation of glands. They usually form networks. They are rather diverse in morphology, but most of them have a basic structure, which is illustrated in Fig. 5.1, showing a pyramidal cell found in the cerebral cortex of a mammal. There is a cell body or *soma* carrying appendages called *dendrites* and *axons*. The dendrites are usually treelike, i.e., multiple branching structures which receive signals and transport them to the soma. However, this is true only for the nerve cells of vertebrates, because in invertebrates the soma does not take part in signal processing. The axons are thin processes which carry signals from their own cell to other nerve cells or to effectors, for instance to glands or muscles. The site of origin of axons is called axon hillock. An axon can be short (a few micrometers) or long (some meters, as in large animals), and can divide into side branches called axon collaterals and eventually make contact with the dendrites, somata or axons of other nerve cells. It is said that neurons "project," mostly via longer axons, to other neurons over some distance, and therefore are called *projection neurons*. Some neurons have more than one axon, while others have no axon at all. Such collaterals may have different targets, and therefore one neuron can project to many other cells. Neurons which have axons that are short and do not leave the immediate vicinity or have no axon at all are called *interneurons*. Often, nerve cells of a certain type or function form cell assemblies, which are visually distinguishable. In invertebrates, and in some cases also in vertebrates, these are called "ganglia" (singular "ganglion"), and in vertebrate brains "nuclei" (singular "nucleus"). Nuclei and ganglia are usually connected via axon bundles of their cells, often in a reciprocal fashion, i.e., nucleus A projects to nucleus B and vice versa, and they mutually influence each other. In most cases one nucleus is connected to several or many other nuclei, yielding a complex connectivity pattern.

Nerve cells have contacts with each other via *synapses*, as shown in Fig. 5.1. Synapses consist of a presynaptic and a postsynaptic part. They can occur between axons and dendrites, axons and cell bodies, axons and other axons, and also between dendrites. There are two types of synapses: electrical and chemical. In electrical synapses, the presynaptic and postsynaptic cell membranes have very close contact and are connected by channels (*gap junctions*), through which electrical current can pass and induce voltage changes in the postsynaptic cell.

5.1 The Structure of a Nerve Cell

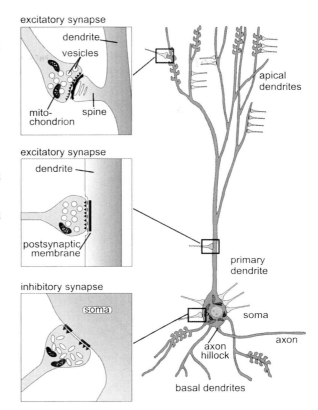

Fig. 5.1 Structure of an idealized nerve cell, i.e., a pyramidal cell of the cerebral cortex of mammals. Apical and basal dendrites serve for the collection of neuronal activity from other nerve cells, and the axon transmits the activity to other nerve cells. To the *left*, three different types of synapses are shown; *above* an excitatory synapse contacting a "dendritic spine," (spine synapse); *middle* an excitatory synapse contacting the primary dendrite; *below*, an inhibitory synapse contacting the cell body (soma). From Roth (2003)

This can happen with almost no delay, but there are only a few mechanisms for the regulation of such signal transmission. In a chemical synapse, the presynaptic and postsynaptic sections are separated by a small extracellular space less than a micrometer wide, the *synaptic cleft*. In the presynaptic part, chemical signaling substances called *neurotransmitters* are released into the cleft that binds to receptors located in the membrane of the postsynaptic part. This binding of the neurotransmitter to a receptor can be either direct or indirect and influence the postsynaptic cell in a wide variety of ways.

5.2 Principles of Membrane Excitability

Communication between cells by means of electrical and chemical substances is universal and not only found in nerve cells (Zimmermann 2013). Therefore, in the nervous system of most organisms, we find a combination of communication via chemical and electric signals. But even electric signaling involves chemical signals in the form of electrically charged atoms or molecules called "ions" (Greek for "wanderers").

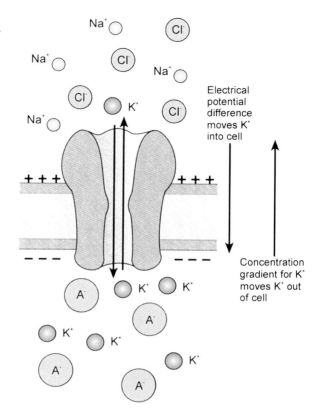

Fig. 5.2 Distribution of ions at a potassium ion channel of a nerve cell membrane. The specific distribution of positively charged sodium (Na$^+$) and potassium (K$^+$) as well as of negatively charged chloride ions (Cl$^-$) and inorganic ions (A$^-$) results from a balance between the electrical potential difference and the concentration gradient of potassium ions. From Roth (2003) after Kandel et al. (2000), modified

These ions are capable of moving across the cell membrane via specific *ion channels*, which can open and close in a reversible and fast manner (i.e., within microseconds) (Egger and Feldmayer 2013). Ions that play an important role in communication between neurons are the positively charged sodium ions (Na$^+$) and potassium ions (K$^+$), the double-positive ("divalent") calcium ions (Ca^{++}), and the negatively charged chloride ions (Cl$^-$) (Fig. 5.2).

Ions possess two peculiar properties. The first one has to do with the fact that ions with opposite charges attract and ions with the same charge repel each other. This is called the *electrotonic force* which builds up a potential difference or voltage across the membrane. The second one is based on the fact that ions of one kind tend to move from sites of high concentration to those of low concentration, i.e., along a *concentration gradient*. The ions are driven by another kind of force, the *osmotic* or *diffusion force*, and this movement happens until the unequal distribution or gradient disappears on both sides of the membrane. However, if we measure the distribution of the ion types mentioned above at the nerve cell membrane, we realize that there is no real equilibrium. Rather, there are many more sodium ions (Na$^+$) at the outer side than at the inner side of the membrane,

5.2 Principles of Membrane Excitability

while the opposite is the case for the potassium ions (K^+). This goes along with a negative potential difference or voltage of 50–75 mV measured inside versus outside the cell. This is called the (negative) *resting membrane potential*, because there is no net movement of ions at that state.

But what causes this unequal distribution of ions, and why is it not abolished by the mentioned capacity of ions to cross the membrane? The reason is that the electrotonic and the osmotic forces counterbalance each other. Let us take the example of the distribution of the K^+ ions. As already mentioned, there are many more of them inside the cell than outside, and we would expect that they tend to move out of the cell until the concentration gradient disappears. This tendency, however, is opposed by the electrotonic force exerted by large negatively charged organic ions inside the cell which, due to their size, are incapable of moving through the membrane channels, but at the same time hold the positively charged K^+ ions back. As a consequence, only a few potassium ions leave the cell until an equilibrium or compromise is achieved between the electrotonic and osmotic force.

If we consider for the moment only potassium ions, then at the equilibrium between electrical and osmotic forces we would measure a voltage of (-75)–(-85) mV which is the potassium equilibrium potential (also often called "potassium reversal potential"). The Cl^- ions play no major role in this situation, but rather distribute themselves according to the potassium distribution. But what about the sodium ions, which are present at the outer side of the membrane at much higher concentration than at the inside? Should they not be attracted by the negatively charged large organic ions in the inside? Here comes into play the idea that at the resting state, most potassium channels are open, and as a consequence potassium ions can permeate the membrane, while most sodium channels are closed and the sodium ions are "locked out." This means that only very few of them can cross the membrane and make the interior less negative. As a consequence, in the resting state the combined potassium and sodium equilibrium potential of the membrane is close to the potential of potassium alone, at about -70 mV.

However, this "resting state" does not mean that nothing happens, because the insulation of the membrane is not ideal and there is a "leakage" of sodium and potassium ions through the membrane, so that a number of sodium ions enter and a number of potassium ions leave the cell. Think of the membrane as a battery, with poles that are insufficiently insulated from each other, the leakage of currents occurs, and the battery needs to be recharged continuously. In the case of the membrane, the recharging is done by the sodium-potassium pump, which continuously exports three sodium ions and imports two potassium ions into the cell and in this way stabilizes the negative voltage of the membrane, because more positively charged ions are exported than imported. This pump requires much energy delivered by ATP and consumes about 2/3 of the entire metabolic energy of a nerve cell. Its breakdown will eventually lead to a loss of excitability of nerve cells.

5.3 Ion Channels and Neural Transmission

The presence of ion channels is fundamental for fast signal transmission through the cell membrane, which as a double-lipid layer constitutes a very efficient barrier against many water-soluble substances. Ions belong to these substances, because they are all embedded in a water shell. The permeation of ions through the ion channels is about 1,000 times faster than the ion transport via specialized pumps (e.g., the sodium-potassium pump), and 100 million times faster than pure diffusion of an ion through the intact cell membrane. Thus, without ion channels a fast control of behavior would be impossible.

It is therefore not surprising that the membranes of all organisms, even of the simplest ones, possess ion channels (Hille 1993; Strong et al. 1993; Anderson et al. 2001; Ghysen 2003). In bacteria and archaea, we find a mechanosensitive ion channel that closely resembles that of eukaryotes. With the help of such a channel, a bacterium can sense that it bumps against an obstacle (Chap. 6), Eukaryotic unicellular organisms like *paramecium* are already equipped with a variety of types of ion channels; besides calcium channels, they already possess four different potassium channels, i.e., the outward rectifier channel (K), the anomalous rectifier (A), the inward rectifier (IR), and the calcium-dependent potassium channel (KCa). In plants and fungi we similarly find a number of types of potassium channels and several calcium channels, while voltage-gated sodium channels are absent in bacteria, archaeans, plants, and fungi. They are first found in a preliminary form at the level of the Cnidaria (jellyfish), and "true" sodium channels are found at the level of flatworms (Platyhelminthes).

The archetype of ion channels appears to be the potassium channel of the inward-rectifier type, which is responsible for inward movement of potassium ions (Fig. 5.3). It consists of a chain of peptides that penetrates the membrane several times. In its simplest form, such a *domain* consists of transmembrane segments, which surround a central pore through which the ions can move. This primitive form of ion channel developed into a domain with six segments, with two of them forming the pore. This form then further developed into a channel consisting of four domains with six transmembrane segments each, which then became the basic type of voltage-dependent sodium and potassium channels (cf. Fig. 5.3) described further below.

The potassium channels constitute the vast majority of ion channels; more than 100 types have been described so far, and for many of them, the function is still unknown. In number, they are followed by calcium channels. There are many fewer types of sodium channels (cf. Fig. 5.3 top) which originated later in evolution, but the reasons for these differences in number are unclear. Potassium channels play a special role in the maintenance and restitution of the resting membrane potential, while sodium channels (supported by potassium channels) are mostly involved in faster changes of the membrane potential, which are responsible for the action potential, as we will learn. Besides their involvement in the formation of certain kinds of action potentials, calcium channels are highly

5.3 Ion Channels and Neural Transmission

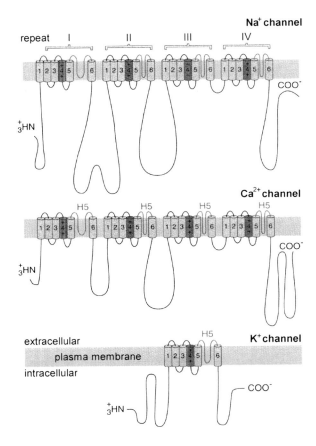

Fig. 5.3 Voltage-sensitive ion channels: *Above*, Na$^+$ channel, *middle*, Ca^{++} channel, *below*, K$^+$ channel. The Na$^+$ and the Ca^{++} channels contain four repetitive domains (I-IV) of six transmembrane segments; the K$^+$ channel contains only one domain composed of six segments. Segment 4 (*red*) functions as a voltage sensor and induces the opening of the channel. The channel pore (H5) is located between segments 5 and 6. From Dudel et al. (1996/2000)

important for the regulation of the intracellular calcium level, and this is in turn very important for neuronal plasticity and memory formation.

In eukaryotic unicellular organisms (protozoans), we mostly find calcium ions as inward carriers of electric charge and as the basis of the nerve impulse, while in cnidarians and ctenophores ("coelenterates") and in flatworms this function is taken over by sodium ions. It is assumed that sodium channels developed from calcium channels. The evolution of voltage-gated sodium channels made a fast conduction of action potentials along the axon possible, and this was a decisive step in the evolution of nervous systems and brains.

5.3.1 The Function of Ion Channels

Ion channels come in three major kinds, i.e., voltage-gated, ligand-gated, and metabotropic (Zimmermann 2013; Egger and Feldmeyer 2013). The first one, also known as *voltage-dependent* ion channels, are channels in which opening and closing is influenced by the membrane potential and its changes (Fig. 5.3). These ion channels possess a sensor for voltage changes, and as soon as the electric signal passes a certain threshold, a change in the conformation of the channel molecule takes place, resulting in an opening (or closing) of the pore. One of the best known and most important members of this group is a type of voltage-gated sodium channel that underlies the generation of action potentials in a way that we will learn in a moment. There are also voltage-gated calcium channels, which play an important role in transmitter release as well as in muscle contraction. Finally, there are many different kinds of voltage-gated potassium channels, some of which are involved in the repolarization of the cell membrane following action potentials (see below).

Ligand-gated ion channels (Fig. 5.4a) are those channels whose permeability is greatly increased when some type of chemical substance, for example a neurotransmitter, binds to the protein structure of the channel. In the nervous system, they are found at postsynaptic sites and open or close by transmitters released by the presynaptic axon terminal. There are many more ligand-gated ion channels than transmitters, because one type of neurotransmitter can bind to more than one type of ion channel, and ion channels can respond to more than one type of transmitter.

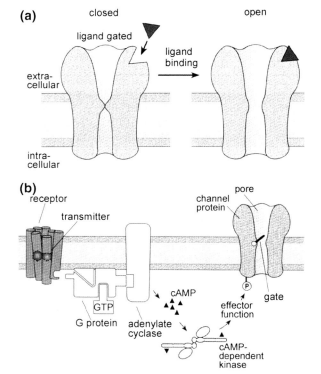

Fig. 5.4 a Ligand-gated ion channel. The channel opens when the transmitter (*black triangle*) binds to a specific site of the channel, the *receptor*. **b** Metabotropic ion channel. Here, the receptor is spatially separated from the channel. The binding of a transmitter molecule to the receptor induces a cascade of chemical processes that eventually leads to the phosphorylation and opening of the channel. Abbreviations: *GTP* guanosine triphosphate; *G-protein* guanine nucleotide-binding protein; *cAMP* cyclic adenosine monophosphate; *P* inorganic phosphate. From Roth (2003), after Kandel et al. (1991), modified

5.3 Ion Channels and Neural Transmission

Finally, *metabotropic* ion channels (Fig. 5.4b) are activated or inactivated by second messengers from inside the neuron rather than from the outside, as is the case for ligand-gated ion channels. Here, the sensor or receptor for a neuroactive substance is not located inside the cell or on the ion channels, but is spatially separated from them. The receptor is coupled to a guanine-nucleotide (GDP or GTP) binding protein ("G-protein"), which may, for example, have an excitatory (G_s) or inhibitory (G_i) effect. This in turn triggers an intracellular "second messenger cascade," e.g., by transmitting a signal to adenylate (or adenylyl) cyclase, which converts adenosine triphosphate into cyclic adenosine monophosphate (cAMP). cAMP is known as a second messenger eventually leading, via phosphorylation, to the opening of ion channels. Other intracellular pathways triggered by G-proteins involve inositol triphosphate (IP_3) and diacylglycerol (DAG) leading to changes in intracellular calcium levels. These channels are called "metabotropic," because a sequence of intracellular metabolic processes is involved.

Ion channels are capable of generating and modulating neuronal signaling. They regulate the membrane potential and make the neurons more sensible (by *depolarization*) or less sensible (by *hyperpolarization*) to an incoming signal, and they control the generation and temporal structure of an action potential or volleys of action potentials.

5.3.2 The Origin of the Action Potential

The action potential, also called "nerve impulse" or "spike," is the most important mechanism of the nervous system for fast propagation of nerve signals. It is based on a very short rise and fall, i.e., depolarization and repolarization plus hyperpolarization, of the negative resting potential, i.e., within the range of a few milliseconds (Fig. 5.5). This process is usually released by the depolarization of a part of the nerve cell membrane, e.g., the axon hillock, where voltage-gated ion channels (mostly Na^+) are concentrated, which rapidly open some of the hitherto

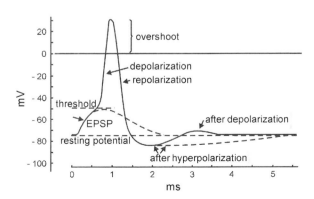

Fig. 5.5 Initiation and course of an action potential. The membrane resting potential is arbitrarily set at − 75 mV. For further explanations, see text. *EPSP* excitatory postsynaptic potential. From Roth (2003)

closed sodium channels. As a consequence, sodium ions flow into the cell, and this makes the negative resting potential less negative (i.e., more positive), *depolarizes* it. When a certain degree of depolarization, the so-called *firing threshold*, is reached (mostly around −50 mV), then the majority of voltage-gated sodium channels open rapidly in a "runaway" fashion, i.e., opening channels excite other channels to open. The following depolarization of the membrane potential may lead to a voltage of +30–40 mV more in the direction of the sodium resting potential (+55 mV; see above). However, the open sodium channels close spontaneously after an open state in the range of microseconds (they "inactivate"), which stops the influx of more sodium ions, and at the same time most of the hitherto closed potassium channels open. As a consequence, potassium ions leave the cell which, together with the closing of the sodium channels, rapidly makes the inside of the cell again negative. This is called *repolarization*. Eventually, for a short time the membrane becomes even more negative in an "overshoot" fashion called *hyperpolarization*, which is then followed by a small positivation which restitutes the resting potential. Only after the end of such a period of insensitivity, called the *refractory phase*, new action potentials can be released. Thus, during an action potential we distinguish a rising phase, a peak phase, a falling phase and a refractory phase, which together may last between 2 and 20 ms, depending on the length of hyperpolarization and repolarization, which in turn critically depends on the properties of the potassium channels involved.

The action potential typically is an *all-or-nothing* signal in the sense that whenever the depolarization of the membrane has passed the "firing threshold", the membrane potential always reaches a maximum depolarization ("peak") of about (+30)–(+40) mV and then falls back to the negative resting potential after hyperpolarization and repolarization. Thus, the amplitude of the action potential remains the same; only during a sequence of many rapidly following action potentials,s the amplitude may decrease slightly. In that sense, the action potential is a "digital signal." What changes, however, is the *frequency* of the action potential, i.e., the number of APs per second. This frequency is in part dependent on the strength of stimulation of the neuron and consequently the amount of transmitter release, which determines how fast the "firing threshold" of the postsynaptic membrane is reached, and in part on the length of the refractory period, which determines the earliest moment at which the next AP can arise. The latter is determined by the temporal dynamics of potassium channels involved in the hyperpolarization and repolarization. If fewer channels are involved, or they do their work more slowly, then—even at maximum stimulation—the postsynaptic cell fires more slowly. Thus, the entire process of the generation of action potentials is a *frequency modulation*, because the strength or amplitude of the stimulation is encoded in the frequency of action potentials.

Action potentials are usually generated at the axon hillock, where the axon originates. Here, numerous voltage-dependent sodium channels are located which open at relatively weak depolarization. Axons propagate APs in a self-generating fashion. In non-myelinated axons this happens in a way that sodium channels that open excite neighboring sodium channels which then open, too. In such a way the

5.3 Ion Channels and Neural Transmission

APs spread over the axon. In principle this could happen in both directions. However, since channels that were open just a millisecond ago are in the refractory period and cannot open immediately, the excitation runs in just one direction, i.e. away from the axon hillock and toward the axon terminals, which is called "orthodromic conduction." The speed with which APs spread along the axons called *axonal conduction velocity* depends on the thickness of the axon. In these fibers the conduction velocity increases with the square root of the diameter. This type of conduction is typical of most axons in the invertebrate nervous systems, but also regularly found in vertebrates, for example in the autonomic nervous system, and also inside the brain, where they constitute the "thin" nerve fibers. It reaches conduction velocities in the range of 0.1–1 m/s.

The situation is different when an axon is covered with a myelin sheath, which is an insulating multilamellar membrane that enwraps the axon. This myelin sheath is found mostly in vertebrates in addition to the just-described unmyelinated thin axons. The myelin sheath is interrupted at regular distances of about 1 mm, and this interruption forms the so-called nodes of Ranvier, where the axon is "naked" for a few micrometers. Here, numerous voltage-gated sodium ion channels are found. The myelin sheath prevents ions from entering or leaving the axon and forces the current to "jump" from one node of Ranvier to the next one. There, it excites (opens) the sodium channels such that a new AP is generated, and so forth. In this way, new APs are generated at every node of Ranvier, and the APs jump from one node to the other along the axon; this mechanism is therefore called "saltatory conduction." This results in a much higher conduction velocity of myelinated compared to unmyelinated axons ranging from 1 to 150 m/s and is believed to increase linearly with the diameter of the myelinated fiber rather than with the square root as in unmyelinated fibers. This is important because this type of axonal conduction is not only much faster, but also much more energy-efficient than the other type. With unmyelinated axons, high-speed conduction of APs can be achieved only with "giant fibers" with enormous diameters up to 1 mm such as are found in some invertebrates, while the same conduction velocity is reached by myelinated axons of much thinner diameters in the range of 3 μm. As we will see, this is of enormous significance in vertebrate brains containing many neurons that need to be efficiently connected.

5.3.3 Neurotransmitters and Other Neuroactive Substances

All the events surrounding the origin of the AP usually occur at the site of the axon hillock or along the axon. The situation is different at the synapse. In the subsynaptic membrane there are no voltage-gated, but ligand-gated sodium, calcium, chloride, and potassium channels that open—as mentioned above—when a certain chemical substance, for example a neurotransmitter, binds to the receptor site of the channel (Lüscher and Petersen 2013).

The most common neurotransmitters, also called "classical transmitters," in the nervous systems of vertebrates and most invertebrates are (1) amino acids, such as glutamate aspartate, gamma-aminobutyric acid (abbreviated "GABA") and glycine, (2) monoamines and other biogenic amines such as dopamine (DA), norepiphrenine (NE, also called noradrenaline, NA), epiphrenine (adrenaline, A), histamine, and serotonin (5-hydroxy-tryptamine, 5-HT) and (3) acetylcholine (ACh). Other neuroactive substances in the nervous systems are peptides (over 50 have been found so far) such as the endogenous opioids, which are often "co-released" together with the transmitters.

Inside the brain, glutamate, GABA and glycine act directly onto the receptor sites within a range of milliseconds and are therefore called "fast" transmitters. In contrast, noradrenaline, serotonin, dopamine, and acetylcholine, besides their ionotropic effect, can work as "neuromodulators," which means that they can change, i.e., enhance or dampen, the effect of the fast transmitters within a range of seconds. Accordingly, they are called "neuromodulators." Often, they are not released at the synapses directly but in their vicinity and then have a more diffuse effect. Usually, neurotransmitters are called "excitatory" (such as glutamate) or "inhibitory" (such as GABA), but this is not correct, because the effect of transmitters depends exclusively on the properties of the receptor they activate. As we will learn, acetylcholine and the other neuromodulators may act either in an excitatory or inhibitory fashion depending on the type of receptors and channels involved.

Neurotransmitters and modulators are synthesized from chemical precursors in the cell body, the axon or the axon terminal and are then packaged into synaptic vesicles in axon terminals which are clustered beneath the presynaptic membrane of a synapse. When a nerve impulse reaches an axon terminal, voltage-sensitive calcium channels open, calcium ions enter the terminal, and a complex chain of chemical reactions causes the vesicles to move toward the presynaptic membrane, fuse with it and release the neurotransmitter into the synaptic cleft (cf. Fig. 5.6). The amount of transmitter released is typically dependent on the electrical activity of the axon terminal, i.e. AP frequency. Thus, here we have another type of neuronal coding, i.e., the translation of AP frequency into amounts of transmitter molecules, which may be considered a digital-analog encoding.

The transmitter crosses the synaptic cleft and binds to its receptor in the subsynaptic membrane, activating it by changing its conformation. In case of a ligand-gated channel, this causes directly and in case of metabotropic receptors indirectly through a cascade of intracellular processes the opening of the channel and the inward or outward diffusion of ions. Many channels can be opened both ways. Depending on the nature of the channel, this will lead to a depolarization or hyperpolarization of the subsynaptic membrane. In case of depolarization due to an influx of sodium ions, this does *not* lead, as is the case at the axon hillock, to the origin of an AP, because voltage-gated ion channels are lacking, but to a *local* and *gradual* (rather than all-or-nothing) *depolarization* in the form of an excitatory postsynaptic potential, abbreviated *EPSP*.

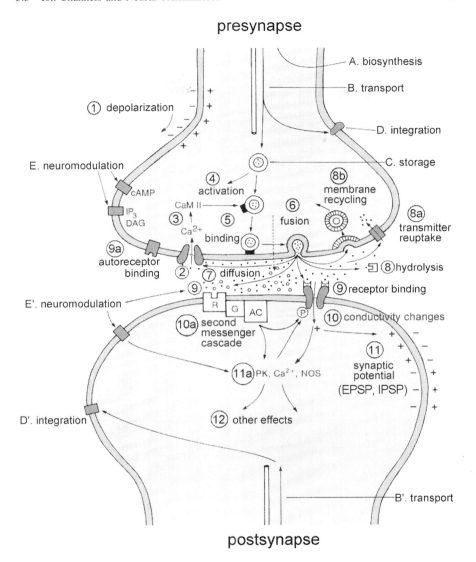

Fig. 5.6 Electrical and biochemical processes at a chemical synapse. Numbers 1–12 denote fast processes in the range of milliseconds, which occur during processing and transmission of electrical signals at the synapse. A-E and A'-E' denote processes in the range of seconds, i.e., synthesis, transport and storage of transmitters and modulators, integration of channel proteins and receptors into the membrane and modulatory effects. Abbreviations: *AC* adenylate cyclase; *cAMP* cyclic adenosine monophosphate; Ca^{2+} calcium ions; *CaMII* calmodulin-dependent protein kinase II; *DAG* diacylglycerine; *EPSP* excitatory postsynaptic potential; *G* guanine nucleotide-binding protein; *IP3* inositol triphosphate; *IPSP* inhibitory postsynaptic potential; *NOS* nitric oxide synthase; *P* inorganic phosphate; *PK* protein kinase; *R* receptor. From Roth (2003), modified

In the case of GABA receptors and channels, the efflux of potassium and the influx of chloride ions leads to a *hyperpolarization* of the membrane and an inhibitory postsynaptic potential, abbreviated *IPSP*, which for a short time makes the membrane less sensitive to subsequent excitations, i.e., inhibits it. The strength of the EPSP or IPSP is proportional to the number of excited subsynaptic receptors, which in turn is proportional to the amount of transmitter released into the synaptic cleft (which in turn is proportional to the AP frequency reaching the axon terminal). As graduated potentials, EPSPs and IPSPs are *analogous* signals as opposed to the digital AP.

This local subsynaptic excitation activates the membrane in the immediate vicinity, which in turn excites their neighbors and so forth, such that the excitation spreads over larger parts of the nerve cell membrane. However, since there are no voltage-gated sodium channels, the excitation is not self-sustaining and weakens after a short distance if it is not corroborated by an immediately following excitation coming from the same or different subsynaptic membrane. Additionally, the dendritic membrane is "leaky," which causes a drop of voltage. Therefore, it may or may not reach the axon hillock of the neuron and release APs (see below).

After a very short time, the transmitter released into the synaptic cleft detaches from the receptor and is removed from the synaptic cleft by specialized transporter mechanisms into the presynaptic terminal (called re-uptake of transmitters). Here, the transmitters are decomposed chemically and eventually resynthesized, carried into the vesicles and ready to be released again.

Let's take a closer look at the most common neuroactive substances and their effects. *Glutamate* is an amino acid and the most common transmitter found in fast excitatory synapses in the nervous systems and brains of animals, including man. Besides ordinary synaptic transmission, glutamate is involved in neuronal plasticity, learning and memory. There are two main groups of glutamate receptors. The first one is an ionotropic receptor characterized by an substance called AMPA (α-amino-3-hydroxy-5-methyl-4-isoxazolepropionic acid) or by quisqualate, because these substances are able to activate this glutamate receptor in the same way as glutamate itself (accordingly they are called "agonists"). This glutamate receptor opens an ion channel that is permeable for Na^+ and K^+, but is relatively insensitive to calcium.

The second type is called NMDA (abbreviation for *N*-methyl *D*-aspartate) receptor, which is voltage- and ligand-gated and is both a a sodium/potassium and calcium channel. Normally, this NMDA channel is blocked by a magnesium ion, and this blockage is removed only after strong depolarization, as happens in *long-term potentiation* (LTP) as one important form of synaptic plasticity (cf. Lüscher and Petersen 2013). Thus, NMDA receptors and channels play an important role in plasticity, learning and memory formation, especially because the opening of the NMDA channels leads to a strong influx of calcium ions into the neuron, which then activates intracellular processes that may lead to functional and structural changes of synaptic coupling.

GABA is an amino acid, too, and is synthesized from glutamate. It is the dominant transmitter in fast inhibitory synaptic transmission throughout the brain

5.3 Ion Channels and Neural Transmission

and is effective via different receptor complexes, of which only $GABA_A$ and $GABA_B$ will be briefly discussed. The binding of GABA to the $GABA_A$ receptor directly opens (ligand-gated) chloride channels and induces a fast hyperpolarization or inhibition. This receptor has a binding site for alcohol, inhaled anesthetics and psychoactive substances like benzodiazepines and barbiturates, which have a sedative and anxiolytic effect, while an overdose may lead to death. The binding to $GABA_B$ opens metabotropic potassium channels and induces a "slower" hyperpolarization or inhibition. Both effects, however, are in the range of milliseconds. GABA receptors are often found on presynaptic terminals, which release the neuromodulators dopamine and serotonin and the neuropeptide substance P and are capable of modulating the release of these modulatory transmitters. While GABA is found in the brain of vertebrates (and invertebrates), the corresponding major hyperpolarizing transmitter in the spinal cord is glycine.

The receptor for the neurotransmitter acetylcholine (ACh) comes in two types that play very different roles in the nervous system. The first one is called the *nicotinic* acetylcholine receptor which is sensitive to the plant alkaloid nicotine. It is a ligand-gated receptor found in the central as well as autonomic nervous system and at the neuromuscular junction, i.e., the connection of motor nerves to muscles. The well-known "arrow poison" curare, the snake venom alpha-bungarotoxin and other animal poisons block ("paralyze") the transmission of ACh at these neuromuscular synapses and lead to complete immobility and death. Inside the brain and in the so-called parasympathetic nervous system, ACh binds to the metabotropic *muscarinergic* acetylcholine receptor, which is sensitive to the substance *muscarine* (produced by the mushroom *Amanita muscaria*). Here it works as a neuromodulatory transmitter, e.g., in the basal forebrain, which is involved in learning and attention.

Dopamine (DA) belongs to the catecholamines. It is synthesized from the amino acid L-tyrosine via the substance L-DOPA and is itself the precursor of the transmitters adrenaline (or epinephrine) and noradrenaline (or norepinephrine). DA is involved in many processes of the brain, including cognitive functions in the context of learning, attention, motivation and the expectation and assessment of reward and punishment as well as in the control of voluntary actions. DA is produced in the *ventral tegmental area* (VTA) and the *substantia nigra pars compacta*, both located in the midbrain tegmentum, and, among other nuclei, in the nucleus arcuatus of the hypothalamus.

DA affects at least five different receptor types, all of which are metabotropic. Only the D_1-and D_2-receptors will be considered here. D_1-receptors are located exclusively postsynaptically. Their activation leads to an increase in the intracellular calcium level and induces excitation. D_2-receptors are found both presynaptically (as autoreceptors) and postsynaptically, and their stimulation leads to an increase in the efflux of potassium ions and as a consequence to an inhibition.

Noradrenaline (or norepinephrine) is a catecholamine like DA and synthesized from it by hydroxylation. Inside the vertebrate brain, it is produced predominantly in the locus coeruleus of the brainstem (cf. Chap. 10). Noradrenaline is absent in the insect nervous system, where the transmitter octopamine plays an equivalent role in many functions including cognition. Noradrenaline, like adrenaline (or epinephrine), binds to α-and β receptors. Binding to the α_1-subtype increases, via a G_q protein and several other chemical processes, the intracellular Ca^{++} concentration from intracellular stores as well as via the opening of voltage-gated Ca^{++}-channels. The dominant effect of binding to the α_1-subtype occurs in the periphery, not in the CNS, and plays a crucial role in the stimulation of heart contraction and the cardiovascular system, but is also involved in the regulation of hormone secretion, e.g., in the context of stress reaction. Binding to the α_2-subtype activates an inhibitory G (G_i) protein and adenylyl cyclase and thereby inhibits the production cAMP. Other subtypes, via G_o, suppress voltage-sensitive calcium channels, and a third subtype stimulates potassium channels, which either directly or indirectly, via a second-messenger cascade, opens K^+ channels leading to inhibition. Stimulation of the β receptor activates a stimulating G (G_s) protein, which either directly activates a Ca^{++} channel and has an excitatory effect, or indirectly via a second-messenger cascade activates K^+ channels having an inhibitory effect. This leads to a modulation of the autonomic system.

Serotonin (5-hydroxytryptamine, 5-HT) is a monoamine neurotransmitter synthesized from the amino acid tryptophan. The vast majority of 5-HT (90 %) is produced in the gut; the rest is produced in the brain, in vertebrates predominantly in the raphe nuclei of the brainstem, from where fibers of serotonergic neurons project to virtually all parts of the brain including the cortex, and here predominantly to the ventral prefrontal areas (cf. Chap. 10). There is a large number of different 5-HT receptors in the brain, which, with the exception of the 5-HT$_3$ receptor (a ligand-gated ion channel), are metabotropic receptors affecting ion channels by coupling to G proteins that activate intracellular messenger cascades. The 5-HT$_1$ receptors have an inhibitory effect via decreasing cellular levels of cAMP, while 5-HT$_2$ receptors have an excitatory effect.

Often, more than just one type of transmitter or neuropeptide is released at a synapse, which is called "co-release." This allows for more complex effects in synaptic transmission. For example, in neurons in the corpus striatum, GABA is co-released with endogenous opioids or substance P. Similarly, GABA and glycine, dopamine and glutamate, ACh and glutamate or the vasoactive intestinal peptide (VIP) are co-released.

5.4 Principles of Neuronal Information Processing

Neurons are the basic elements of neuronal information processing (Druckmann et al. 2013). They receive and generate signals, filter, amplify and diminish them, and control their spatial and temporal properties and propagation. Most neurons have an extended dendritic tree and an axon (or more than one) and are connected with thousands or tens of thousands of other neurons via synapses forming *neuronal networks*. Thus, the activity of each neuron is directly influenced by the activity of many other neurons in a highly complex fashion.

5.4 Principles of Neuronal Information Processing

Two important principles are *spatial* and *temporal summation* in the case of depolarizing synapses. Each synapse of a neuron depolarizes, via an EPSP, the postsynaptic membrane of the contacted neuron by much less than 1 mV, in motor neurons, for example, by 0.2–0.4 mV. Normally, such a tiny EPSP would not make it to the axon hillock in order to release an AP, for which at least a depolarization of 10 mV is necessary. How far a decaying depolarization *passively* spreads over the cell membrane before dissipating depends critically on the *length constant* (e.g., signified by the Greek letter "lambda") of the membrane, which in turn depends on the resistance of the membrane as well as of the inner structure. The larger the lambda, the farther the spread of an EPSP over the membrane toward the axon hillock and the greater its contribution to the origin of an action potential.

As just mentioned, normally one depolarization is not sufficient to induce an action potential. In case of a motor neuron, at least 50 synapses must fire simultaneously in order to sufficiently depolarize the membrane of the axon hillock. The contribution of one synapse to the origin of an AP likewise depends on its distance from the axon hillock: the farther away from the axon hillock the synapse impinges on the neuron, e.g., on so-called distal dendrites, the less effective the EPSP generated by it even at the same length constant. This means that under normal conditions, either one neuron must impinge on that neuron with many synapses in order to fire another neuron, or that many neurons must connect to that neuron that fires simultaneously. This effect is called *spatial summation*.

Temporal summation is based on the fact that EPSPs from one synapse that follow each other sufficiently fast, i.e., before each EPSP decays completely may add up. This is determined by the membrane *time constant* (e.g., signified by the Greek letter "tau"), which is critically dependent on the membrane resistance and capacitance. If the time constant is large enough, EPSPs build up sufficiently high depolarization that makes it to the axon hillock. This effect of temporal summation can occur, at least in principle, through just one synapse, or—under normal conditions—via many synapses from several neurons. Obviously, neurons with a long membrane length constant and a large membrane time constant carry EPSPs more readily toward the axon hillock and increase the probability of neuron firing, while in those with a short length and a small time constant, there is strong resistance against stimulation by other neurons. In this way, neurons may exert *strong filter functions* with respect to the inflow of excitation.

Besides excitation, synapses can also have an inhibitory effect on postsynaptic neurons. These synapses can block excitation induced by excitatory synapses from running to the axon hillock, and this effect is the stronger the closer inhibitory synapses are to the axon hillock. At the bifurcation point of primary dendrites they can block the entire excitation of that part of the dendritic tree, and at least in vertebrates they can exert the strongest influence when they impinge at the soma, i.e., very close to the axon hillock. Thus it is not surprising that on average, excitatory synapses are found more at "distal" parts of the dendritic tree i.e., away from the axon hillock, and inhibitory synapses more "proximally," closer to the axon hillock. Thus, the ratio between excitatory and inhibitory synapses, their individual strength and the site, where they impinge on the nerve cell is one of the most important factors for the integrative ability of a nerve cell. Depending on their contact site, inhibitory synapses can finely tune the excitatory work of neurons. Final summation takes place at the axon hillock and determines whether or not APs are released at all and with what frequency.

At a finer level, there are numerous other factors that control the flow of excitation throughout nerve networks, most of which concern synaptic transmission. The most important of these factors are (1) the amount of transmitter released at the presynaptic site, (2) the sensitivity of the subsynaptic membrane to the transmitter, and (3) the time of presence of the transmitter within the synaptic cleft. The first factor depends on the strength of excitation of the presynapse by incoming APs and the availability of transmitters, which in turn depends on the rate of its synthesis and the number of transmitter vesicles that move toward the presynaptic membrane, fuse with it, and release the transmitter into the synaptic cleft.

The second factor depends on the number and sensitivity of transmitter-specific receptors and ion channels in the subsynaptic membrane: the higher their number and their sensitivity, the more effective the transmitter molecules that bind to them and open the ion channels, and the stronger the EPSP or IPSP. These two processes may vary independently, i.e., the number of receptors may decrease, while the sensitivity of the receptors increases, or vice versa.

The third factor depends on the efficacy of the specific re-uptake systems, because they determine how long a transmitter remains in the synaptic cleft and binds to their receptors. A final important factor is the speed of decomposition and resynthesis of the transmitters. Many psychopharmacological drugs are targeting these two latter processes, i.e., they prolong the presence of certain transmitters, e.g., serotonin, in the synaptic cleft by slowing down the work of the transporter mechanism, or they speed up the re-synthesis of the transmitter.

Inside the nervous system and the brain, from several to many neurons usually form smaller or larger networks or *assemblies* by often reciprocal synaptic contacts. These networks have inputs from other networks (e.g., sensory input) and an output via projection neurons. There is *convergence* and *divergence* inside these networks in the way that one neuron collects excitatory and/or inhibitory input from many (sometimes thousands) of other neurons and spreads, in turn, its activity via synaptic contacts to other neurons. The combination of convergence and divergence is one decisive basis for information processing inside the nervous system and brain, because on the one hand different kinds of information can converge onto one neuron and are thus *integrated* in different ways depending, among others, on temporal and spatial summation properties of that neuron. On the other hand, neurons can *distribute* their activity to thousand or hundreds of thousands of other neurons. The proportion between the degree of convergence and divergence can vary enormously. In tiny brains (e.g., those of small worms), the nervous system is mostly dominated by convergence, e.g., of information from sense organs, onto motor centers, while large brains usually reveal a much higher degree of divergence than of convergence, which means that the information processing between sensory and motor parts is much more complex. In the human brain, divergence is at least five orders of magnitude larger than convergence.

Equally decisive for information processing in the nervous system and brain is the ability of at least some neurons to undergo short-term and/or long-term changes of their processing properties mostly regarding the properties of synaptic transmission, but also propagation of EPSPs along dendrites (i.e., temporal and spatial summation). This *neuronal plasticity* is the basis for learning and memory and is a prime example for the phenomenon discussed in the previous chapter that components of a system may undergo changes while interacting with other components. This has the consequence that nervous systems and brains continuously change with each interaction of their components.

5.5 What Does All This Tell Us?

Organisms are highly complex electrical-chemical machines. Accordingly, inside the nervous system and brain, there is always an interaction between chemical and electrical transmission. The basis of signal generation, processing and transmission is the fact that nerve cell membranes, like all biological membranes, exhibit an electrical potential or voltage because of asymmetrical distribution of ions at the inside and the outside of the membrane. Due to changes in the distribution of ions, either a local potential originates that moves relatively slowly over the membrane, or an action potential that is transported relatively fast via an axon to another nerve cell or to an effector organ such as a muscle or a gland.

Particularly impressive is the large variety of chemical substances subserving chemical transmission and of receptors and ion channels and related intracellular signaling pathways. Transmitters can either work locally, like the "fast" transmitters glutamate and GABA, or more globally, like the neuromodulators (e.g., acetylcholine, serotonin, dopamine), which are often released in large parts of the brain by a widely distributed pattern of axons. In this way, a given information can be sent to many parts simultaneously, and local specificity is then reached via specific receptors and their spatial distribution. At the same time, most neuromodulators can have either an excitatory or an inhibitory effect, depending on the receptor types and ion channels affected, and in this way modulate the activity of the fast transmitters.

Usually neurons do not occur singly inside the brain, but form smaller or larger assemblies that can often be distinguished anatomically and are called *ganglia* (mostly in the nervous systems and brains of invertebrates and in the peripheral nervous system of vertebrates) and *nuclei* (mostly in the brains of vertebrates). In some parts of laminar structures, such as the cerebral cortex of mammals, *areas* are distinguished. These structures act as a unity, because their neurons are connected more densely with other neurons inside than outside the ganglia, nuclei, or areas. The latter, in turn, are connected with other (often many) ganglia and nuclei, and quite frequently in a reciprocal manner, and this happens not only (although more often) with ganglia, nuclei and areas located in one part of the brain, but also with those located in different parts, which requires long "ascending" (mostly sensory) or "descending" (mostly premotor or motor) fiber pathways.

Thus, we recognize an anatomical and functional hierarchy beginning with single cells, then cell assemblies in ganglia, nuclei or areas within major parts (e.g., midbrain) or subparts of the brain (e.g., tectum) and eventually the brain as a whole. The processing of information within and between such hierarchical levels proceeds in the divergent-convergent and parallel fashion, as described above, plus many recurrent pathways. While the information processing abilities of single neurons are relatively well known, the interaction of cell assemblies within and among nuclei, ganglia and areas is not sufficiently understood. Even small populations consisting of only a few hundred densely connected neurons may produce such highly complex activity, which at present can neither be elucidated nor mathematically described in detail, because there are no adequate mathematical tools for that task. At a gross level, we are able to simulate fairly well the work of small assemblies, but *how* a certain pattern of activity is generated by the interaction of single cells is not well understood.

This, however, is common to all complex systems (one may think of the weather) and leads to phenomena that philosophers tend to call "emergent" in the sense of something unexpected or enigmatic, and the larger the neuronal networks, especially when forming information processing hierarchies, the more unexpected and "enigmatic" do the results of their interaction look like. In this context, the ability of neurons for *plasticity*, i.e., short- and long-term changes of its information processing and anatomical properties, plays a fundamental role. Neural plasticity is an excellent example of the phenomenon that within a system the components modify each other through their interaction such that the properties of the entire system change continuously. Given the incredible complexity even of small brains, like that of the honeybee, we tend to greatly underestimate their achievements and hasten to regard them as something inexplicable or even mystical.

Chapter 6
Bacteria, Archaea, Protozoa: Successful Life without a Nervous System

Keywords Bacteria · Archaea · *E. coli* · Flagellar motor · Protozoans · *Paramecium* · *Chlamydomonas* · Origin of multicellular organisms

In biology it is commonly assumed that for a successful life it is advantageous for an organism to have a complex nervous system, because an increase in neural complexity appears to be adaptive. How should we otherwise explain a seemingly universal increase in complexity during evolution? That this assumption is not generally valid is easily demonstrated by the fact that the most successful organisms in terms of duration of its existence, number of species, and diversity in ecology and physiology, i.e., the unicellular organisms, by definition have no nervous system at all. In this chapter, we will ask how they manage to survive despite this fact—and perhaps better than any other organism.

6.1 Bacteria and Archaea

About 3.6 billion years ago (or even earlier), the evolution of living beings began with unicellular organisms without a cell nucleus—in Greek "karyon," and therefore called "prokaryotes." These organisms most probably resembled the extant bacteria. Bacteria are not only the simplest, but also the most diverse organisms. About 9,300 bacterial and archaean species have been described, but estimates of the species number range from 10 million to 1 billion. Like all living beings, they supply themselves with matter and energy and for that they need to recognize and localize potential sources of food, identify adverse events and protect themselves against them, avoid or pass obstacles, etc. Trivially, bacteria as unicellular organisms have no nervous system, but they, of course, possess mechanisms for stimulus recognition and information processing, which already function according to the principles of more complex organisms (cf. preceding chapter). They even have a sort of memory, as we will learn in this chapter. This contradicts the common view that these very simple unicellular organisms are nothing but "reflex machines." True reflex machines would not be able to survive, because at least short-range changes of behavior as adaptation to changing environmental conditions are a necessary prerequisite for survival.

A well-studied bacterium is *Escherichia coli*, which populates our gut in incredible numbers. *E. coli*, as it is mostly called, measures only a few micrometers and is therefore invisible to the naked eye and weighs only 1 picogram (Berg 1999; Alberts 2002). It has a 30 nm thick lipopolysaccharide cell membrane surrounding a cytoplasm, in which as genetic substance only one strand of DNA is found that is not enclosed in a cell nucleus. The membrane of *E. coli* carries more than a dozen types of chemoreceptors, which serve to recognize food and other substances such as sugar or amino acids, but also toxic substances like heavy metals, and they possess mechanoreceptors by which the bacterium can detect obstacles.

The information gathered about the environment through these receptors guides the behavior of the organism, i.e., its movement pattern. Like many other bacteria, *E. coli* moves by means of rotating *flagella*, while others such as the spirochetes "screw" themselves through their environments, and still others move by creeping over surfaces. The flagella, of which *E. coli* has six, are 15–20 nm thick protein filaments, which can rotate either clockwise or counterclockwise. They are driven by a flagellar motor located inside the membrane with a diameter of only 45 nm. This motor is composed of 20 different kinds of parts. It is driven by proton current like a turbine and can spin at a maximum of several hundred hertz. To move forward, the flagella join to form a "superflagellum" that rotates counterclockwise in a joint fashion, but as soon as it disintegrates and the single flagella rotate clockwise independently, the bacterium "tumbles" in place for a fraction of a second and randomly changes its orientation. In the next moment, the bacterium starts swimming again, because the flagella re-integrate into a superflagellum that begins to rotate counterclockwise synchronously, but now swims in another direction.

Movement direction is controlled by chemoreceptors like one for glucose or aspartate signaling "food." The receptors "test" whether food concentration increases or decreases, and in the former case the bacterium keeps swimming in the same direction, while in the latter it "tumbles" and eventually swims in another direction. This is repeated, until the chemoreceptors again signal an increase in food concentration. In this way, *E. coli* is able to move along food gradients. In a similar manner, gradients of toxic substances are detected, and the bacterium moves away from them. Mechanoreceptors are stimulated by touching an obstacle, and this induces tumbling, such that the bacterium moves in another direction and eventually swims away from the obstacle.

E. coli does not possess spatial orientation, which means that it cannot tell front from back, above from below, or sense the direction of its movement, let alone measure distances or its own velocity. Its orientation is merely based on temporal information in the sense that the receptor activity at a given moment is compared with that of three seconds ago. This yields information about possible changes within this short period of time, e.g., about changes in the concentration of certain substances. In order to do so, *E. coli* possesses a short-term memory, which lasts only for a few seconds, but is sufficient to orient the organism in its environment.

6.1 Bacteria and Archaea

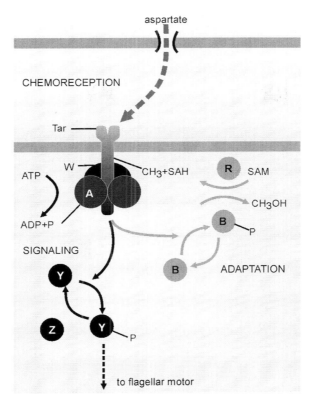

Fig. 6.1 Control of chemotaxis in the bacterium *Escherichia coli*. The figure depicts the major processes of chemotaxis elicited by the amino acid aspartate. The receptor complex consists of two molecules, W and A, with Tar spanning the cytoplasmic membrane. When an aspartate molecule binds to the Tar chemoreceptor, the Y molecule is phosphorylated (Y-P), and this activates the flagellar motor. Adaptation occurs via a methylation-de-methylation process (*right*). For further explanations, see text. Abbreviations: *ATP* adenosine triphosphate (phosphate donor); *SAM* S-adenosylmethionine a (methyl donor <), *ADP* adenosine diphosphate, *SAH* S-adenosyl homocysteine; *CH₃* methyl group, *CH₃OH* methanol, *P* inorganic phosphate. After Berg (1999), modified

In *E. coli*, we already find a separation of *sensorium* and *motorium*—a separation found in all multicellular animals At the same time, both subsystems need to communicate, at least in one direction, i.e., the sensory parts have to tell the motor parts what to do next. In multicellular organisms (except for sponges), this happens through nerve cells, which trivially are absent in bacteria. So how does the information about the environment reach the flagellar motor in *E. coli*?

As illustrated in Fig. 6.1, this happens by means of an intracellular communication pathway, which has been studied in greater detail in the context of stimulation by aspartate. Aspartate molecules in the environment of the bacterium bind to a specific "Tar" receptor. This receptor complex consists of two pairs of molecules, *CheA* and *CheW*, and can assume two states, i.e., one that eventually leads to clockwise rotation flagellar motor inducing *tumbling*, and the other one leading to counterclockwise rotation inducing *forward swimming*. The more aspartate molecules activate the receptor, the longer the period of forward swimming, while a decrease in aspartate concentration and stimulation of the receptor induces tumbling. Chemically, this happens in the way that the protein *CheY* is in a *phosphorylated* configuration (Y-P), binds to a protein complex of the flagellar motor, and induces the clockwise rotation and with this tumbling, while the absence or termination of such a *CheY-P*-binding (Y) induces the counterclockwise rotation and forward swimming. *CheA* now influences this phosphorylation-dephosphorylation process of *CheY* and inforces dephosphorylation. This causes a prolonged forward swimming along the aspartate gradient.

But how does *E. coli* know that it is swimming in the right direction? In order to do so, it must test whether the aspartate concentration increases or decreases. This requires the measurement of a change in concentration. At the same time, the receptor must always remain optimally sensitive, even at major changes of the concentration of the substance under question. Both problems are solved by an ingenuous chemical process which is illustrated in Fig. 6.1. It is based on the fact that the degree of methylation of the glutamate resting at the receptor (the methyl group CH_3) is modified by attaching SAH. SAM serves here as the "donor" of SAH (Fig. 6.1, *right*). The Tar receptor is the more susceptible to aspartate, the higher the degree of methylation of the glutamate. The receptor can control the methylation process via an inhibition of a methyl transferase B (i.e., an enzyme that transfers a methyl group) and this happens the stronger, the more the receptor is activated. At increasing aspartate concentration, the receptor becomes less and at decreasing concentration more susceptible. Thus, the receptor possesses its own negative feedback or adaptation mechanism. Since this feedback of methylation and demethylation takes a few seconds, it creates a "short-term memory" of what happened a moment ago. Similar things happen at the recognition of other nutrients or toxic substances.

We can interpret this entire process as the simplest example of goal-directed behavior known in nature. It is goal-directed in the sense that it guarantees—at least for a short time—survival of the organism by enabling it to approach that which promotes survival (i.e., food) and to avoid that which is aversive (i.e., toxic substances). *E. coli* has neither a nervous system nor reason nor insight, and its behavioral repertoire is of the greatest simplicity. But it has a short-term memory, although it is—in contrast to protozoans—unable to "keep in mind" something for more than a few seconds and it cannot learn, i.e., acquire a new behavior based on individual experience. If there are more substantial changes in behavior, this happens by changes over generations, and not during the lifetime of an individual. Yet *E. coli* is one of the most successful organisms on earth.

Other prokaryotes have developed more complicated sensorimotor mechanisms. One of these is *Halobacterium salinarum*, which lives in salt meadows or ponds. Despite its name, it is now considered to be an archaean and not a bacterium. It possesses a rudimentary visual system and absorbs light in the orange part of the spectrum. The molecular structure of the light-sensitive pigment, the bacteriorhodopsin, is similar to the rhodopsin found in the retina of vertebrates, and, as in the vertebrate retina, its retinal part undergoes a conformation change upon absorption of a photon. This photosensitive mechanism acts on the motor apparatus as the chemoreceptors do in *E. coli* enabling the organism to detect and orient toward light, i.e., show phototaxis. The archaea, to which *Halobacterium* belongs, are interesting because many of them live under extreme conditions, e.g., at very high or very low temperatures, at high pressures or, like *Halobacterium*, at strong salt concentration or in a very acidic environment.

6.2 Protozoa

A large step in orientation to the environment and with this, in cognitive and executive functions, is found at the transition from prokaryotes to eukaryotes. Eukaryotes have a nucleus and specialized cell organelles such as mitochondria and—in plants—chloroplasts. While chloroplasts have the important ability to gain energy from sunlight which is then used for the splitting of water molecules into hydrogen and oxygen, mitochondria are the "power plants" of the eukaryotic cells, because they oxidize glucose to produce the energy-rich molecule adenosine triphosphate (ATP) (see Chap. 4).

Protozoans are much larger than the bacteria or archaea, but still so small that only few of them can be seen with the naked eye. They live a single life or form colonies or pseudo-multicellular organisms. They reproduce either asexually by mitotic cell division or by sexual reproduction with a fusion of the two nuclei. Many of them are autotrophic, mostly phototrophic, while others are heterotrophic, i.e., live on organic matter from plants or animals. Again, others conduct a predatory lifestyle, i.e., they "eat" other animals. Among protozoans we find more plant-like and more animal-like organisms and lifestyles. They move either by means of cilia, flagella, or pseudopodia.

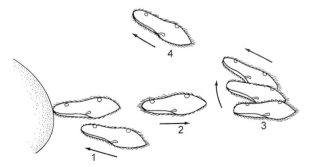

Fig. 6.2 Behavior of the protozoan *Paramecium* encountering an obstacle. Numbers 1-4 indicate the different behavioral responses: *1* hitting an obstacle, *2* backward swimming, *3* slight turn, and *4* forward swimming in another direction. After Hille (1992), modified

The system for energy acquisition found in eukaryotes is much more efficient than that of prokaryotes and has been retained in all multicellular organisms. This allows a more complex behavior based on novel sensory and motor mechanisms (Armus 2006). Many protozoans, such as the ciliate *Paramecium* (Fig. 6.2), possess new mechanisms for movement in the form of cilia, with a construction fundamentally different from the flagella of bacteria. Cilia have a surface (pellicle) that is equivalent to the plasma membrane of eukaryotes and encloses a fiber bundle consisting of nine pairs of microtubuli surrounding a central pair of microtubuli. This $9 + 2$ structure of motile cilia is found in all multicellular organisms. In *Paramecium*, cilia cover the entire surface of the egg-shaped body. In contrast to the flagella of bacteria, which function as a propellor and rotate in two directions as described above, the cilia work like a row, beating forward and backward. Wave-like synchronized backward beating of the cilia results in forward movement and synchronized forward beating induces backward movement. However, since the cilia beat at a 120° (instead of 180°) angle, *Paramecium* moves by spiraling through the water on an invisible axis.

Paramecium has an oral groove at its front end that collects food, mostly bacteria, algae and fungi, and transports it by means of cilia to the "mouth," and an anal pore as well as a contractile vacuole for osmoregulation, which is connected via tiny canals to the surface. Upon stimulation, *Paramecium* ejects trichocysts, which are organelles that release long filamentous proteins that may capture prey or serve to anchor the organism on a surface.

When *Paramecium* hits an obstacle, it moves back, turns slightly, and moves forward again. It will repeat this process, until it can get past the object (Fig. 6.2). For this strategy, the *Paramecium* needs an appropriate sensory mechanism based on the electric potential of its membrane. If the membrane has a negative potential due to the work of open potassium channels, the *Paramecium* moves forward. When it hits an obstacle, mechanosensitive calcium channels open which depolarize the membrane and reverse the beat of the cilia inducing backward swimming. Here, a calcium action potential is generated. After about one second, the calcium channels close, the potassium channels open again, and the membrane is repolarized. As a consequence, the cilia beat backwards and the *Paramecium* swims forward again. If it hits an object with its rear end, potassium channels located there open, inducing a hyperpolarization and forward movement (Hille 1992).

The protozoan *Chlamydomonas* belongs to the green algae and possesses a simple visual system based on light-gated ion channels, called channelrhodopsins, which are most sensitive in the blue-green wavelength range around 488 nm. Absorption of light of that wavelength induces a change in retinal conformation as happens in the vertebrate retina. *Chlamydomonas* has two cilia located at the front end with the same "$9 + 2$" double-microtubuli structure like that found in *Paramecium*. They beat forward and backward and are under the control of the "visual system" enabling phototaxis, i.e., movement toward a light source. As all unicellular organisms, *Chlamydomonas* likewise possesses chemoreceptors enabling *chemotaxis*, which, however, differs in function from that of bacteria. While bacteria are too small to make use of chemical gradients directly, but measure them by means of the short-term memory as described above, eukaryotic cells are much larger and can make use of chemoreceptors that are distributed all over their surface and are stimulated differently by a chemical gradient. In this way, protozoans can move along the gradient via direct perception and not by trial and error as does *E. coli*.

Not all protozoans use cilia for movement, but many of them, like the amoeba, creep forward by extending finger-like structures called pseudopodia. Chemical gradients are detected via chemoreceptors, which induce, via intracellular signal cascades, the contraction of small actin filaments inside the organism and the "ameboid" movement patter.

It has been discussed to what extent protozoans can learn—in contrast to bacteria and archaea. It appears that they are capable of non-associative learning such as habituation and sensitization, but alleged evidence for associative learning in the context of tube-escape behavior could not be verified.

6.3 Why Did Multicellular Organisms Evolve?

As discussed above, prokaryotic and eukaryotic unicellular organisms were and still are extremely successful. This leads to the question of why multicellularity evolved at all. Interestingly, multicellular organisms apparently have evolved several times independently, i.e., within the prokaryotes in the Actinobacteria, Cyanobacteria, and Myxobacteria, and within the eukaryotes with the origin of plants, fungi and animals, and volvocine green algae (Rokas 2008). As advantages of multicellularity, authors have suggested (1) reducing predation, (2) improving the efficiency of food consumption, (3) facilitating more effective means of dispersal, (4) limiting interactions with noncooperative individuals, and (5) functional specializations. First cases of multicellularity may have been filamentous Cyanobacteria (previously called "blue algae"), which already had distinct cell types and appeared about 2.5-2.1 bya, multicellular Actinobacteria which may have originated 2.0-1.9 bya, and multicellular Myxobacteria which originated much later, around 1 bya. The first unicellular eukaryotes appeared 1.8-1.2 bya, and the first multicellular eukaryotes no later than 1.2 bya, and the ancestors of plants, animals, and fungi presumably appeared around 1 bya.

As the basis of animal multicellularity and development, a "genetic toolkit" is believed to have evolved regarding three key processes, i.e., cell differentiation, cell communication, and cell adhesion. These include (1) the Hox *gene transcription factors* determining the basic structure of the organisms, e.g., segmentation, (2) the *Wnt signaling pathway*, which is a network of proteins that passes signals from receptors on the cell surface to DNA in the cell nucleus in the context of cell-cell communication in the embryo and adult and receptor tyrosine kinases working as cellular "switches," and (3) gene families of cadherins and integrins involved in cell adhesion. The genomes of bilaterally symmetric animals are characterized by rather similar toolkit sets. Cell adhesion molecules were probably present before the origin of animals, while cell signaling and cell differentiation genes evolved at the same time or soon after (Rokas 2008).

As to the origin of multicellular animals, it is now believed that this happed via "evolutionary radiation" in the sense that sponges, cnidarians-ctenophorans, and bilaterians originated within a very short time, making the exact relationship between the early branches of animals difficult to determine (Rokas 2008).

6.4 What Does All This Tell Us?

Prokaryotes, i.e., bacteria and archaeans, are the simplest organisms, and they exist since the beginning of biological evolution. Despite their apparent simplicity, they are equipped with relatively complex mechanisms for orientation in their environment and consequent survival. We find sensory receptors for nutrients and toxic substances, a separation between sensorium and motorium and finally a short-term memory serving the detection of chemical gradients. At the level of eukaryotic protozoans, many of the mechanisms of cellular signal recognition and processing (including action potentials) as well as cellular movement mechanisms (ciliary, ameboid), which are found at the level of multicellular organisms, are already present. No wonder, because these metazoans are nothing but assemblies of eukaryotic unicellular organisms. As we saw in the previous chapter, the "language of neurons" developed much earlier than the nervous systems and brains.

Multicellularity has evolved continually among bacteria and eukaryotes, the latter giving rise to multicellular plants, fungi, and animals. The key events for multicellularity were the evolution of genes responsible for cell-cell signaling and communication, cell differentiation, and cell adhesion, which remained more or less the same since one billion years ago.

Prokaryotes, as living beings, clearly exhibit a behavior which some philosophers as well as biologists tend to describe as "teleonomic" in the sense of goal-directedness or purposefulness of behavior. They hasten to distinguish "teleonomy" from "teleology," which implies actions that are planned or intended (cf. Mayr 1974), but even the former notion of "teleonomy" is based on the attribution of certain internal states that "drive" an organism to do something. Since at least in the case of *E. coli* we can nicely identify (most of) the molecular mechanisms underlying the alleged goal-directed behavior, there is no such "drive" beyond the interaction of molecules. The proponents of "teleonomy" either have to consider such kinds of behavior as "strongly emergent" (which some authors do indeed) or accept that, in a panpsychistic sense, the constituent molecules are already goal-directed. At least in the case of *E. coli*, both assumptions clearly violate the parsimony principle, i.e., they assume something that is superfluous for the description of behavior of the bacterium.

There remains, of course, the question of at which level of the organization of animals and their nervous systems or brains we are allowed to speak of "intentionality" in a non-metaphysical sense. If we do not want to restrict ourselves to a

6.4 What Does All This Tell Us?

radically behavioristic position (i.e., reject any speculation about internal states), we could argue that for intentionality we would need the presence of long-term memory, evaluation, and motivational systems and internal representations of future events. Following these lines, protozoans would not be regarded as being "intentional," honeybees perhaps, and birds and mammals for certain. Intentionality, therefore, could be regarded as a slowly emerging phenomenon.

Chapter 7
The "Invertebrates" and Their Nervous Systems

Keywords Invertebrate-protostome nervous systems · Sponges · Coelenterates · Lophotrochozoa · Annelids · Mollusks · Cephalopods · *Octopus* brain · Ecdysozoa · Nematods · Arthropods · Chelicerates · Insects · Honeybee · Mushroom bodies

In the preceding chapter, we have dealt with prokaryotic and eukaryotic unicellular organisms and their "struggle for survival." In this chapter, we will have a look at multicellular organisms, the *Metazoa*, which originated about 1 bya. The most widely accepted phylogenetic relationship of metazoans today is given in Fig. 7.1. Metazoans are fundamentally divided into those that *do not* exhibit a bilateral organization and those that *do* so. Representatives of the former group, the non-bilaterians, are sponges and "coelenterates," while all other animals are *bilaterians*. Bilaterians and non-bilaterians, therefore, have to be regarded as sister groups taxonomically. The bilaterians are basically divided into *protostomes* and *deuterostomes*, including the *vertebrates*. There is a common division of all metazoan animals into "invertebrates" and "vertebrates," which from a strict taxonomic point of view is incorrect, because "invertebrates" simply means "animals without a backbone." Strictly speaking, this is a "paraphyletic group" (like the group "reptilian;" cf. Chap. 3), because besides sponges, "coelenterates" and all protostomes, it would include deuterostome taxa as well, such as echinoderms, hemichordates, cephalochordates, urochordates, and even myxinoids, because the latter likewise have no backbone. However, the term "invertebrates" for the protostomes plus (at least) cnidarians and ctenophorans is used quite often, and I will do that occasionally in this book.

In this chapter, we will first deal with sponges and "coelenterates" (i.e., cnidarians and ctenophorans) and then with the largest group of animals, the protostomes.

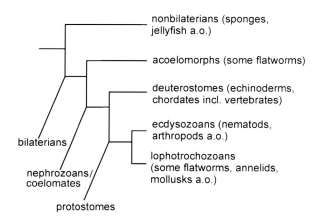

Fig. 7.1 Phylogeny of metazoans, i.e., multicellular organisms. Metazoans are divided into non-bilaterally organized organisms such as sponges and coelenterates, and bilaterally organized ones, which include organisms *without* a secondary body cavity or "coelom" (*Acoelomorpha*) and those *with* a coelom (*Coelomata*)

7.1 Non-bilaterians

7.1.1 Sponges

Sponges (*Porifera*, 8–9,000 species), together with the enigmatic one-species genus *Placozoa*, are considered the most primitive form of metazoans. According to the present view first formulated by the German evolutionary biologist Ernst Haeckel (1834–1919), multicellular organisms have evolved from flagella-bearing ancestors, which, like the extant choanoflagellates (cf. Rokas 2008), aggregated to colonies, e.g., in the form of hollow spheres. In later evolution, single cells specialized in different ways, remaining together and forming a multicellular organism.

Sponges can be from a few millimeters to 3 m in size, possess a free-swimming larva, but are sessile in the adult stage, i.e., they are anchored onto a surface. They feed by water filtration and can live for thousands of years. Their body tissue differs from that of all other metazoans by having only two "germ layers," called *ectoderm* and *endoderm*, rather than three germ layers, i.e., with a *mesoderm* in addition. These two layers "sandwich" a jelly-like substance called *mesohyl*. The surface of sponges is covered with many pores (hence the name "porifera" meaning pore-bearers), which are openings of channel-like structures, the "spongocoel," that penetrate the body. Water flows through these channels, carrying oxygen and nutrients (bacteria or other water particles), and in most sponges is ejected, together with waste, through an "osculum" ("little mouth"). The water stream is enhanced by choanocytes ("collar cells") sitting in the wall of the channels and carrying one central flagellum that can move the water in a coordinated fashion. How they coordinate themselves is unknown, but most probably electric excitation jumps from one cell to the other. Sponges respond to their environment by opening, closing, or changing the shape of the pores, and thus are able, together with the choanocytes, to control the water stream. The presence of true nerve cells is debated. Some authors believe that bipolar and multipolar nerve cells exist, while others deny it. Undoubted, however, is the existence of "independent effector cells" called myocytes, which have both sensory and motor functions and are capable of local responses to stimulation, but are insensitive to electrical stimulation. There are larger changes in body shape, but how these are achieved is unclear. Conduction of electric signals between cells, if present, may occur only over a distance of a few millimeters and cannot be responsible for this phenomenon. It has been suggested by the Austrian zoologist von Lendenfeld that mobile cells carrying neuroactive substances move from the stimulated site to the responsive sites and thus transmit information between sensors and effectors.

7.1.2 Coelenterates

The term "coelenterates" means animals with a hollow body cavity. Today, this term has become obsolete, because the two groups that previously formed the phylum *Coelenterata*, the *Cnidaria* and the *Ctenophora*, are now considered to represent two independent phyla. Members of both phyla evolved more than 700 mya from sponge-like ancestors and represent the first true metazoans. Like sponges, the bodies of most members of both phyla consist of an outer and inner layer, i.e., ectoderm and endoderm, with a mesogloea in between, but they possess muscle-like tissue which in other invertebrates and vertebrates animals originates from the mesoderm as the middle layer; it has been assumed that cnidarians evolved from ancestors with three layers and lost the middle layer. However, recent data by Steinmetz et al. (2012) suggest an independent evolution of striated muscles in cnidarians and bilaterians.

The phylum *Cnidaria* comprises up to 11,000 species belonging to the anthozoans (sea anemones, corals, sea pens), scyphozoans (true jellyfish), cubozoans (box jellies) and hydrozoans (freshwater cnidarians, *Hydra*, Portuguese *Man o'War* and others). The reproduction of cnidarians often involves a complex life cycle with both polyp and medusa stages. In the jellyfish and in box jellies, a freely swimming larva exists which eventually becomes sessile and transforms into a

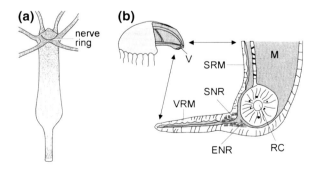

Fig. 7.2 (a) The nervous system of the polyp *Hydra* with nerve ring. (b) Radial section through the umbrella of a hydromedusa. *ENR* exumbrellar nerve ring, *M* mesogloea, *RC* ring canal, *SNR* subumbrellar nerve ring, *SRM* subumbrellar ring muscle, *V* velum, *VRM* velar ring muscle. After Roth and Wullimann (1996/2000), modified

polyp (Fig. 7.2a). The polyp grows, absorbs its tentacles, and splits into a series of disks (a process called strobilation) that become juvenile medusae. The juveniles swim off and slowly grow to maturity, while the polyp regrows and may keep strobilating periodically. By sexual reproduction, medusae produce larvae which transform into polyps and so forth. However, some hydrozoans, like the freshwater polyp *Hydra* (see below), and all anthozoans, have lost their medusa stage.

The jellyfish medusae have an umbrella- or bell-like shape with a central stomach tube and a mouth at its end (Fig. 7.2b). Most of them have fringes of tentacles at the bell equipped with cnidocytes or nematocysts, and a ring of tentacles around the mouth. They move in a jet propulsion-like fashion by contraction of the bell squeezing out water. The cnidocytes are weapons that function either as harpoons or by the injection of venom, which may be very painful or even deadly.

The exclusively marine and free-swimming ctenophores are a small group with less than 100 species. They lack cnidocytes as well as a polyp stage, because the freely swimming larva develops directly (i.e., without metamorphosis) into the adult form. The phylogenetic relationship between the cnidarians and ctenophores is unresolved, but it is believed that the former are more closely related to the bilaterians.

Cnidarians as well as ctenophorans are the first metazoans that possess true nerve cells (Bullock and Horridge 1965). The evolutionary origin of first nerve cells is debated. One theory postulates that sensory cells and nerve cells originated from neuromuscular cells, while other authors assume an independent origin of sensory, nerve, and muscle cells from epithelial cells. The "paraneuron" concept developed in the early twentieth century by Parker and colleagues proposes the evolution of nerve cells from secretory cells. As already mentioned, many features of nerve cells, such as membrane potential, transmitters and other neuroactive substances, membrane receptors, ion channels, many chemical processes relevant for neuronal information processing, and even action potentials, are older than nerve cells and nervous systems and are already found in protozoans, plants, and non-neuronal cells, and are thus more than one billion years old.

Cnidarians have no central nervous system or even brain, rather a decentralized nerve net or nerve rings. They comprise both the simplest types of nervous systems (nerve nets) and relatively complex forms, i.e., radially symmetric nervous systems (Fig. 7.2). Epidermal nerve nets are found in sessile hydrozoans, like the freshwater polyp *Hydra*. There is a concentration in the form of nerve rings around the mouth and the peduncle of this animal. This makes sense, because the tentacles need to carry food toward the mouth. Complex sense organs are absent, but *Hydra* responds to mechanical, chemical, visual, and temperature stimuli.

The freely swimming medusa forms of scyphozoans (the "jellyfish"), in contrast, possess complex nervous systems that appear to have evolved independently from other complex nervous systems in the animal kingdom. The medusae have a complex ring nervous system at the rim of the umbrella consisting of an outer or *exumbrellar* nerve ring composed of fine nerve fibers with low conduction velocity interrupted by ganglia containing small multipolar sensory cells, which are in contact with light-sensitive cells, the mouth and tentacles, and an inner or

7.1 Non-bilaterians

subumbrellar nerve ring with thick nerve fibers and consequently high conduction velocity and many large bipolar "swim motor neurons" for synchronous umbrella contraction. This ring receives information from statocysts, i.e., balance cells (Fig. 7.2b). Both ring nervous systems are interconnected.

Signal transmission in this nerve ring system differs from that in other animals, because electrical synapses dominate over chemical ones. This makes fast signal conduction possible, but as in all electric synapses, restricts the modulation of signal transmission. There are chemical synapses, the presence of which was debated for a long time, but chemical transmission is mostly exerted by a number of neuropeptides (e.g., FMRF-amides and RF-amides; cf. Grimmelikhuijzen et al. 1992), although there is evidence of cholinergic, serotonergic, dopaminergic, and glutamatergic transmissions in different cnidarians species (cf. Anderson et al. 2001).

Sense organs found in the medusae are visual organs, i.e., pigment spots, cup ocelli, or even "eyes" with biconvex lenses, statocystes (i.e., organs of balance), and "rhopalia," i.e., complex club-like balance organs, often combined with photo- and chemoreceptors, which initiate the rhythmic contraction of the medusae. In essence, in the cnidarians and ctenophorans, we already find a highly developed nervous system, which represents a striking alternative to all other complex nervous systems.

7.2 Bilaterians

Animals with bilateral symmetry comprise the three major groups of phyla, the *Acoelomorpha*, the *Protostomia*, and the *Deuterostomia*. All of them have three germ layers, an endoderm, an ectoderm, and, in-between, a mesoderm. The *Acoelomorpha* have only a primary body cavity in the form of a digestive tract with or without an opening, while protostomes and deuterostomes comprise all organisms with a secondary body cavity and are therefore called "Coelomata" (sometimes also called "Nephrozoa," because they possess kidneys). The coelom is partially or fully covered with mesodermal tissue. However, it may be that a secondary body cavity has evolved and was lost several times independently among eumetazoans. The currently most accepted phylogenetic relationship is given in Fig. 7.1.

The coelomates are divided into protostomes and deuterostomes, which differ in embryonic development. In the former—at least in the common view that has been challenged several times—the embryonic "mouth" remains the entrance to the secondary body cavity (the *coelom*), and an anus is formed secondarily on the opposite side of the embryo. In deuterostomes, the embryonic mouth becomes the anus, and a secondary mouth is formed as an entrance to the coelom. Another important character distinguishing protostomes and deuterostomes is the position of the central nervous system (CNS).

7.2.1 Acoelomorpha

The *Acoelomorpha* are considered the simplest bilaterally organized animals and include very small bilateral animals resembling flatworms. For that reason, until recently they were included into the phylum *Platyhelminthes* (see below), but are now considered a separate phylum. They possess a diffuse subepidermal nerve net resembling that of *Hydra* representing the simplest form of a bilateral nervous system. Since such diffuse subepidermal nerve nets are likewise found in other flatworm-like organisms, it is debated whether this type represents the ancestral form of all bilaterial nervous systems or has to be regarded as a product of secondary simplification. This would presuppose that the ancestral state of the brains of all bilaterians (i.e., of all "invertebrates" and "vertebrates") was already relatively complex, exhibiting a tripartite organization, and that the nervous systems and brains of many invertebrates underwent secondary simplification (Hirth and Reichert 2007). Other authors, however, maintain that more complex types have developed from simpler ones independently (Moroz 2009).

7.2.2 Protostomia

According to recent taxonomy, mostly based on genetic data ("phylogenomics"), protostome phyla are grouped into the *Lophotrochozoa*, i.e., animals carrying a lophophor (i.e., a tentacle "crown") or possessing a trochophora larva, and the *Ecdysozoa*, i.e., showing ecdysis (molt, see below). The phylogenetic and taxonomic status of the phylum *Platyhelminthes* is unclear. While some authors put them into the Lophotrochozoa, others consider them to be a sister group of the Lophotrochozoa, which together form the super-group *Spiralia*.

Lophotrochozoa

As indicated by their taxonomic naming, the lophotrochozoans are divided into the lophophorates, i.e., those carrying a tentacle crown around the mouth, called lophophor, and the trochozoans, having a trochophora larva of a characteristic shape. The *Lophophorata* include the phyla *Bryozoa*, *Phoronida*, *Brachiopoda*, and *Entoprocta*, either including or excluding the *Platyhelminthes* and *Rotatoria* (*Rotifera*), while the *Trochozoa* include the phyla *Nemertea* (also called *Nemertini*), *Mollusca*, *Sipuncula*, *Echiura*, and *Annelida*. However, this taxonomy varies among authors.

7.2 Bilaterians

Platyhelminthes

Platyhelminthes (or *Plathelminthes*, flatworms; 25–30,000 species), of unclear taxonomic status (see above), comprise a number of species previously called "Turbellaria" (other "turbellarians" are now included in the *Acoela*), the endoparasitic *Cestoda* (tapeworms, 3,500 species) and *Trematoda* (flukes, about 20,000 species).

Some of the "turbellarian" platyhelminthes (about 3,000 species) possess very simple nervous systems resembling the subepidermal diffuse nerve net found in the *Acoela*. In other forms, in addition to a subepidermal nerve net, there is a supraesophageal ganglion giving rise to dorsal and ventral longitudinal nerve cords connected by commissural tracts as illustrated in Fig. 7.3. The longitudinal nerve cords consist either entirely of fibers or of fibers forming regularly arranged ganglia as local densities of nerve cells. This type of CNS can either be regarded as the evolutionary starting point of all more complex CNS found in bilaterian animals or as the product of independent secondary reduction in many taxa (cf. Hirth and Reichert 2007).

More complex forms of the brain are found in some flatworm taxa, which either represent the ancestral state of bilaterian brains (while others underwent secondary simplification, see above) or have evolved independently. The most complex ones are found in predatory planarians such as *Notoplana* and *Stylochoplana*, with cerebral ganglia consisting of five different brain masses. Their sensory equipment is impressive: flatworms possess a variety of sense organs such as touch or chemoreceptors on the head and all over the body, statocysts, and inverse or everse eyes containing several hundred photoreceptors enabling the animals with motion and the necessary input for object perception (Fig. 7.4). Some of them are found as a pair of ocelli on the head; other terrestrial flatworms have more than 1,000 ocelli.

Cestodes and *trematodes* are endoparasites, and due to their lifestyle, they have a simple or *simplified* nervous system consisting of a simple supraesophageal ganglion or "brain" and a varying number of longitudinal (mostly three) fiber tracts. The mouth region is heavily equipped with nerve ends. There are a few photo- and mechanoreceptors, while chemoreceptors are apparently absent.

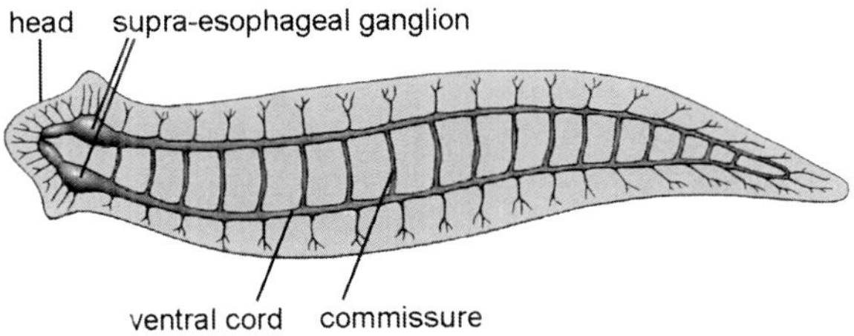

Fig. 7.3 Nervous system and brain of a flatworm. For further explanations, see text

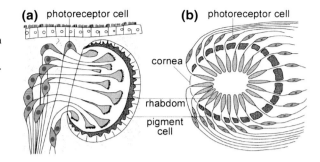

Fig. 7.4 Eyes of "turbellarian" flatworms. (a) Inverse pigment cup eye of a freshwater planaria, (b) everse eye of a land planaria. After Roth and Wullimann (1996/2000), modified

Lophophorata

Brachiopods (340 species) are sessile, marine animals that have a striking resemblance to bivalve mollusks (clams) because of a hard shell, but in contrast to clams have a pedicle fixing them to the surface and a lophophore that filters particles out of the water. Their nervous system is very simple (or simplified) and consists of a nerve ring around the esophagus with a smaller dorsal and a larger ventral ganglion, or in the adults of some species, only a ventral ganglion. Ventral cords originate from the ganglia or the nerve ring and supply the body, the tentacles, mantle, and valve muscles. There are touch receptors and statocysts.

Bryozoans (moss animals, also called *Ectoprocta*; 4–5,000 species) and *entoproctans* (meaning "anus inside," about 150 species) are tiny filter feeders using a lophophor. *Phoronids* (10 species) live inside self-built tubes and are likewise filter-feeders, but can grow up to 50 cm long. All of them have simple nervous systems with or without ganglia like the brachiopods, with an esophageal nerve ring and nerve cords, some of them with "giant" axons for fast retraction.

Trochozoa

Besides the nemerteans, the trochozoan group includes annelids and mollusks as larger groups as well as small groups of marine worm-like animals like *sipunculids* ("peanut worms," about 300 species) and *echiurids* ("spoonworms," 150 species). All of them live relatively simple lives and accordingly have very simple nervous systems—either primitively or as a consequence of secondary simplification. Their CNS includes a small supra- and sub-esophageal ganglion connected with a sub-epidermal nerve net and a small number of sensory cells such as statocysts, eye spots, and mechanoreceptors on the tentacles.

The *nemerteans* (also called "Nemertini," "ribbon worms," or "proboscis worms," about 1,200 species) are considered the relatives of flatworms, while they are not parasites but marine and often colorful predators, ranging in size from a few centimeters to 30 m (the maximum is 54 m—the longest animal ever found!). They feed on annelids, clams, and crustaceans by means of a highly protrusible structure called proboscis, which is everted from the "rhynchodeum" above the mouth and captures prey with poison.

In the context of a predatory lifestyle, their nervous system is more complex than the forms mentioned before and consists of a brain forming a ring of four ganglia (one pair of dorsal and one pair of ventral ganglia) positioned around the rhynchocoel, a body cavity containing the rhynchodeum, rather than around the pharynx, as is the case in most other worm-like invertebrates. At least one pair of ventral nerve cords originates from the brain and runs along the length of the body without segmentation into ganglia. Most nemerteans have chemoreceptors and pigment cup ocelli on their heads, but some forms have lens eyes.

- Annelids

Annelids (ringed or segmented worms, 16-18,000 species) are one of the largest trochozoan groups. They are divided into the *Polychaeta* ("worms with many hairs" called "chetae," and leglike appendages called "parapodia") and the *Clitellata* comprising oligochaetes (earthworms) and hirudineans (leeches), both without hairs and parapodia. The body of annelids is *segmentally* organized into a number of identical parts with the same sets of organs including segments of the ventral nervous system and in most polychaetes a pair of parapodia used for locomotion.

Until some years ago, a close phylogenetic relationship was assumed between the annelids and the largest animal group, the arthropods, especially because all members of the latter group, too, have a segmented body plan, and the nervous system of both resembles each other in a number of details. However, today the phylum Annelida is included in the Lophotrochozoa and in closer relationship with the platyhelminthes and mollusks, while the phylum Arthropoda now belongs to the Ecdysozoa (see below). With respect to body segmentation, we are confronted with the same question as with the brain, i.e., a segmented body either represents the ancestral condition of all protostomes and perhaps of all bilaterians and was lost in all phyla without a segmented body, or body segmentation evolved several times independently in lophotrochozoans, ecdysozoans, and in chordates (perhaps all deuterostomes).

In annelids, we find—together with a segmented body plan—a "rope-ladder" CNS (Fig. 7.5). In its simplest (either ancestral or derived) state, this structure consists of a supra-esophageal or cerebral ganglion located in the "head" ("prostomium") of the animal and a nerve ring around the esophagus, giving rise to paired ventral nerve cords with a pair of ganglia per body segment connected by transverse connectives or "anastomoses." Some of the cords are very large in diameter, forming "giant fibers" which—due to their large diameters—have high conduction velocities carrying information at high speed from the head to the hind end of the animal.

In some polychaete taxa, the cerebral ganglion reveals a complex organization. In predatory species there is a three-partite brain strikingly resembling the proto-, deuto-, and tritocerebrum of the insect brain (see below), which might indicate that

Fig. 7.5 "Rope-ladder" central nervous system of annelids. After Roth and Wullimann (1996/2000), modified

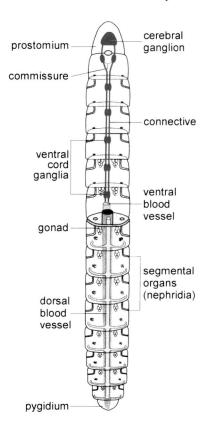

all protostomes and even all bilaterians indeed had such a tripartite organization. Here, but also in some oligochaetes, the first segments of the ventral nerve cords are often fused into a sub-esophageal ganglion. Predatory polychaetes have protocerebra that contain structures resembling the mushroom bodies (MB) of insects (see below), but it is debated whether these structures are homologous (at least in the sense of "deep homologies") or have evolved independently. In the annelids, these "MB" are connected with centers that receive information from the tentacle-like palps, the optic centers and the antennal centers of the deutocerebrum, and apparently represent multi-sensory integration centers as is the case with the MB of insects. In the oligochaetes, we find a modest, and in hirudineans a massive simplification of this basic organization.

Annelids possess a large variety of tactile and chemosensory organs, feelers or antennae, tentacle-like palps, or ciliated "nuchal organs" possibly involved in chemoreception for food and/or light detection. Other light-sensitive organs range from very simple pigment spots and eye pits to compound eyes and lens eyes with accommodation mechanism in some predatory polychaetes, and have evolved independently of similar eye types in other animal groups (Fig. 7.6).

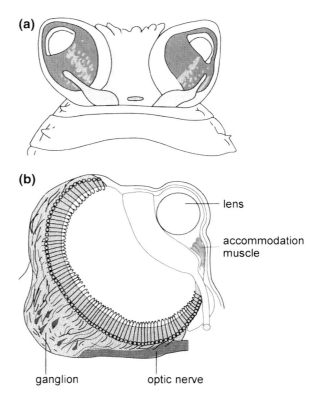

Fig. 7.6 Camera eye of the polychaete *Alciope* with lens accommodation mechanism. (a) ventral view, (b) cross section through the optical axis. After Roth and Wullimann (1996/2000), modified

- Mollusks

Mollusks are the largest lophotrochozoan group (about 100,000 described and probably many more existing species). Their phylogenetic relationship is debated. Besides several smaller groups, there are three large taxa, i.e., the *Gastropoda* (snails and slugs, about 70,000 described species), *Bivalvia* (clams, oysters, mussels, scallops, about 20,000 described species), and *Cephalopoda* (inkfish, about 800 species).

The molluskan nervous systems range from relatively simple (or simplified) forms resembling those found in acoelans to the most complex ones found among invertebrates, in the cephalopods. The basic pattern is a *tetraneural* nervous system consisting of a cerebral ganglion, which gives rise to two dorsal pleurovisceral and two ventral pedal nerve cords (Fig. 7.7a). In the ancestral state, nerve cell bodies are not concentrated in ganglia, but are dispersed throughout the cords. Therefore, in mollusks the formation of ganglia may be considered either an ancestral or a derived state that happened independent of the formation of ganglia in other forms such as annelids and arthropods and maybe even several times independently in mollusks (Moroz 2009).

Fig. 7.7 Central nervous system of mollusks. (**a**) nervous system of the sea slug *Aplysia*, (**b**) site of the brain of *Octopus*, (**c**) *Octopus* brain and nerves. After Roth and Wullimann (1996/2000), modified

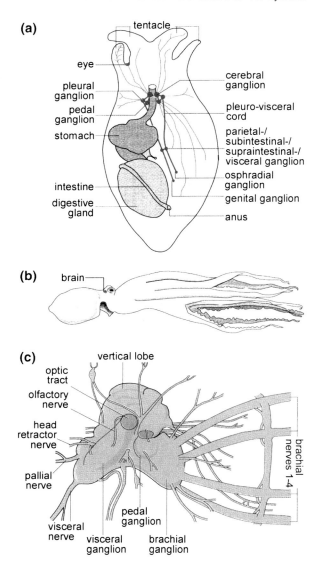

Gastropods: The tetraneural system of snails and slugs exhibits maximally six pairs of ganglia and mostly one unpaired visceral ganglion (Fig. 7.7a). The cords are mostly linked by commissures. The paired cerebral ganglia connected by a commissure are located around the esophagus and process information from and to the eyes, statocysts, head tentacles, skin and muscles of the lip, head, and sometimes penis region. One pair of buccal ganglia with a commissure is situated below the esophagus and innervates the pharynx, salivatory glands, a nerve plexus of the esophagus and the stomach. One pair of pleural ganglia without a commissure is connected by cords with the cerebral, buccal, and parietal-visceral ganglia.

The pedal ganglia innervate foot muscles and skin. The cerebral, pleural, and pedal ganglia together form the brain mass. The supra- and sub-intestinal ganglia innervate the gills, the "osphradium" (an olfactory organ), and parts of the mantle and skin; one pair of parietal ganglion (not present in all gastropods) innervates the lateral walls of the body. Finally, the unpaired visceral ganglion supplies the caudal region of the gut, anus, and neighboring regions of the skin and body wall, sexual organs, kidney, liver, and heart. It completes the "visceral loop," i.e., the chain of ganglia and cords from the pleural to the visceral ganglion.

A fusion of ganglia, mostly of the "visceral loop," is observed in many gastropods, e.g., in air-breathing landsnails. The most highly developed gastropod brain is found in the Burgundy or Roman snail *Helix pomatia*. It consists of a protocerebrum with globuli and dense neuropils, a mesocerebrum and a postcerebrum with pleural and pedal lobes. This organization is remarkably similar to that of other invertebrates with complex brains, but has probably evolved independently.

Gastropods have chemoreceptive and mechanoreceptive sense organs distributed all over the body. Complex sense organs comprise statocysts, eyes ranging from widely open pit eyes (*Patella*), pinhole eyes (*Trochus*) to lens eyes (*Helix*), and chemosensitive osphradia in the mantle near the gills.

Some sea slugs have gained fame in modern neurobiology, e.g., the Californian "sea hare" *Aplysia californica*, which can be 75 cm long. It possesses some very large nerve cells that can be detected with the naked eye and are well-suited for studies of neuronal information processing and learning processes, as did the American Nobel Laureate Eric Kandel (cf. Kandel 1976).

Bivalves: Bivalves (about 20,000 species) form the second largest molluskan group. In accordance with their sessile lifestyle, all of them have very simple or simplified nervous systems with only three pairs of ganglia with an emphasis on the visceral ganglion, which is often fused with the parietal ganglia. In most species, the rostralmost ganglion is a fused cerebral, pleural, and buccal ganglion. Some bivalves, e.g., the scallop *Pecten*, have eyes on the rim of the mantle, which can have a complex anatomy (e.g., a distal and proximal retina, the latter including a crystalline argentea), but their special functions are unclear.

Cephalopods: Cephalopods (10,000 extinct and a little more than 700 extant species) are a phylogenetically old group that originated about 500 mya in Cambrian times. Some groups, like the well-known ammonites, which strongly resemble the extant *Nautilus* (see below) were very specious from the Ordovician to the end of the Cretaceous, i.e., until 65 mya. Today, only two subclasses of the class Cephalopoda exist, i.e., the *Nautiloidea*, with two genera *Nautilus* and *Allonautilus* consisting of six species, and the *Coleoidea*, which comprise all other cephalopods with the *Dekabrachia*, i.e., cephalopods with ten arms, to which the sepiida or cuttle fish and the squids (*Theutida*) belong, and the *Octobrachia*, i.e., cephalopods with eight arms, with the *Octopoda* as the main group.

Cephalopods are exclusively marine animals and live at various depths from the abyssal plane to the surface. They all can move relatively fast by water jet propulsion. The head gives rise to tentacles—hence their name "cephalopods," i.e., head with legs, and these tentacles can lengthen rapidly and are used for capturing prey and drawing it to the mouth as well as for slow movement. The tentacles often terminate in sucker-coated clubs. In squids, the tentacles may reach a length of eight meters. All cephalopods have a parrot-like beak with an upper and lower jaw, and most of them have a tongue-like radula. Cephalopods are strikingly short-lived; while *Nautilus*, with about 20 years, becomes relatively old, most species of the supraorder Coleoidea barely reach one year, and the maximum is five years. These animals grow rapidly and reach sexual maturity early.

Cephalopods generally have highly developed nervous systems characterized by the fusion of ganglia and subsequent development into lobes forming a complex brain around the esophagus. There are brain lobes that correspond to the cerebral, buccal, labial, pleural, and visceral ganglia of other mollusks, while other structures like the central optic-visual, olfactory, and peduncular ganglia as well as peripheral branchial and stellar ganglia are newly formed.

The well-known pearl boat *Nautilus* is a living fossil, because it probably represents the ancestral form of cephalopods living in the Cambrian about 500 mya. Under the perspective of evolutionary biology, this group should be considered extremely successful, although—somewhat paradoxically—only one genus with a few species survived mass extinction at the end of the Cretaceous 65 mya.

Nautilus possesses an outer chambered shell, i.e., the beautiful pearl boat that grows life-long by adding new chambers to the end. The animal lives in the last formed chamber. It can regulate the gas content in the chambers and in this way adjust its buoyancy. However, gas exchange limits the range of depth, where it can live, to 100–400 m. For forward swimming, *Nautilus* draws water into and out of the last chamber. It has between 60 (males) and up to 90 (females) short, yet strong tentacles, which bear no suckers, but are sticky. It forages mostly on small crustaceans.

Nautilus possesses a relatively simple brain without externally visible supra-esophageal lobes and with unfused sub-esophageal lobes, which probably represents the ancestral state of cephalopods (Grasso and Basil 2009). However, according to Young (1971) and Nixon and Young (2003), this ring-like cord structure has a complex inner structure with five different regions. A cerebral cord connects the laterally situated cerebral lobes, which receive and process information from the tentacles, gills, eyes, and olfactory system. There are separate upper and lower buccal ganglia innervating the pharynx and the mouthparts.

The dekabrachian squids (*Theutida*, about 250 species including the common squid, *Loligo vulgaris*) live in the open ocean, partly in abyssal regions. Squids may become very large, e.g., the giant squids (*Architeuthis*), with tentacles that span 20 m and weigh half a ton or more (there are many horror stories about them!). Squids, like cuttlefish and cephalopods, and unlike nautiloids, have no external shell, but a horny inner strip for body stabilization. They have eight shorter and two longer arms covered with suckers, which they use for capturing prey, which is then crushed with the beak and the radula. The main body is covered by the mantle, with a siphon at the front of the mantle cavity. Like all cephalopods, squids use the mantle cavity and siphon for movement by jet propulsion. Squids can move very fast and even jump over the water surface. Some species have enormously large eyes—up to 20 cm in diameter—the largest eyes in the animal kingdom.

The cuttlefish (*Sepiida*, 120 species) live in tropical and subtropical oceans close to the coast, from surface to 400 m deep. They can grow up to 60 cm and weigh up to 10 kg. They possess an internal shell, the cuttlebone, for buoyancy. They eat small mollusks, crabs, shrimp, fish, octopuses, worms, and other cuttlefish, but are not fast swimmers like squids, but rather ambush feeders. They catch their prey with eight sucker-covered arms and bring them to the beak and mouth with the two shorter tentacles.

The brains of squids and cuttlefish exhibit an organization typical of all *Coleoidea* with an increasing fusion of ganglia and concentration around the esophagus. The sub-esophageal region consists of three parts, an anterior one supplying the tentacles, a middle multi-lobed one innervating the chromatophores responsible for the color change, for which the animals are famous, and the siphon, and a caudal one supplying, among others, the mantle and the gills. These sub-esophageal parts interact with the magnocellular lobes located lateral of the esophagus and partly surrounding it.

The brain or supra-esophageal ganglion is composed of many externally visible lobes including the buccal ganglia, which innervate the gills. There are basal lobes for the control of movement and a dorsal part enclosing a chemo-tactile and a visual system, which are composed of four lobes each and process information from the tentacles and the eyes. Squids have a vertical lobe like *Octopus*, which here, too, is the site of higher cognitive functions (see below). Relative to the entire brain volume, albeit not in absolute terms, it is even larger than in the *Octopus*.

Cuttlefish eyes, like those of the *octopus*, are among the most developed in the animal kingdom. They have a W-shaped pupil and two foveae, one for forward, and one for backward view. The eye changes focus by reshaping itself entirely, instead of reshaping the lens, as in mammals. The cuttlefish eye is an everted eye, which means that the photoreceptors point toward the lens and light.

The *Octopoda* (about 300 species; Fig. 7.7b) are considered by many experts to be most highly developed mollusks and most intelligent invertebrates. They possess neither an external nor an internal shell, and this enables them to creep even through small openings. Like squids and cuttlefish, they exhibit a spectacular color change either serving for camouflage or signaling emotional arousal, e.g., during reproductive behavior. All octopods are poisonous, but only the poison of the Australian blue-ringed octopod *Hapalochlaena maculosa* is deadly for humans.

The most prominent feature of octopods is the eight long arms covered with suckers that originate from the head and are used to capture prey (mostly crustaceans) and for slow movement over the surface. Fast movements occur by jet propulsion with the head ahead. Octopods weigh between 15 and 75 kg, but very large specimens can weigh 270 kg or more; an arm span width of 9 m has been found.

Octopods, like all coleoids, are short-lived—from a few months in small species—and up to five years in larger ones. Usually, they die after reproduction, i.e., spawning and egg deposition, because they stop feeding. This apparently is caused by the release of a substance that has to do with the optic glands, because removal of these glands after spawning leads to resumption of feeding and greatly prolonged life. The short life expectancy of octopods is interesting because it is an exception to the rule that animals with large and complex brains and high intelligence are long-lived. Female octopods exhibit only brief parental care, which is another exception to the rule and prevents them from transferring experience to their offspring. They live solitary lives and produce, like most invertebrates, many eggs (up to 200,000).

The nervous system and brain of the *Octopus* (Figs. 7.7c and 7.8) has been studied in detail thanks to the pioneering work of J. Z. Young and his collaborators (cf. Young 1971). The nervous system contains about 550 million neurons, 350 million of which are located inside the eight arms, about 160 million neurons in the giant optic lobes, and 42 million neurons inside the brain. The arm nervous system exhibits a great degree of autonomy and is capable of exerting stereotyped movements without the help of the brain.

The entire brain mass encircling the esophagus has been formed by the fusion of numerous ganglia, and according to the classical description by Young and colleagues (Young 1971), is composed of 38 lobes (Fig. 7.8). The supra-esophageal part, or brain proper, is divided into 16 lobes and contains the mass of neurons. It has a ventral motor portion consisting of several lobes that are involved

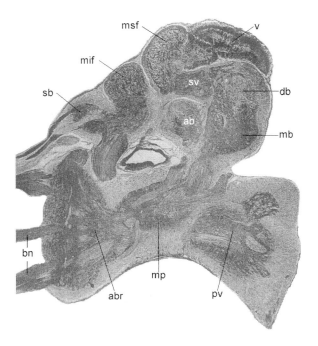

Fig. 7.8 Cross section through the brain of *Octopus vulgaris*. The figure depicts the different lobes which together form the supra- and sub-esophageal mass. *ab* Anterior basal lobe, *abr* anterior brachial lobe, *db* dorsal basal lobe, *mb* median basal lobe, *mif* median inferior lobe, *mp* median pedal lobe, *msf* median superior frontal lobe, *pv* palliovisceral lobe, *sb* superior buccal lobe, *sv* subvertical lobe, *v* vertical lobe. From Nixon and Young (2003), modified

in the control of feeding, locomotion, and color change, and a dorsal portion exerting sensory information processing and higher cognitive functions. It receives visual afferents from the eyes and the optic lobes as well as tactile-chemosensory information from taste and touch receptors of the arms. Each of these sensory-cognitive systems are divided into four lobes each, forming a lower and an upper row. The visual and the tactile chemosensory system are closely interconnected.

The vertical lobe is considered the most complex part of the *Octopus* brain (Young 1979; Hochner et al. 2006). It is composed of five lobuli, similar to the gyri of the cortex of mammals, and contains about 26 million neurons (more than half of the neurons inside the brain). It consists only of two major types of neurons, i.e., nearly 26 million tiny interneurons, the smallest inside the *Octopus* brain, and 65,000 large projection neurons, with the former converging on the latter. As we will see, the mammalian cerebral cortex likewise consists of only two major types of neurons, large projection neurons (the pyramidal cells) and small interneurons, but with the difference that in the mammalian cerebral cortex the ratio between them is inverse with respect to the vertical lobe of the *Octopus*, i.e., in the cortex there are 80 % projection neurons and 20 % interneurons.

The vertical lobe receives afferents predominantly from the so-called median superior frontal lobe, which belongs to the upper row of the visual system. These afferents form a distinct tract composed of 1.8 million fibers, which terminates in the rind of the vertical lobe (Hochner et al. 2006). The processes of the nearly 26 million interneurons located there penetrate that tract in a rectangular fashion and form "*en passant*" contacts with them (cf. Fig. 17.2). According to Shomrat et al. (2008), this is the site of long-term potentiation and formation of long-term memory. The vertical lobe as a major center for "higher" cognitive abilities of *Octopus* (cf. Chap. 8) is closely connected, via the projection neurons, to the subvertical lobe which contains about 800,000 neurons, and the interaction of both lobes is based on the work of an impressively regular network of millions of crossing fibers. The subvertical lobe then sends numerous fibers back to the optic lobes.

Likewise complex are the giant optic lobes exhibiting a five-fold laminar neuropil resembling the retina and cortex of mammals. They process visual information arriving from the large lens eyes. The eyes are the main sense organs of *Octopus*. They possess external muscles for eye movement and inner muscles for accommodation and pupil control, and there is pigment migration between the photoreceptors for light and dark adaptation. These eyes have a striking resemblance to the vertebrate eye, although they are built of different embryological material. Also, the *Octopus* eye is *everted*, which means that the photoreceptors point toward the lens and light. Accordingly, it has no "blind spot," because the optic nerve originates behind the retina. This difference has to do with differences in eye formation: while in the *Octopus*, the eye is formed via an invagination of the head surface, in vertebrates it originates as an extension of the brain, more precisely of the diencephalon.

A magnocellular lobe is found between the brain and the sub-esophageal ganglion, where giant fibers originate, mediating fast defense and flight reactions.

The sub-esophageal ganglion is the "motor brain" of the *Octopus* and is divided into an anterior part consisting of a pre-brachial and a brachial lobe, a medial part (pedal lobe), and a posterior part, which may be considered to be part of the brain proper. In the sub-esophageal ganglion, the brachial nerves of the arms originate and at the same time afferents from the arms terminate there.

Ecdysozoa

The *Ecdysozoa* comprise all invertebrate animals that possess a rigid body surface called *cuticula*, which does not grow together with the body. Accordingly, these animals regularly shed or molt their exoskeleton under the influence of a hormone, in insects called *ecdysone*. This process is called "ecdysis." The term "Ecdysozoa" partially replaces the previous term "Articulata," because it is based on the annelids that have been removed and are now part of the Lophotrochozoa, as mentioned above. The new superphylum "Ecdysozoa" comprises the *Cycloneuralia*, which, according to most authors, include the nematods (*Nematoda*), horsehair worms (*Nematomorpha*), penis worms (*Priapulida*), mud dragons (*Kinorhyncha*), and brush-heads (*Loricifera*), and the *Panarthropoda* including the arthropods (*Arthropoda*), velvet worms (*Onychophora*), and tardigrades (*Tardigrada*). Tardigrades and onychophorans together are sometimes called "lobopods" and both, or at least the onychophorans are considered the stem group of the arthropods.

The *kinorhynchans* (150 species) are a few-millimeter-long marine worms with a segmented body without limbs. They live in the mud, but also on algae and sponges. They have a nervous system consisting of a multilobed nerve ring surrounding the esophagus and a ventral nerve cord with one ganglion per body segment. There are tiny bristles as touch receptors all over the body as well as eye spots or ocelli. Horsehair worms (*Nematomorpha*, about 320 species described, although probably many more exist) are thin and extremely long (up to 1 m) parasitic worms, which in the juvenile stage live inside arthropods, but as adults are free-living. As in most parasites, their nervous system is very simple, probably simplified, and consists of a nerve ring around the esophagus and a ventral cord with one ganglion close to the anus. There are simple sensilla and a light-sensitive pit below the cuticula.

The *nematods* (roundworms, 25–28,000 species) are among the most numerous multicellular animals on earth and are found in almost all marine or terrestrial biotopes. They are mostly slender and a few millimeters long, but some are microscopic. More than half of the species are parasites; others are scavengers or predators. Many of them are dangerous endoparasites like hookworms, filarias, and pinworms or whipworms. In accordance with their predominantly parasitic lifestyle, they have very simple nervous systems (Fig. 7.9) consisting of a nerve ring around the esophagus and a number of ganglia connected to this ring. Four to twelve ventral cords originate from the ring and are connected by half-sided commissures in an irregular fashion. Local ganglia and nerves are found in the

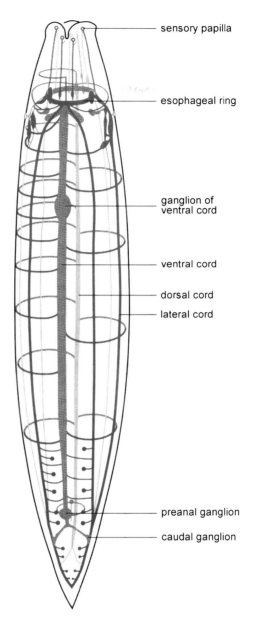

Fig. 7.9 Central nervous system of the nematode *Ascaris*, ventral view. After Roth and Wullimann (1996/2000), modified

caudal gut and anal region. Some nerves extend from the esophageal nerve ring to the sense organs in the head region such as sensory papillae and bristles. Other sense organs are chemoreceptive organs called "amphidia." Some free-living nematodes have paired eyes in the form of pigment cup ocelli, which sometimes have a lens.

The tiny nematode *Caenorhabditis elegans*, like the "sea hare" *Aplysia*, has gained fame in molecular and developmental neurobiology by the work of the South-African molecular neurobiologist Sydney Brenner and colleagues, because it has a very simple nervous system composed of exactly 302 neurons. The connectivity of this nervous system has been completely mapped (Brenner 1974). Subsequent studies explored the neural mechanisms responsible for a variety of behaviors shown by *C. elegans*, such as chemo- and thermo-taxis, and mating behavior. *C. elegans* has played an important role as a model organism for molecular and developmental genetics.

Tardigrades (also called "water bears" or "moss piglets," 800–1,000 species), considered by some authors to be closely related to the arthropods, are tiny animals 1 mm or less long. They live inside the water film on mosses or lichens, but are also found in dunes, beaches, and in marine water or freshwater in large numbers. They are famous for their resistance to extreme environmental conditions such as very high or very low temperatures, pressure, dehydration, toxic environment, UV radiation, and even vacuum. They have a relatively large brain consisting of a larger supra-esophageal and a smaller sub-esophageal ganglion connected by a nerve ring. The brain gives rise to two ventral cords forming a chain of four ganglia that control the four pairs of legs.

Arthropoda

With about 1.2 million described, and more than 10 million estimated species, arthropods are by far the largest and most diverse group of animals. They are divided by many authors into *proto-arthropods* (onychophorans, perhaps also tardigrades) and *eu-arthropods* (chelicerates, crustaceans, myriapods, and insects/hexapods; the latter three taxa together are called "mandibulates"), while other authors put the onychophorans closer to the chelicerates (cf. Withington 2007).

Like annelids, arthropods have a "rope-ladder" nervous system, i.e., regularly segmented ventral nerve cords. Based on the new taxonomy of protostomes mentioned above, this organization either was already present at an "ur-bilaterian" nervous system (Hirth and Reichert 2007) or has evolved independently in the lophotrochozoans and ecdysozoans from an unsegmented ancestral state.

In all arthropods, the first ganglia have fused into a complex brain. In mandibulates, there are three major brain divisions, i.e., a proto-, deuto-, and tritocerebrum. The protocerebrum is associated with the paired optic lobes, the deutocerebrum with the first and the tritocerebrum with the second pair of antennae, if present. In mandibulates, a sub-esophageal ganglion has formed by fusion of the three first ventral ganglia. It supplies the mouth region and mandibles, in crustaceans the first and second maxillae, and in insects the maxillae, the mandibles, and the labium. Caudal ganglia of the ventral cords exhibit a strong tendency to fuse and to form specialized abdominal structures.

7.2 Bilaterians

- Onychophora

Onychophorans (about 150 species) have a segmented body with multiple pairs of legs, are 0.2–20 cm long, and are nocturnal ambush predators, living mostly in tropical zones and subtropical zones of the southern hemisphere. A well-known species is *Peripatus*, a "living fossil," because it remained unchanged for 570 million years (until today). The onychophoran brain resembles that of the (other) euarthropods, especially because in its anterior part (protocerebrum) mushroom bodies (MB) are found, which in the eyes of some authors closely resemble those of the MB of chelicerates, but differ from those of the mandibulates (crustaceans and insects) (Strausfeld et al. 2006). There are numerous papillae covering the entire body and carrying mechanoreceptive bristles, and sensilla as chemoreceptors, which are found on the mouth and the two antennae as well as eyes with a lens and cornea, which some authors assume to be homologous to the median eyes (ocelli) of arthropods.

- Chelicerata

Extant *Chelicerata* (about 100,000 described species, probably many more) comprise the *Arachnida* (spiders, scorpions, mites and others), *Pantopoda* (sea spiders), and *Xiphosura* (horseshoe crabs). They all possess specialized feeding appendages called chelicers originating from the second head segment, with fangs, by which they inject venom into their prey. Antennae are lacking (for an overview, see Foelix 2010).

The CNS of the chelicerates (Fig. 7.10) is characterized by the absence of a deutocerebrum because of the lack of antennae; the tritocerebrum supplies the chelicerae. In xiphosurans, scorpions, and araneans (i.e., spiders), we find an increasing tendency toward fusion of ganglia during ontogeny. In many species of these groups, the entire chain of ventral ganglia forms a compact mass around the mouth, in the araneans below the brain.

The largest group of chelicerates, the *Arachnida* (about 100,000 species), comprises, among several other groups, the spiders in the classical sense (*Aranea*), scorpions (*Scorpiones*), harvestmen (*Opiliones*), pseudoscorpions (*Pseudoscorpiones*), and mites (*Acari*). Their bodies are divided into two segments, the prosoma or cephalothorax carrying the four pairs of legs, and the opistosoma or abdomen (Fig. 7.10a).

The brain (supra-esophageal ganglion) of arachnids consists of a protocerebrum and tritocerebrum. In the anterior median part of the protocerebrum, mushroom bodies are found, which—in contrast to insects—are exclusively visual neuropils associated with the secondary eyes. A central body is found in the posterodorsal part and is probably an integrative center for visual information from the main eyes. The homology of both the MB and the central body of arachnids with those of insects is doubted, but recent studies suggest a "deep homology" possibly present in the last common ancestor of all protostomes or even bilaterians (Strausfeld and Hirth 2013). The tritocerebrum is the ganglion linked with the

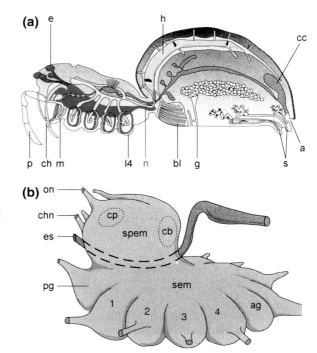

Fig. 7.10 Central nervous system of arachnids. (a) Site of the CNS (*blue*) inside the body of the house spider *Tegenaria*, side view. (b) Closer view of the CNS. *1-4* leg ganglia, *a* anus, *ag* abdominal ganglia, *bl* book lung, *cb* central body, *ch* cheliceres, *cp* corpora pedunculata, *cc* cloacal chamber, *chn* cheliceral nerve, *e* esophagus, *g* aperture of gonads, *l4* insertion of leg 4, *m* mouth, *n* nerve to abdomen, *on* optic nerve, *p* pedipalp, *pg* pedipalp ganglion, *s* spinneret, *spem* supra-esophageal mass, *sem* sub-esophageal mass. After Roth and Wullimann (1996/2000), modiied

chelicerates and is often fused with the sub-esophageal mass supplying the legs. This mass is found below the brain. It consists of a highly variable number of fused ventral ganglia (16 in araneans).

Arachnids have a large variety of sense organs. There are lyriform organs involved in the detection of vibration and in proprioception and hair sensilla called "trichobothria" on the legs and lateral and dorsal parts of the body, which are involved in the detection of airborne vibration and air currents (cf. Chap. 11). Species differ in the number of primary and secondary eyes. The primary eyes are considered homologous to the ocelli, and the secondary eyes to the compound eyes of insects.

Mites (*Acari*, about 50,000 species) are mostly very small or even microscopic, and are either free-living in soil or water or are parasites on plants or animals. Their brains exhibit the highest concentration of ganglia among invertebrates in the form of one compact "synganglion" around the esophagus.

Araneans (spiders, about 40,000 species) are mostly predatory. By means of their claws, they inject venom into their prey and some pump digestive enzymes into the prey and then suck the liquified tissues into the gut, while others grind the prey to pulp. Most of them, but not all, build sticky webs to trap insects; others catch their prey using a "bola" made of a single thread, tipped with a large ball of very wet sticky silk. These bola spiders emit chemicals that resemble the phero-mones of moths to attract their prey. Jumping spiders (*Salticidae*, more than 5,000 species) have excellent visual abilities due to their four pairs of eyes, of which the anterior median eyes are most prominent.

The brains of araneans consist of a fused supra- and sub-esophageal ganglion forming a compact mass around the esophagus (Fig. 7.10b), very much in the same way as found in *Octopus*. The brain is relatively large and represents up to 10 % of the cephalothorax. Like the brain, the ventral nerve cord ganglia in the opistosoma are fused into a compact mass.

- Crustacea

Crustaceans (crabs, lobsters, crayfish, shrimp, krill, and barnacles, more than 50,000 species) again exhibit an enormous diversity of forms and lifestyles. Together with the insects and myriapods, they form the group of mandibulates, i.e., arthropods carrying mandibles (rather than cheliceres), but differ from the myriapods and insects by the presence of gills rather than tracheae for air respiration and the possession of two pairs of antennae rather than one pair. The largest crustacean group are the *Malacostraca* or "higher" crustaceans such as the *Decapoda* (crabs, lobsters, crayfish), the *Isopoda* (woodlice, pill bugs, etc., about 10,000 species), *Amphipoda* (7,000 species), *Euphausiacea* ("krill"), *Remipedia*, *Branchiopoda, Ostracoda* (seed shrimp), *and Cirripedia* (barnacles).

Crustaceans have a typical rope-ladder nervous system as their ancestral form. The brain (supra-esophageal ganglion) is linked via two connectives with the ventral nerve cords. The protocerebrum consists of two lateral optic lobes and the median protocerebrum containing the anterior and posterior optic neuropils, the protocerebral bridge and the central body. Neuropils of the optic lobes are highly variable, but always possess a distal lamina ganglionaris. In decapod crustaceans (e.g. crabs), there are additional visual neuropils within the optic lobes, viz., a terminal medulla and the so-called hemi-ellipsoid bodies. Both include a varying number of complex neuropils; most of them contain glomeruli. The hemi-ellipsoid bodies and some of the other neuropils have connections with the accessory and olfactory lobes of the deutocerebrum. The deutocerebrum contains the medial and lateral neuropils receiving vestibular and mechanosensory input from the first antennae, the olfactory and parolfactory lobes (the latter with unknown input), and the lateral glomeruli. The tritocerebrum receives information from the second antennae and sends motor nerves to them. There are strong differences in the degree of fusion of ventral cord ganglia. A sub-esophageal ganglion controlling mouth appendages is found in many malacostracans, but is very small or even absent in many other crustaceans. There is a strong tendency for fusion of the ventral cord ganglia, which is maximal in crabs.

Crustaceans have a large number of sense organs, of which eyes and antennae predominate. An unpaired nauplius eye, frontal simple eyes, and compound eyes are found, the latter are located either immobile in the head, like the other types of eyes, or on movable stalks. The compound eyes may consist of a few or several thousand ommatidia (see below). The body surface including distal limbs and antennae is covered with mechano- and chemoreceptors possessing sensillae or setae. There are two pairs of antennae; their information is processed in the deutocerebrum (first pair) and the tritocerebrum (second pair). Only malacostracans have vestibular organs. There are proprioceptive mechanoreceptors of leg joints, the chordotonal organs.

- Insecta (Hexapoda)

Insects are by far the most specious group of eumetazoan animals, with more than a million species described and up to 10 million estimated species. Most of them are terrestrial as opposed to crustaceans. There are large differences in size among taxa: The smallest insect, the parasitoid wasp *Dicomorpha echmepterygis*, has a body length of 139 μm; the largest one, the walking stick *Phobaeticus chani*, a total length of nearly 57 cm. The insect body is segmented into head, thorax, and abdomen. The head carries one pair of antennae, eyes, and mouthparts (maxillae, mandibles, labrum, and labium). The thorax is divided into pro-, meso-, and metathorax, where the three pairs of legs originate (hence the alternative name "Hexapoda," i.e., animals with six legs), and wings—when present—originate at the meso- and meta-thorax. In dipteres with only one pair of wings, only the mesothorax carries wings, and the metathorax drumstick-like organs called halteres. Terrestrial insects breathe through tracheae; in aquatic insects and their larvae, gills have developed.

The insect CNS consists of a brain (supra-esophageal ganglion) and ventral nerve cords with ganglia (Fig. 7.11a; for an overview see Mobbs 1985). The brain, formed by fusion of the first three ganglia, consists of a large protocerebrum, a smaller deutocerebrum and a very small tritocerebrum. Fiber tracts connect the brain with the sub-esophageal ganglion, formed by fusion of the first three ventral cord ganglia. The protocerebrum consists of two hemispheres, which are continuous with the lateral optic lobes processing the input from the compound eyes. In the median protocerebrum, the *central body* (CB) or *central complex* and the mushroom bodies (MB) are found. Terminal fields of the nerves from the ocelli are found in the posterior median protocerebrum.

The nearly hemispherical calyces (cups) of the MB and the so-called aglomerular protocerebrum (also called lateral horn) receive olfactory input from the antennae via the antennal lobes situated in the deutocerebrum and the antennocerebral tract (ACT) to the MB. In flies and dragonflies, which heavily depend on vision, the MB are largely reduced, while in hymenopterans (bees, wasps, ants), the MB are very large. Their sensory input is organized according to modalities: the lip ring region receives olfactory input, the collar visual input, and the basal ring mixed mechanosensory and olfactory input (cf. Chap. 11). In the median protocerebrum, the central complex and the optic tubercle are found, which likewise receive visual input from the optic lobes. These structures are connected with the ventral cords via descending tracts. The smaller deutocerebrum receives mechanoreceptive fibers from the antennae terminating in its dorsal lobe. Here, the antennal lobe is likewise found representing the terminal fields of olfactory afferents from the antennae organized in multiple glomeruli. Projection neurons of

Fig. 7.11 Insect brain. (a) *Lateral view* and (b) *ventral view* of the brain and nerves in the scorpionfly *Panorpa*, (c) schematic drawing of the brain of a honey bee. After Roth and Wullimann (1996/2000), modified

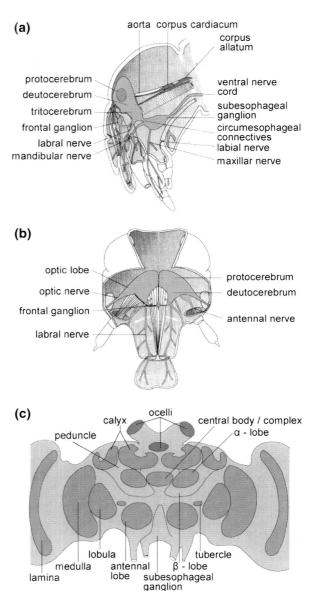

the antennal lobe send axons to the MB and to the protocerebral lobe via the ACT. The deutocerebrum gives rise to the sensory and motor antennal nerves. The small tritocerebrum is the origin of the frontal connectives and the labral nerves.

The chain of ventral cord ganglia consists of sub-esophageal, thoracic, and abdominal ganglia. The sub-esophageal ganglion innervates the mandibles, maxillae and labium as well as the neck musculature. It is also involved in the innervations of the salivary glands, the *corpora allata* (endocrine glands producing the juvenile hormone), and the frontal ganglion, and is considered a higher motor center for the initiation and control of behavior. Most insects have three thoracic ganglia, viz., a pro-, meso-, and meta-thoracic ganglion supplying legs and wings, if present, with sensory and motor nerves. Abdominal ganglia (11 in the embryonic stage) are reduced and fused during development.

The visual system of insects comprises the retina of the compound eye (cf. Chap. 11) and three optic neuropils, the lamina, medulla, and lobula complex, the last of which in flies and butterflies is divided into a lobula and lobula plate. In addition to the compound eyes, insects have dorsal eyes, ocelli, which are simple lens eyes and thought to exert steering functions during walking and flight.

Antennae carry mechanosensitive, olfactory, hygroreceptive, and temperature-sensitive receptors. The neuropil of the antennal lobe in the deutocerebrum contains a species-specific number of glomeruli, in which sensory afferents and interneurons make contacts. Macroglomeruli are found in some male insects related to the excretion of sexual pheromones. Axons of projection neurons form the ACT, which runs to the protocerebral lobe and the MB.

The paired MB are the most prominent structures in the insect brain and are composed of one calyx in many insects or two calyces in hymenopterans (a medial and a lateral one), and a peduncle consisting of two lobes, in hymenopterans and the cockroach, the α and the β lobe (Figs. 7.11b, 17.13). In the honeybee, the MB occupy about half of the volume of the brain (without optic lobes; Mobbs 1985). The neurons, called "Kenyon cells" (KC) (in the honeybee about 300,000—R. Menzel, pers. comm.) forming with their axons ("Kenyon fibers") the peduncle, are the smallest ones found among insects, and their packing density is 15 times higher than the highest one found in the vertebrate brain. The KC get input from about 800 projection neurons of the antennal lobe via about one million presynaptic contacts, plus about 10 postsynaptic contacts (Menzel 2012). KC are located mostly within a cup-shaped indentation of the calyx, but some somata lie also around the outer rim of the calyx. Axons of KC divide in the peduncle, and one collateral enters the α, and another one the β lobe. In the honeybee, the calyces exhibit three vertically arranged regions, the lip ring region processing olfactory input, the collar ring region processing visual input, and the basal ring region processing mixed olfactory and mechanosensory input (Menzel 2012).

The α and β lobes send fibers to about 400 projection or output neurons, which in turn project to the median protocerebrum between the two MB, to the protocerebral lobe lateral to the MB, the contralateral MB, the optic tubercle and back to their own calyces. In hymenopterans, the MB represents a highly complex multimodal center that forms the neural basis of processing and integrating olfactory and visual information and enables learning (mostly olfactory and visual) complex cognitive functions and complex behavior such as navigation (see Chap. 8). Their output recodes the sensory input representing the learning-based evaluation of sensory information.

The central body or central complex of insects consists of four neuropils, i.e., the protocerebral bridge, an upper division (in *Drosophila* called fan-shaped body), a lower division (in *Drosophila* called ellipsoid body), and the paired nodules. It receives strong visual as well as mechanosensory input, but only weak input from the MB. It is probably involved in sensory coding related to navigation (e.g., representing the time-compensated polarization pattern of the sky as a reference for the sun compass), the initiation and control including inhibition and disinhibition of motor sequences and habit formation. The functional similarity between the central body/complex of insects and the basal ganglia of vertebrates is striking, e.g., concerning the role of dopamine, and some authors assume a "deep homology" between them (Strausfeld and Hirth 2013).

The homology of the MB and the central bodies (CB) or central complexes in arthropods is unclear. The CB of insects and crustaceans are probably homologous, whereas a homology with CB of chelicerates and other deuterostomes is debated (see above). Likewise debated is the homology between MB in insects and crustaceans (here called a hemiellipsoid body) on the one hand, and of chelicerates on the other, because in the latter, the CB receive only visual input. Nevertheless, this could be another case of "deep homology."

7.3 What Does All This Tell Us?

A diffuse nervous system found in polyps is usually regarded as the most primitive state of nervous systems found in all true metazoans. From this starting point, two fundamentally different evolutionary pathways departed. One pathway led to the complex ring nervous system of the cnidarians and ctenophorans, which exhibit a dominance of peptidergic neural signal transmission. The other led to the bilaterians. Here, diffuse nerve nets are found in the acoelans, simple bilateral nervous system in planarians consisting of head ganglia and three to six pairs of ventral cords connected by commissures at irregular distances, and complex, tripartite brains like those in some planarians, in polychaete annelids, in mollusks, and in arthropods. In the lophotrochozoans as one of the large groups of bilateral invertebrates, we find an enormous variety of lifestyles, from sessile or parasitic animals to active predators, and accordingly we find large differences in the structure and function of nervous systems and brains, from a simple supraesophageal ganglion and a simple system of ventral cords to the most complex invertebrate brain found in cephalopods like *Octopus*.

In ecdysozoans we again find simple nervous systems in many worm-like, mostly sessile or parasitic taxa as well as highly complex tripartite brains in combination with highly sophisticated sense organs, such as compound eyes. This tripartite brain exhibits considerable variability in the diverse groups of arthropods, i.e., the chelicerates, crustaceans, and insects, which—besides the cephalopods—exhibit the most complex brains found in invertebrates. High-level learning abilities and other well-developed cognitive functions are closely correlated with specialized brain centers, i.e., the vertical lobe in the brains of squids and octopods and the MB found in insects as centers for integration of multisensory information, learning, memory formation, and abstraction, as we will learn in Chap. 8.

Whether these complex brains of lophotrochozoans and ecdysozoans evolved independently from simple ancestral states in the earliest bilaterians, or whether the last common ancestor of all bilaterians already had a relatively complex tripartite brains is hotly debated at present (see above). If the latter was the case, then we would have to assume many independent cases of secondary simplification among protostome taxa. I will come back to that problem in Chap. 9.

Chapter 8
Invertebrate Cognition and Intelligence

Keywords Insect intelligence · Honeybee learning · *Drosophila* learning · Spatial navigation · Abstraction · Waggle dance honeybees · Learning parasitoid wasps · *Octopus* intelligence · *Octopus* learning by observation

After having dealt with the nervous systems and brains of invertebrates, we will ask how intelligent these animals are. This is of particular interest, because even in the behavioral sciences many authors tend to attribute intelligence only to the "higher animals," i.e., vertebrates or even only mammals or primates. Invertebrate animals are—perhaps with the exception of the *Octopus*—considered by many biologists to be pure "reflex machines" and guided by instinct rather than learning. However, if we conceive intelligence as the general ability to respond in a flexible manner to changes in the environment in a way that is favorable for survival, then we have to view even bacteria, archaea, and protozoans as intelligent, because they can modify their behavior on the basis of short-term learning and memory, as we have seen in Chap. 6. The same holds true for the simple, often sessile or slowly moving animals that show habituation and sensitization, but also Pavlovian (classical) conditioning, as has been demonstrated in planarians, earthworms, and slugs, but not in cnidarians or sponges. Operant conditioning as well as "higher" forms of learning such as context learning, is well demonstrated in vertebrates, but is rare in invertebrates and has been convincingly shown only in insects and cephalopods. In the following, I will concentrate on these two groups of animals.

8.1 Learning, Cognitive Abilities, and Intelligence in Insects

Some insects, above all honeybees, exhibit an impressive repertoire of behavior in the domain of feeding, spatial orientation ("navigation"), and social and communicative behavior, and can learn very quickly, especially the association between the color and odor of flowers. This indicates a high degree of behavioral flexibility (for overviews, see Menzel et al. 2007; De Marco and Menzel 2008; Pahl et al. 2010).

Honeybees can be conditioned efficiently to odors by reward, even under artificial conditions, e.g., when a honeybee is immobilized in a test apparatus and only able to move its antennae and mouthparts, i.e., mandibles and proboscis. The antennae are the main olfactory organs of the bee, while they suck nectar or other sugar solutions with their proboscis. When the antennae of a hungry bee come into contact with a sugar solution, the proboscis is reflexively protruded to suck the sucrose. In naïve animals, unfamiliar odors or other stimuli applied to the antennae do not release such a proboscis extension response (PER). If, however, in the course of a classical conditioning experiment, an odor A is presented *immediately before* the presentation of sucrose (which is called "forward pairing"), an association is formed between odor and sucrose such that the odor alone releases the PER, whereas another odor B, which had not been paired with the sucrose before, will not elicit the response. In terms of classical conditioning, odor A is the conditioned (paired) stimulus (CS+), and sucrose the unconditioned, reinforcing stimulus (US), and odor B is the unpaired stimulus (CS-). Thus, the bee has learned to respond to odor A (CS+) and not to odor B (CS-). It is important to note that in the honeybee such classical conditioning is successful only as "forward pairing" or "forward conditioning," i.e., the conditioned stimulus CS+ must precede the US in time (see Chap. 2).

In a series of spectacular experiments, the two German neurophysiologists Martin Hammer (who unfortunately died at a young age) and Randolf Menzel from the Free University of Berlin demonstrated that the activity of one specialized neuron located in the subesophageal ganglion of the bee called "VUMmx1" (which means "ventral unpaired median neuron of the maxillary neuromere 1") represents *an evaluation system* like the limbic system inside the vertebrate brain (cf. Hammer 1993; Menzel and Giurfa 2001). The activity of the VUMmx1 neuron functions as the *neuronal representation* of the food reward in appetitive associative olfactory learning. This type of neuron is connected only with the olfactory pathway and not with other sensory pathways in the brain of the honeybee, and intracellular electric stimulation of this neuron fully substitutes the US (i.e., the sucrose) as reward.

Such a simple conditioning process can be made more complicated in the context of *configural* learning, for example, *negative patterning discrimination*. Here, two stimuli, A+ and B+, are separately reinforced, while the combination of both stimuli, AB, remains *un*reinforced. Normally, a bee responds more intensely to the stimulus combination, but in this case it learns to suppress exactly that response because of lack of reinforcement. Furthermore, a bee can be brought to *contextual learning*, i.e., to exert different kinds of behavior depending on the site and the conditions. Finally, honeybees are capable of *categorical learning*, i.e., they learn to assign differently shaped objects to certain basic forms (oval or rectangular, symmetric or asymmetric) or to group together objects with the same pattern (e.g., vertical or horizontal stripes) or assign novel objects to one of the two categories—tasks that animals with giant brains, such as elephants, master only with difficulty or not at all (see Chap. 12). In the same way, bees are able to learn the category "same–different" and to transfer this concept to novel stimulus arrangements. Giurfa, Menzel, and their colleagues demonstrated that in the process of such categorical learning, a prominent "eureka effect" happens, i.e., after an initially slow learning effect, there is a sudden leap to high learning success (cf. Giurfa 2003).

Experiments by the Menzel group also reveal that honeybees show selective attention for a stimulus, i.e., they are able to "concentrate" on a particular stimulus and actively process this information while ignoring irrelevant stimuli. They can be trained to focus on certain colors against innate preferences, and this significantly increases their sensory color discrimination abilities. Also, honeybees master so-called *delayed match-to-sample* or its opposite, the *delayed non-match-to-sample* (cf. Chap. 2). Here, an animal has to keep in mind for a few seconds a briefly presented target stimulus, before the next series of stimuli is presented, and the animal has to decide whether the target stimulus reappeared or not—and is rewarded for a correct answer. Bees are also able to select a novel stimulus from a series of stimuli. The working memory needed for mastering such tasks had a span up to 8 s (Pahl et al. 2010). This seems to be short, but memory spans of 5-15 s are typical even of vertebrates, including mammals, and are wider only in humans using the "phonological loop" of their working memory (see Chaps. 2 and 15).

"Counting" or numerical abilities are much-studied tasks in cognitive behavioral experiments, most of which are being carried out in vertebrates (cf. Chap. 12). In bees, numerical abilities were tested using the delayed match-to-sample method. Bees first had to learn by reward a sample stimulus carrying a certain number of objects and then were placed into a Y-maze with visual stimuli differing in a number of objects shown at the end of each arm. The animals had to take the left or right course according to the number of visual objects shown on the stimulus that matched the number of objects of the sample stimulus. Objects could be dots, stars, or lemons. The animals learned the abstract number of the objects shown and to transfer this knowledge to new types of objects. They were able to discriminate well between two and three objects and could identify the "three objects" in the three versus four choice, but failed when confronted with higher numbers (Pahl et al. 2010).

It is interesting to test what the "intelligent" honeybees did *not* learn. This includes the well-known law of transitivity—a form of logical reasoning saying, for example: if A is larger than B and B is larger than C, then A is likewise larger than C. Bees were confronted with a sequence of visual stimuli $A > B$, $B > C$, $C > D$, and $D > E$ and then tested with the hitherto unfamiliar pair B versus D. The animals failed in this task. They mastered this task pairwise and only in a pair familiar to them, but not nonadjacent pairs like B and D, and they showed a preference for the pairs shown last (*recency effect*). Apparently, the working memory of bees is incapable of remembering and comparing longer chains of stimuli. In Chap. 12, we will ask whether vertebrates do better.

Spatial orientation capabilities of honeybees have always fascinated experts as well as laymen. Pioneer bees leave the hive and search for food sources, usually flowers. Once they have found a good source, they return to the hive and communicate this to their sisters in the dark beehive by executing the famous waggle dance, as described first by the Austrian-German zoologist and Nobel Laureate Karl von Frisch (1886–1982) (cf. von Frisch 1923, 1965). By dancing in the dark on the vertically oriented comb, they indicate the direction, distance, and attractiveness of the source. In the waggle dance, the dancing bee executes fast and short forward movements straight ahead, returns in a half-cycle in the opposite direction, and starts the cycle again with regular alternations, which means that each waggle dance includes several cycles (von Frisch 1967). The straight portion of this clockwise and counterclockwise movement, called "waggle-run," consists of a single stride expressed by lateral waggling of the abdomen. The length of the single waggle-run represents the *distance* that a bee has to fly to reach the source, while the angle of the run relative to gravity represents the *direction* of the foraging flight relative to the azimuth of the sun in the open field and sun-linked patterns of polarized skylight. In this way, distance and direction of the source are communicated to the colony members. The duration of the overall waggling performance apparently encodes the *attractiveness* of the source—and perhaps other parameters.

Important in this context is the fact that the dances can be used by the bees for communicating very differing kinds of information. The animals may dance for desirable sources of nectar and pollen, but also for water, which is essential for downregulation of the nest temperature, when the hive is in danger to become overheated, and during swarming, for informing about a potential new nest site. This may indicate a certain degree of semantics of the bee dance.

During their exploration flights, bees return safely to the hive, using path integration and landmark learning as guiding cues. They are able to use egocentric and allocentric information in reference to the sun—a mechanism that is called "sun compass." The bees recognize the sun's azimuth (i.e., the direction in which the sun is standing) by the sun itself or by the pattern of polarized light in the sky, but also make use of visual landmarks learned in relation to the sun compass. The components of navigation are thought by some authors to be stored in different "modules" of the bee brain. The question now is whether these modules of navigation by the sun compass, egocentric information as derived from path integration, goal-directed information (beacon orientation), memory of the panorama as seen at the respective sites or geometric relations between landmarks work independently or interact. This can be tested by multiple training procedures, e.g., by training bees to search for food at two foraging spots, i.e., at spot A in the morning, which is situated 115° from the north at a distance of 630 m, and in the afternoon at spot B, situated 40° from the north at a distance of 700 m. Consequently, the bees learn two different paths home to their hive. Now, if the experimenter releases the successfully trained animals at the "wrong" time, i.e., at spot B in the morning or spot A in the afternoon, then the animals fly directly back to the hive, which means that they remember the correct return flight for a given site. If the animals are released halfway between A and B, then half of them fly back to the hive. These results, and many others (see below), appear to indicate that the bees use an "internal" topographic or cognitive map.

The existence of such cognitive maps for spatial orientation has been hotly debated for years (cf. Wehner and Menzel 1999), but could be investigated seriously only after it became possible to track the flight of single bees over longer distances, e.g., 1 km, using a so-called harmonic radar, which detects a radar transponder carried by the bee. Using these methods, Menzel and his collaborators demonstrated that bees reliably find their way back to the hive from all directions, if they are released in an area known to them from exploration flights. In such experiments, bees that had been trained to return to a stationary feeder are used. As soon as they had sufficiently fed at that spot and were prepared to fly back to the hive, they are displaced to a different site of their area of orientation. The animals then first fly in the direction of the hive according to their memory, which has become incorrect. After some searching, they fly back straight to the hive without using any special landmarks or other information coming from the hive. This means that they possess sufficient, "maplike" information about the ground structure and use this map for return. In general, bees appear to use spatial memories "opportunistically" in three different contexts, i.e., (1) a general landscape memory acquired via initial orienting flights, (2) route memory while flying repeatedly to and from and to a specific field location, and (3) the dance memory while following dances in the beehive.

In the context of navigation, *context learning* can likewise be found in bees. These animals are capable of associating different stimulus configurations, e.g., flowers differing in color or shape or a feeder, with a different site and a different daytime. Thus, they can learn to fly to one type of flower in the morning and to another type in the afternoon (Pahl et al. 2010). This shows that they can remember rewarded visual patterns separately regarding spatial and temporal context information.

Other insects, for example the fruit fly *Drosophila* or parasitoid wasps, likewise exhibit good learning abilities. This is especially interesting because these insects are even much smaller than the bees, and their brains contain a considerably lower number of neurons—the brain of *Drosophila* contains roughly 200,000 neurons compared to 1 million in bees. Some parasitoid wasps have become very tiny, i.e., they are about 200 μm long and thus smaller than the protozoan *Paramaecium*. Their brains contain about 4,000 nerve cells, of which only 5 % (i.e., 200) possess a soma, while the somata of the others have lysed, which means an enormous saving of space (Niven and Farris 2012). In addition, these tiny insects are very short-lived. It was generally assumed that small-brained and short-lived animals exhibit no or very little learning abilities, because "investments" into learning appear worthless. Rather, the behavior of such animals should be guided predominantly by instinct. This, however, is an error.

On the one hand, *Drosophila*, like all other insects, dispose of a large repertoire of behaviors or parts of behaviors guided by instinct, and the animals can switch between them in a context-dependent fashion. This is an important basis of their behavioral plasticity. At the same time, however, there is a clear genetic disposition toward learning and memory formation as has been demonstrated by Frederic Mery and Tadeusz Kawecki from the Fribourg University in Switzerland. They were able to select mutants in *Drosophila melanogaster* mutants with significantly higher abilities in learning and memory formation in the context of aversive conditioning at egg deposition (Mery and Kawecki 2002).

In this context, the question of the mechanisms arise that underlie long-term memory (LTM). As already mentioned in the first chapter, one type of LTM depends on gene expression or protein synthesis and can be impaired by administering antibiotics, while other forms of LTM are not affected by the administration of antibiotics and are regarded as being independent of protein synthesis and gene expression. Learning experiments that address this problem, among others, were undertaken with parasitoid wasps, of which about 100,000 species exist—apparently an evolutionary successful group of animals. These animals deposit their eggs inside the larvae of different insects, e.g., of butterflies or fruit flies. For them it is advantageous to learn at which sites or on which substrates (mostly plants) host animals, mostly larvae of other insects, are found more frequently, and they can learn this by associating the odor and aspect of the substrate.

Different species of the parasitoid wasp family *Braconidae* have developed different strategies. The species *Cotesia rubecola* deposits the majority of their eggs in its host animal, the Large White or Cabbage Butterfly (*Pieris brassicae*), on one single plant, while the species *Cotesia glomerata* deposits only one egg onto one host animal, the Small White (*Pieris rapae*), per plant, and therefore visits many plants. The Dutch biologist Smid and colleagues (cf. Smid et al. 2007) trained these two species to deposit their eggs in *Pieris* larvae living on watercress, which is not much "liked" by naïve wasps. In order to elucidate the type of LTM consolidation, the authors treated one half of the wasps either with the antibiotics Actinomycin, which inhibits gene transcription, or Anisomycin, which inhibits protein synthesis, and both groups were compared with control animals. The studies demonstrated that *C. glomerata* needs only one single learning trial to learn, in order to learn the association between plant and host, while *C. rubecola* needed three separate learning trials. In addition, in *C. glomerata* a protein-synthesis dependence of LTM consolidation was found, while in *C. rubecola* two parallel processes of LTM consolidation were observed, one dependent and the other independent of protein syntheses, and this parallelism lasted for about three days. This suggests that various types of formation of LTM exist in insects that may even be found in closely related species.

Investigations of my Bremen colleagues Andra Thiel and Thomas Hoffmeister (Thiel and Hoffmeister 2009) of egg deposition behavior in parasitoid wasps reveal an astonishingly "rational" decision making process, in which information, for example, about the distribution and the residence of a host, the most favorable larval stage of the host and the parasitization status of the host are taken into account. During egg deposition, the wasps have to make complex "decisions" about whether to continue depositing eggs in a given host or switching to another one, which come close to abstract optimization models. This demonstrates that these animals are capable of highly flexible egg deposition behavior despite their extremely small brains.

Bees can learn both via classical and operant conditioning, while in fruit flies, interestingly, classical conditioning is difficult to demonstrate. This reveals that both conditioning procedures represent, in fact, different types of associative learning, and that operant conditioning is not just a more "complex" or "higher" form of learning compared to classical conditioning. The neurobiologist Martin Heisenberg and his colleagues from Würzburg University gave evidence of operant conditioning in the fruit fly *Drosophila*. A fly was fixed on an apparatus in the center of a cylinder in such a way that it could not fly away but, could still bend its body and move legs and wings. These movements caused the cylinder, or parts of it, to move in one or the other direction around the fly. A heat source was coupled with the cylinder or its parts, such that the heat source was brought close to the fly or moved away from it, depending on how the fly was behaving. If now the fruit fly experienced a dangerous increase in temperature, it first fidgeted around wildly, until it executed, by chance, a certain movement that moved the heat source away, and the fruit fly quickly learned to execute these favorable movements following the scheme of operant conditioning, here negative conditioning (Brembs and Heisenberg 2000; Menzel et al. 2007).

8.2 Learning, Cognitive Abilities, and Intelligence in Cephalopods

As we have already learned, some cephalopod taxa, predominantly squids and octopods, possess large and complicated brains. This correlates well with their predatory lifestyle—predators in general tend to have larger and more complicated brains. Accordingly, in the *Octopus*, with a brain containing 42 million neurons and the most complex neuroanatomy among invertebrates, we expect a high degree of intelligence.

The lifelong studies of the eminent British zoologist J. Z. Young (1907–1997) not only supplied us with a wealth of knowledge about the anatomy of the nervous system of cephalopods in general and of the *Octopus* in particular, which Young obtained while working in the famous Stazione Zoologica in Naples, Italy, but gave evidence of the high cognitive abilities that led Young to put the *Octopus* into the neighborhood group of the most intelligent vertebrates, i.e., primates. In following these lines, the *Octopus* was regarded as a genuine "egghead." Indeed, astonishing achievements of *Octopus* are reported in the domain of spatial orientation. The animal not only remembers well where tasty food can be found, but upon returning to its home site after long travels, it often takes the shortest route, which it had never taken before. Such behavior shows that the *Octopus* has good spatial memory, but it is unclear whether this allows us to conclude from that the existence of a "mental map," because some experts argue here that the animal applies path integration, which is found in many other animals, such as crabs or ants (cf. Menzel et al. 2007). Other observations and experiments demonstrate that the *Octopus* uses its siphon for cleaning sand and garbage from its cave and its surroundings. Also, it could be observed that the animal collects little stones and piles them up at the entrance to its cave for protection against predators. Some *Octopus* specialists interpret this as evidence of tool use. Finally, these animals became famous for playing with plastic bottles and being able to unscrew lids from jars filled with prawn (which can be watched in Internet videos).

In experiments by the American zoologist Jean Boal, the *Octopus* learned relatively quickly how to escape from a complicated maze, and remembered this experience for one week. It could also master some tasks in the domain of object and pattern recognition, but its achievements were modest and did not exceed those of teleost fishes. Over the years of intense *Octopus* studies, the general impression was that the more one investigates the behavior of this animal, the more modest its cognitive abilities appear (which is similar in dolphins, as we will learn in Chap. 12).

In 1992, the Italian authors Fiorito and Scotto published a study showing that the *Octopus* is capable of learning by pure observation of the behavior of its conspecifics (Fiorito and Scotto 1992). Their experiment proceeded as follows: First, a group of octopods (the "demonstrators") were trained by reward and punishment to select from a pile of red and white balls either the red or the white ones. During the second phase of the experiment, untrained ("naïve") animals (the "observers") were allowed to watch the demonstrators choosing either the red or white balls, while no reward or punishment was applied to them during that choice. Finally, the observers were confronted with the choice between the red and white balls, and they preferred—something that is very significant—the type of ball that the demonstrators had chosen while being watched. The authors reported that this preference acquired by observation was stable for at least five days after the observation (cf. Fiorito and Chichery 1995; Fiorito et al. 1998).

What irritated the skeptics in particular, besides the methodological question, was the fact that such an ability makes sense in highly social big-brained animals, but not in the *Octopus*, which conducts a solitary life and interacts with conspecifics only during mating and never meets it parents or children. Thus, the *Octopus* would reveal a behavior which apparently is not part of its behavioral repertoire. But—as we know today—this is nothing unusual in intelligent animals: chimpanzees, for example, can learn to use sign language or keyboards for communication with humans and even conspecific, although, somewhat astonishingly, they do not make as much use of such "practical things" as we would expect for humans.

However, the already mentioned *Octopus* specialist Boal was unable to reproduce these results, but in 1995 Fioriti and Chichery showed that removal of the vertical lobe (cf. preceding chapter) abolished the ability for "learning by observation" in the *Octopus*. Until today, it is unclear how the findings by Scotto and Fiorito should be interpreted, i.e., whether or not they give clear evidence of observational learning. Some years after the appearance of the article, Fiorito and the American expert on learning, Biederman (together with two other colleagues), published an article in the context of the mentioned capability of the *Octopus* to unscrew a jar and mentioned a weak pre-exposure effect (Fiorito et al. 1998).

In any case, there can be no doubt that the site for learning and memory formation in octopods is the vertical lobe. In neurophysiological experiments, the vertical lobe was stimulated via the nerve bundle running from the median superior frontal lobe (the MSF, see preceding chapter and Fig. 17.2) to the vertical lobe. Processes of the type of long-term potentiation (LTP) as one important mechanism for learning were found, which, however, did not involve NMDA receptors (cf. Chap. 5). Interestingly, sectioning the mentioned nerve bundle impairs only the long-term, but not the short-term memory (Boycott and Young 1955; Hochner et al. 2006; Shomrat et al. 2008).

8.3 What Does All This Tell Us?

Invertebrates—comprising more than 95 % of all animals described so far—exhibit in their vast majority a relatively simple behavior mostly based on reflexes and instinct behavior. However, simple forms of nonassociative learning, like habituation and sensitization, are universal. Widely distributed are effects of classical conditioning in mollusks (e.g. *Aplysia*) and arthropods. Operant conditioning has been demonstrated in insects and here predominantly in hymenopterans like bees and parasitoid wasps, dipterans (*Drosophila*), as well as in the *Octopus*. Possibly, the latter are capable of observational learning, as are honeybees with respect to the waggle dance. Honeybees exhibit categorical learning, the existence of "mental maps" and a "eureka effect" in learning. They fail in tests for logical reasoning, e.g., in the form of the law of transitivity, but here many vertebrates (e.g., pigeons) are no better.

While the capabilities of the honeybee are truly impressive, the findings in the *Octopus* do not meet popular expectations, although they appear to exceed all the cognitive abilities found in other invertebrates except honeybees. However, these two groups of animals are not directly comparable because *Octopus* has much higher manipulatory abilities compared to the bee due to its long and flexible arms.

Insects and octopods possess the most complex brains among invertebrates, but at the same time, differences in brain size and number of neurons are dramatic between these two groups: The brain of the *Drosophila* contains 200,000 neurons, the honeybee about 1 million, and parasitoid wasps 100,000 or less. This strongly contrasts with the 42 million neurons found in the brain of the *Octopus*. We have to ask ourselves later what that means for the question of the relationship between brain properties and cognitive abilities.

Chapter 9
The Deuterostomia

Keywords Origin deuterostomes · Myxinoids · Petromyzontids · Chondrichthyans · Osteichthyans · Amphibians · Reptiles · Birds · Mammals

9.1 The Origin of Deuterostomes and Their Nervous Systems

Deuterostomes are the sister group of the protostomes, as illustrated in Fig. 3.1. They are characterized by the fact that their central nervous system is located in the dorsal part of the body, whereas in protostomes it is found ventrally forming the ventral nerve cords and their ganglia. An exception to that rule seems to be the echinoderms, which have a radially organized body like the cnidarians, and accordingly have a radially organized nervous system. However, adult echinoderms develop from bilateral larvae which in the eyes of many authors indicates that radial symmetry of the adults is secondary and that the ancestors of echinoderms were bilaterally symmetric. The presently accepted taxonomy of deuterostomes is given in Fig. 9.1.

The basic organization of protostomes and deuterostomes appears to be fundamentally different, which may suggest that both groups of bilaterians developed independently from a sponge-like ancestor. However, already early in the theory of evolution there was the hypothesis that they have a common ancestor and are more similar than they appear. The French biologist Etienne Geoffroy St. Hilaire (1772–1844) had the idea that the only thing one has to do is to turn the elongated invertebrates (the "worms") by 180° around their long axis, and their ventral nerve cords become the dorsal "spinal cord" of vertebrates (cf. Hirth and Reichert 2007).

There is the idea that a long time ago it was more favorable for some invertebrates to swim upside down—and there is the lancelet, one of the most primitive chordates (see below) that swims both ways. Furthermore, some years ago, together with the impressive progress of developmental genetics, St. Hilaire's idea

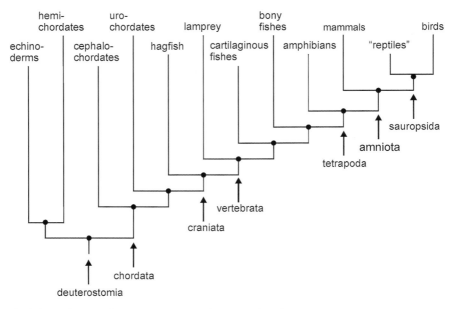

Fig. 9.1 Phylogeny of deuterostomes. For further explanations, see text

became attractive again, when it could be demonstrated that very distantly related species, i.e., the fruit fly *Drosophila* and the clawed toad *Xenopus*, possess the same developmental genes that are responsible for the basic organization of the body, including the brain. These genes determine the body axes (head and tail, dorsal and ventral), the basic segmentation of the body, the nervous system and brain, the formation of nerve cells and their processes, and even the formation of the sense organs, including the eyes (Martinez et al. 2013).

Within the central nervous system of the *Drosophila*, Hox genes determine the formation of the ventral cord and the tritocerebrum, and in the *Xenopus* they influence the formation of the spinal cord and hindbrain. Furthermore, while in insects non-hox genes like *otd/Otx* control the formation of the more anterior parts of the brain, i.e., the proto- and deutocerebrum (see above) and in the chordates and vertebrates the development of the mesencephalon, diencephalon, and telencephalon (see below). In between there is a so-called *Pax region*, which in the vertebrate brain defines the isthmic region of the brainstem (Farris 2008; Martinez et al. 2013). The same pattern of developmental genes has been found in all bilaterians investigated so far. Accordingly, some authors believe that in all bilaterians there is a basic organization for a tripartite brain, despite the great differences observed (cf. Hirt and Reichert 2007 and above).

Such a view solves many problems regarding the evolution of nervous systems and brains, but at the same time creates new ones. For example, despite the fact that there is a homology of genes in protostomes and deuterostomes, the regions that they control in the nervous systems and brains could have evolved independently of each other. The same or similar genes could have undergone a functional change in the two groups and accordingly could do very different things. But it may also be that even the structures, and not only the genes, are homologous. In this case, a relatively complex central nervous system and brain would have developed once 600 million years ago in the so-called "ur-bilateria" (Lichtneckert and Reichert 2007), and all differences found would be nothing but evolutionary modifications of that basic organization.

Other authors do not deny the existence of common developmental or "organizer" genes for the nervous systems of bilaterians and maybe even for all eumetazoans, but reject the idea of a common origin of their nervous systems (Moroz 2009). In their eyes, central nervous systems evolved at least three times independently from ancestral diffuse nerve nets, for the first time in the lophotrochozoans, for the second time in the ecdysozoans, and for the third time in the chordates. This would imply that the radially symmetric nervous system of the echinoderms is primitive and not a product of secondary simplification of a bilateral nervous system.

In this context, there is the already discussed question of how the same genes can lead to very different structures. In most details, the insect brain differs from the vertebrate brain, and the compound eye of insects has no resemblance to the vertebrate eye despite the great similarity of the underlying developmental genes. If we accept the existence of such genes, then we have to accept that these genes do not control a precise structure, but rather give more general commands, such as "Build a tripartite brain!" or "Form a light-sensitive organ!" and that other and more specific genes, together with epigenetic mechanisms, determine which structures are precisely formed, for example, a pigment spot, a simple pit eye, a compound eye, or a vertebrate lens eye. Every student of biology learns that the lens eyes of the *Octopus* and of the vertebrates, despite their striking resemblance, are not homologous, but products of convergent evolution. In fact, the embryonic tissues from which the cornea, lens, and photoreceptors develop are different, and the *Octopus* has an *everted* and vertebrates an *inverted* eye. But how to interpret the finding that the ancestral genes of both eyes are the same and even in the case of the compound eyes of insects? The situation becomes even more complicated by the fact that in invertebrates we find many cases of inverted as well as everted brains, and that compound eyes appear to have evolved several to many times independently.

In Fig. 9.1, the presently most accepted taxonomy of the deuterostomes is shown. On the one hand, we find the echinoderms (*Echinodermata*), hemichordates (*Hemichordata*), and the enigmatic xenoturbellarians (not shown in Fig. 9.1) with only one genus. On the other hand, there are the chordates (*Chordata*), which include the cephalochordates (*Cephalochordata*), the tunicates (*Tunicata*), and the craniates (*Craniata*, i.e., animals possessing a cranium or skull. The craniates comprise the hagfish (*Myxinoidea*) and the vertebrates (*Vertebrata*) with the Petromyzontidae (*Petromyzontida*, lamprey) and all other classes of vertebrates, i.e., cartilaginous fishes, bony fishes, amphibians, mammals, "reptiles," and birds.

9.2 Echinoderms

The phylum *Echinodermata* (meaning "animals with spiny skin;" up to 7,000 species described) comprises, among others, starfish (*Asteroidea*), sea lilies (*Crinoidea*), brittle stars (*Ophiuroidea*), sea urchins (*Echinoidea*), and sea cucumbers (*Holothuroidea*), all of which are marine animals living from intertidal to abyssal zones. The echinoderms originated in late Precambrian times early after the split between protostomes and deuterostomes, i.e., about 560 mya. All echinoderms exhibit a five-fold (pentameric) radial symmetry as basic organization, which during ontogeny develops, as mentioned, from a bilaterally organized free-swimming larva resembling embryonic chordates.

Echinoderms greatly differ in their feeding modes. Crinoids and some brittle stars are mostly passive filter-feeders, sea urchins are grazers, and sea stars extract organic particles from the mud or are active hunters.

Similar to the cnidarians, echinoderms have no brain, but two decentralized nervous systems. The first one is called the *ectoneural* nervous system that surrounds the mouth. From there nerve cords radiate into the arms (if present) or along the body. There is a nerve plexus, which is closely connected to the skin, and chemo-, mechano-, and photoreceptive cells located there. The second one is called the *hyponeural* nervous system and has purely motor functions. Experts believe that the ectoneural system is of ectodermal origin, like the central nervous system of all chordates, while the hyponeural system derives from the mesoderm, which in chordates gives rise to muscles (among others). The similarity with the double nerve ring system of cnidarian medusae (cf. Chap. 7) is striking, although there is little doubt that both developed independently.

9.3 Hemichordates

The hemichordates (70–100 species) are wormlike or sessile marine animals. They have an unsegmented body ranging from a few millimeters to 2.5 m in length. The main group are the acorn worms (*Enteropneusta*). They live in sand burrows and either extract organic material from the sand or are filter feeders. They are an ancient deuterostome group originating in the middle to late Cambrian Age. They are characterized by a diverticulum of the foregut called stomochord, which for a long time was interpreted as a forerunner of the "notochord" (or *chorda dorsalis*) of chordates (see below), and therefore were viewed as direct ancestors of the chordates (hence the name "hemichordates"). However, these two structures are not considered homologous anymore, and now the hemichordates are believed to be more closely related to the echinoderms than to the chordates. The smaller group of hemichordates, the pterobranchs (*Pterobranchia*) are small, sessile marine filter feeders that live in colonies.

The nervous system of the hemichordates essentially consists of a ventral and a dorsal nerve cord, which are connected by nerve rings in the head lobe and around the gut. In the acorn worms, the dorsal cord is a hollow tube and considered by some authors as homologous to the spinal cord of the chordates, while others view it as having independently evolved from a diffuse nerve net (Moroz 2009).

9.4 Chordates-Craniates-Vertebrates

The phylum *Chordata* (about 65,000 species) comprise the urochordates or tunicates (*Urochordata* or *Tunicata*), the cephalochordates (*Cephalochordata*) and the craniates (*Craniata*). They are characterized by the presence (at least in embryonic stages) of a notochord, i.e., a fairly stiff cartilaginous rod extending along the dorsal part of the body and stabilizing it, a hollow dorsal nerve cord or "neural tube," pharyngeal slits as part of the throat immediately behind the mouth, and a tail extending behind the anus. Accordingly, chordates mostly have elongated bodies. While there can be no doubt that the chordates are related to the echinoderms, the hemichordates, and the xenoturbellarians, the detailed phylogenetic relationship is unclear, and accordingly there is no clear picture of the common ancestor of these groups. The first chordates already existed in the Cambrian Age.

The *Urochordata* (2–3,000 species) have a free-living larva, but are sessile as adults and are marine filter feeders. The notochord is present only in the tail region. The nervous system is very simple, like in all sessile animals. In contrast, the *Cephalochordata* (about 30 species, including the well-known lancelet, *Branchiostoma,* previously called *Amphioxus*) are lifelong freely swimming animals with a translucent, fish-like body 5–7 cm long, without a true skeleton, paired fins or other limbs. Paired sense organs are likewise absent. As all chordates, they have a hollow nerve cord running along the back, pharyngeal slits, and a post-anal tail. Also, they have muscles arranged in blocks called myomeres. Lancelets live in shallow sea water, mostly burrowed in the sand and moving water and larger particles with "cirri," i.e., tentacle-like strands at the mouth, into the pharynx, where organic particles are trapped by the mucous surface of the gill slits.

The nervous system of *Branchiostoma* consists of a neural tube and a modestly developed cerebral vesicle, which together contain about 20,000 neurons. Despite its simple appearance, recent studies reveal that it already possesses most of the developmental genes that are required for the formation of the craniate-vertebrate brain (cf. Holland and Short 2008). There is a region homologous to the craniate hindbrain (*rhombencephalon*), because the typical Hox and ParaHox genes are expressed here, although a segmentation into rhombomeres (see Chap. 10) is not visible, while segmentally arranged motor neurons exist. In addition, according to most authors, there is a midbrain (*mesencephalon*) and part of a forebrain (*prosencephalon*), at least in the form of an in-between-brain (*diencephalon*), while the presence of a true endbrain (*telencephalon*) or parts of it is debated. Some experts believe that there are at least ventral regions of the prospective telencephalon, because genes like Pax6 and Otx are expressed. *Branchiostoma*, like all craniates, possess a neural plate, but no neural crest typical of craniates, although, somewhat paradoxically, neural crest genes are present.

The subphylum *Craniata*, i.e., chordates with a *cranium* (skull), comprise the hagfish (*Myxinoidea*) and the vertebrates (*Vertebrata*) including the petromyzontids. According to Northcutt and Gans (1983), the evolution of a neural crest of placodes and in this context of a true head, together with sense organs carried by the head, was a key event in the evolution of craniates. The multipotent cells of the newly formed neural crest invade the embryonic body and—among others—transform into the branchial skeleton, cranium, peripheral nervous system, pigment cells (melanocytes), and adrenal medulla. Placodes are thickenings in the embryonic epithelial layer, from which, among others, sensory epithelia are formed. These include the otic placode that forms the otic pit and the otic vesicle, eventually giving rise to organs of hearing and equilibrium, the lens placode which, under the influence of the optic vesicle, gives rise to the eye lens, the olfactory or nasal placode, which gives rise to the olfactory epithelium, the trigeminal placode, which gives rise to some of the sensory ganglia of the head and in mammals leads to the formation of the ophthalmic and maxillo-mandibular branch of the trigeminal nerve, the epibranchial placode forming other sensory ganglia of the head, the adenohypophyseal placode giving rise to the anterior part of the pituitary (adenohypophysis), and the lateral-line placode which in aquatic vertebrates give rise to the lateral line system (see Chap. 10).

Thus, the placodes are essential for the sensory epithelia of the ear, eye, nose, and the electro- and mechanoreceptive lateral-line system, the formation of a branchial apparatus, which later in terrestrial vertebrates transforms into the jaws, making new ways of feeding and prey capture possible. Some authors assume that these novelties evolved in early vertebrates in the context of competition with dominant forms of invertebrates, e.g., cephalopods.

In the following, I will briefly describe the myxinoids and the vertebrates. The description of their brains will be given in Chap. 10.

9.4.1 Myxinoids

Myxinoids (*Myxinoidea*), or hagfish (about 60 species), are eel-like exo- and endoparasitic animals of about 50 cm in length. They were previously, together with the lampreys (*Petromyzontida*), included into the taxon *Agnatha* (which means jaw-less animals) or *Cyclostomata* (which means animals with round mouths), while today they form a craniate taxon of their own as a sister group of the vertebrates, including lampreys. Hagfish live in coastal regions and can produce enormous quantities of slime when captured and held and therefore are also called "slime eels." Their ancestors evolved about 530 mya in the early-middle Cambrian age. The anterior end of the animals is marked by tentacles, mouth, and nose openings, and eyes that are covered with skin and have no lenses. It is unclear whether this is a primitive or derived feature, because such reductions of the eyes often occur in parasitic or cave-dwelling animals. Photoreceptive cells are found all over the body, but olfaction and touch are the major senses. A lateral-line system is absent, and their vestibular apparatus has only two, rather than three canals. Hagfish feed on small aquatic organisms, are scavengers or conduct a parasitic lifestyle by attaching their mouths to dead or dying fishes, forming openings through which they can enter their prey and eat them from the inside.

9.4.2 Vertebrates

The subphylum *Vertebrata* comprises the class *Petromyzontida* (lampreys, 40–50 species) and the superclass (or infraphylum) *Gnathostomata*) (i.e., jawed animals). The latter comprise the classes *Chondrichthyes* (about 1,100 species), *Osteichthyes* (bony fishes, about 30,000 species) *Amphibia* (amphibians, i.e., frogs, salamanders, caecilians, about 6,000 species), *Mammalia* (about 5,700 species), and the group *Sauropsida*, which include the former class "Reptilia" in the classical sense (i.e., chelonians, rhynchocephalians, squamates, and crocodilians; about 9,500 species) and the class *Aves* (birds, about 9,500 species). Lampreys, cartilaginous, and bony fishes as well as amphibians together are called "anamniotes," because their eggs have no amnion (i.e., a membrane surrounding and protecting an embryo), while mammals and sauropsids are called "amniotes."

The ancestors of extant vertebrates were the *Ostracodermi* ("shell-skinned animals") living in the Upper Silurian period (around 430 mya) and covered with a bony shell, a pair of pectoral fins, but lacking jaws and a bony inner skeleton. At the end of the Devonian, i.e., 359 mya and after the appearance of jawed fish, the entire group became extinct.

Petromyzontids

Petromyzontids (or *Petromyzontida*) are jawless, eel-like animals (hence the common name "lamprey eels") and considered the most primitive group of vertebrates. They are often called "nine-eyed" fish, because the three pairs of gill slits were taken for eyes, in addition to the pair of eyes and the unpaired parietal eye. Adult animals such as the "sea lamprey" *Petromyzon* mostly live in the open ocean or in coastal regions, but for reproduction they invade streams with suitable habitats. The "ammocoetes" larva develops there and then migrates toward the ocean. Many species are parasitic, like *Petromyzon*, which has a toothed, funnel-like mouth with which it attaches to larger fish and sucks their blood. In contrast to hagfish, lampreys have a number of well-developed sense organs, i.e., an unpaired nostril and olfactory epithelium, paired eyes, an unpaired pineal body (parietal eye), a paired auditory and vestibular inner ear with three canals, a mechanoreceptive system with relatively simple epidermal neuromasts and an electroreceptive system (see Chap. 11).

Chondrichthyans

Chondrichthyans (*Chondrichthyes*), i.e., cartilaginous fishes, comprise the elasmobranchs with sharks (*Selachii*, about 500 species), rays and skates (*Batoidea*, about 600 species), and the chimaeras (*Holocephali*, 34 species). The chimaeras are considered more ancient, from which about 350 mya the elasmobranchs diverged. Chondrichthyans are characterized by a skeleton made of cartilage rather than bone. Since their Silurian ancestors, the *Placodermi*, had a bony skeleton, the presence of a cartilaginous skeleton must be interpreted as a derived feature, which perhaps has evolved in order to reduce weight, because chondrichthyans have no swim bladder like the bony fish and need to move constantly in order not to sink.

Sharks and chimaeras have fish-like bodies and mostly live in the ocean, with the exception of the "bull shark" *Carcharhinus leucas*, which besides coastal regions, lives in great lakes and rivers. Sharks can achieve considerable lengths up to 13 m (e.g., the whale shark *Rincodon typus*) and can, therefore, be longer than any bony fish. They are mostly predators or scavengers, but the largest sharks are herbivores and not dangerous to humans. As an adaptation to life on the seafloor, most rays have flattened bodies and enlarged pectoral fins that are fused to the head, and their nostrils, eyes, and gill slits are located on the ventral surface of their heads. Rays and skates feed on small, ground-dwelling invertebrates.

Some groups of sharks, the galeomorphs, as well as myliobatid rays, independently evolved large and complex brains (see Chap. 10). Sharks and rays are well equipped with sense organs, e.g., an excellent olfactory sense which is connected with large and often protruding olfactory bulbs, a likewise excellent sense of taste with receptors in the mouth and gills, large movable eyes well suited for vision in the dark (i.e., with dominance of rod photoreceptors), a well-developed inner ear, and two lateral-line systems, i.e., a mechanoreceptive and an electroreceptive one (see Chap. 11). Interestingly, cartilaginous fishes, although possessing electroreception, have no electric echolocation, as do the so-called weakly electric fish (see below). However, the electric rays (*Torpediniformes*) can generate electric discharges up to 220 V to stun prey and for defense.

Osteichthyans

Osteichthyans (*Osteichthyes*), or bony fish, form the largest class of vertebrates. This class comprises as its largest group the actinopterygians (ray-finned fish, *Actinopterygii*), then the brachiopterygians (arm-finned fish, *Brachiopterygii*), and the sarcopterygians (lobe-finned fish, *Sarcopterygii*), which are divided into the dipnoans (lungfish, *Dipnoi*) and the coelacanth (the only member of the *Coelacanthimorpha*).

The extant brachiopterygians consist only of one family, the birchirs (*Polypteridae*) living in freshwater habitats in tropical Africa and the Nile river system. They are considered to resemble archaic Osteichthyans. They have two thick pectoral fins which they can use for forward swimming. They also have lungs and can breathe air. This enables them to live in muddy and oxygen-poor environments of tropical Africa.

The actinopterygians are divided into three superorders, i.e., chondrosteans (*Chondrostei*), holosteans (*Holostei*), and teleosts (*Teleostei*). The chondrosteans (sturgeons and paddlefish, 25 species) are considered the most ancient group of actinopterygians. Sturgeons have an elongated body with a long nose and an extended tail. Similar to cartilaginous fish, their bony skeleton was replaced almost completely by cartilaginous material. They have small eyes, but an excellent olfactory and gustatory system. In addition, they have both a mechanoreceptive and electroreceptive system with many ampullary organs covering the entire body surface. Their brain is relatively simple with the exception of a highly developed cerebellum in combination with the mechanoreceptive and the electroreceptive system.

The holosteans are likewise considered primitive actinopterygians and comprise two orders, the *Lepisosteiformes* (gars) and the *Amiiformes* (bowfins). The body of gars is covered with ganoid scales like in sturgeons, while bowfins have cycloid and ctenoid scales like the teleosts. Therefore, gars are considered more primitive and more closely related to chondrosteans than the bowfins.

The "true" bony fish or teleosts (*Teleostei*, about half of vertebrate species) are characterized by a gaseous swim bladder for the control of buoyancy, a movable maxilla and premaxilla that makes it possible for teleosts to protrude their jaws outwards from the mouth, and a specialized body musculature for fast movements. They originated in the Triassic period. Major groups (superorders) are the *Osteoglossomorpha* (i.e., bony-tongued fish—probably the most primitive teleost group), the *Elopomorpha* (eels and their relatives), *the Clupeomorpha* (herrings, etc.), the *Ostariophysi* (carps, catfish, electric eels, etc.), and the *Acanthopterygii* with the largest teleost group, the *Perciformes* (or *Percomorpha*, about 7,000 species) including the largest teleost family *Cichlidae* (cichlids), considered the most modern group of teleosts. Many cichlid species have evolved only 100,000 years ago in the large lakes of Africa or Central and South America.

Lungfish (*Dipnoi*) are considered an ancient group of vertebrates originating in the lower Devonian, with a peak in the upper Devonian and Carboniferous. They once had a worldwide freshwater distribution, and the current distribution of the surviving lungfish in South America, Africa, and Australia is considered a consequence of the breakup of the ancient continent Pangaea and afterwards, Gondwana. They have the ability to breathe air and possess lobed fins with a well-developed internal skeleton. Only six species survived, i.e., the Australian *Neoceratodus forsteri*, the African genus *Protopterus* with two species (*P. dolloi* and *P. annectens*), and the South American *Lepidosiren paradoxa*. Like their ancestors, lungfish live primarily in rivers and are capable of surviving longer periods of seasonal drying out of their habitats by burrowing into mud. They have small eyes, but a well-developed olfactory system as well as a mechano- and electroreceptive lateral-line system.

The group of coelacanthimorphs or coelacanths (meaning fish with "hollow spines") includes only two species, *Latimeria chalumnae* and *L. menadoensis*. It was long believed that the coelacanths disappeared 70 mya, i.e., toward the end of the Mesozoic, until in 1938 one specimen, later called *Latimeria chalumnae*, was discovered off the Chalumn river at the coast of South Africa by Marjorie Courtenay-Latimer. This evoked worldwide interest because experts believed that with this, a direct version of the ancestor of terrestrial vertebrates had been discovered. Since then about 100 specimens have been found and studied in their natural habitats. *Latimeria* is characterized by its "lobed," fleshy pectoral fins, which are used to stabilize their movement through the water, but not for forward movement on the ground, as was previously believed. Thus, they are not a model for an ancestral pattern for locomotion of terrestrial vertebrates.

Amphibians

Modern amphibians (*Amphibia*) are represented by the subclass *Lissamphibia* (i.e., smooth-skinned amphibians), which comprises three orders: frogs (*Anura*), with 29 families and about 5,100 described species, salamanders and newts (*Urodela* or *Caudata*, with ten families and about 545 described species), and caecilians (*Gymnophiona*), with six families and about 170 described species. Anurans are found worldwide with the exception of Arctic and Antarctic regions, while salamanders are found only in the Northern Hemisphere of Eurasia as well as in Northern and Central America and including northern parts of South America. Caecilians are found only in the tropics and subtropics of Eurasia, Africa, and America. Most experts now believe that lissamphibians form a monophyletic group and that their closest relatives are the lungfish. Therefore, the extant dipnoans are the sister groups of extant amphibians and of all tetrapods. The first terrestrial vertebrates, the labyrinthodonts (which means "maze-toothed animals"), to which the crocodile-like *Ichthyostega* belonged, lived from mid-Devonian until early Mesozoic (390-210 mya) and developed in a still unknown way to the modern amphibians, with the again crocodile-like *Temnospondyli* as the more immediate ancestors of the lissamphibians. The phylogenetic relationship of the three amphibian orders is still debated. On the ground of morphological as well as molecular data, most authors believe that salamanders and caecilians are more closely related to each other than both with frogs. While the elongated bodies of salamanders resemble the ancestral amphibian condition, frogs underwent a thorough reorganization of their bodies by strong reduction of their long body axis and equally strong elongation of their hindlimbs, which enabled them to make large leaps. Caecilians adapted to underground life by developing a wormlike body shape, loss of limbs, and strong ossification of the skull.

Reptiles

The traditional vertebrate class "Reptilia" comprises four groups: turtles (*Chelonia*, 290 species), tuataras (*Rhynchocephalia*, two species, *Sphenodon punctatus*, and *S. guentheri*), squamates (*Squamata*, i.e., lizards, snakes, gekkos, and amphisbaena, together more than 9,000 species), and the group of crocodiles and alligators (*Crocodilia*, about 20 species). However, according to new taxonomy, this class is not a monophyletic, but a paraphyletic group, because the crocodiles are more closely related to birds than to the other "reptiles" (cf. Chap. 3 and Fig. 3.1). The crocodiles, together with the extinct dinosaurs and the birds (*Aves*) are now grouped into the *Archosauria*, as opposed to the *Lepidosauria*, which include the tuataras and the squamates. Together, they form the *Diapsida* (animals with a skull with two openings in addition to the eye). Turtles are now considered an outgroup of the diapsids and are called *Anapsida* (animals with a skull without an opening) and regarded as relatively evolved. All groups of former "reptiles" plus birds are called "sauropsids" and form the sister group to the mammals. In the following, I will use the traditional term "reptiles" in quotation marks.

The first reptile-like terrestrial vertebrates appeared in the lower Carboniferous (about 320 mya). Their most characteristic traits, compared to the amphibians, is that their skin is covered with scales (which perhaps was a heritage from extinct amphibians—as opposed to the extant smooth-skinned ones), a higher position of the limbs and body above the ground enabling a more effective locomotion, and strictly terrestrial egg deposition and development inside an egg that possesses an amnion, hence the term "amniotes" for both mammals and sauropsids. Lizards and crocodiles resemble reptilian ancestors, while turtles and snakes underwent a thorough modification of their bodies. Turtles are characterized by a flattened body covered with a special bony or cartilaginous shell (carapace and plastron) developed from the backbone and the ribs. Snakes originated from lizard-like ancestors and developed, as an adaptation to burrowing lifestyle, a highly elongated body while losing limbs, eyelids, and external ears (still present in limbless lizards). Eyes were strongly reduced or lost at all and later newly developed, which explains many differences in their eyes to those of extant lizards. A similar reduction took place with the inner ear, which is capable of perceiving low frequencies only.

Birds

Birds (*Aves*) represent, with about 9,500 species, the second largest group of vertebrates and forms, together with the crocodiles and the extinct dinosaurs, the sauropsid group *Archosauria*. Their evolution from crocodile-like ancestors lies in the dark. The well-known *Archaeopteryx* ("first bird" or "Urvogel") lived in the upper Jurassic about 150 mya and is considered a transitorial form from reptiles to modern birds. It still had a reptile-like body and no ossified wishbone, possessed teeth in the upper and lower jaws and a long tail, but already carried feathers, which probably served predominantly for insulation. In addition, the forelimbs were already turned into wings, which are believed by experts to serve mostly for catching insects and making a gliding flight possible.

Extant birds are divided into 28 orders, including *Struthioformes* (ostriches, emus, kiwis), *Pelecaniformes* (pelicans and allies), *Ciconiiformes* (storks and allies), *Anseriformes* (waterfowl), *Galliformes* (fowl), *Falconiformes* (falks, eagles, hawks, and allies), *Columbiformes* (doves and pigeons), *Psittaciformes* (parrots and allies), *Strigiformes* (owls) and *Piciformes* (woodpeckers and allies), and by far the largest group, *Passeriformes* (passerines, about 5,700 species).

The most striking feature of birds, besides endothermy (warm-bloodedness), which perhaps was already present in dinosaurs, is their feathers and (in most birds) the ability to fly. Another evolutionary innovation, which made a metabolically expensive enduring flight possible, is the lung, which is based on the principle of simultaneous inhalation and exhalation, thus making a constant supply of oxygen possible. Birds possess a highly developed visual system, and some birds, like owls, have an equally well-developed auditory system, while others, like ducks, have a highly sensitive touch organ at the tips of their beaks. Pigeons possess a well-developed olfactory sense, and migratory birds have developed a magnetic sense for navigation along the magnetic field of the Earth.

Mammals

Mammals (*Mammalia*, about 5,700 species) are a surprisingly ancient group of vertebrates. The first ancestors of mammals evolved from cotylosaurian (or stem) reptiles and appeared about 224 mya in the Triassic, but conducted an inconspicuous life until the end of the Mesozoic. Important intermediate reptile-mammals were the therapsids, which existed from 275 to 180 mya. In contrast to other archosaurs, they showed no tendency toward bipedalism, but retained a quadruped locomotion and showed hair and lactation. There was a mass extinction of the precursors of mammals at the Permian-Triassic transition, and in the course of the Mesozoic, around 170 mya, the modern type of mammals, the multituberculates, evolved; they had small bodies and conducted a nocturnal and/or arboreal life. The split between the *Prototheria* and the *Theria* is believed to have occurred era 150 mya, and that among the *Theria*, between *Metatheria* and *Eutheria* 125 mya or earlier. The great time of mammals began near the end of the Cretaceous, around 70 mya, and particularly with the extinction of the dinosaurs 65 mya.

Distinguishing features of mammals are a hairy skin, the internal development of the embryo and—with the exception of egg-laying monotremes—birth of very small, but more or less fully developed young, a heterodont dentition with different types of teeth, lactation by mammary glands, the development of a new lower jaw (dentale), and, in this context, of a new inner ear.

The first infraclass of mammals comprises the *Prototheria* with only one order, the monotremes (*Monotremata*) with the platypus (*Ornithorhynchus*) and echidna (*Echidna*, 4 species). These animals lay eggs, but this may be a derived and not a primitive trait. The hatchlings are lactated, as in all other mammals. The second infraclass, the *Metatheria* or *Marsupialia*, comprises seven extant superorders with a total of about 334 species. The larger group, the *Australidelphia* (five superorders

9.4 Chordates-Craniates-Vertebrates

with 234 species, among them the kangaroos), live in Australia and New Guinea, whereas a smaller group, the *Ameridelphia* (about 100 species, among them the opossum, *Didelphis*) live in North, Central, and South America. Marsupials originally had a worldwide distribution and were later rolled back by the *Eutheria/ Placentalia*. The name-giving trait is the marsupium or front pouch. Marsupials give birth at a very early stage of development, about 4–5 weeks, and the newborn and relatively undeveloped marsupials have to crawl up the bodies of their mothers and attach themselves to a nipple located inside the marsupium.

The *Eutheria* or *Placentalia* originated about 100 mya and conducted a modest life as insect-eaters parallel to the marsupials. All exant major mammalian taxa were already present towards the end of the Mesozoic. A distinguishing feature is a uterus with a placenta as a nourishing organ connecting the embryo/fetus to the uterine wall. There is huge variability in body shapes and sizes as well as in shape, length, and function of limbs, which are used for fast running in many ungulates, for flying (primarily the forelegs) in bats and flying foxes, for digging, as in the moles, for swimming in aquatic mammals such as seals or whales or for manipulation, as in primates.

Placental mammals are divided into four groups or superorders. The first group, *Afrotheria* (39 species) includes, among others, the orders *Afrosoricida* (tenrecs and golden moles), *Tubulidentata* (aardvarks), *Hyracoidea* (hyraxes and allies), *Proboscidea* (with elephants, *Elephantidae*, a single family with three species), and *Sirenia* (dugong and manatees). The second group comprises the *Xenarthra* (sloths, anteaters, and armadillos, 30 species). The third group, *Euarchontoglires*, includes the orders *Scandentia* (treeshrews), *Dermoptera* (colugos or "flying lemurs"), *Primates* (lemurs, bushbabies, monkeys, and apes, including humans, about 440 species; see below), *Rodentia* (rodents, about 2,300 species), and *Lagomorpha* (pikas, rabbits, hares). Finally, the fourth group, *Laurasiatheria*, comprises the orders *Eulipotyphla* (hedgehogs, moles, shrews, previously called "insectivores"), *Chiroptera* (bats, about 1,100 species), *Pholidota* (scaly anteaters), *Carnivora* (carnivores, about 290 species), *Perissodactyla* (odd-toed ungulates like horses, zebras, tapirs; 16 species), *Artiodactyla* (even-toed ungulates, cattle, pigs, sheep, deer, camels, antilopes; 315 species), and *Cetacea* (whales, dolphins, porpoises, about 80 species).

In this taxonomy, primates are put into the neighborhood of the colugos and treeshrew, and rodents and the cetaceans (whales) into the neighborhood of the ungulates. The traditional order "insectivores" is now split into two new orders, *Afrosoricida* and *Eulipotyphla*, because it was shown that the tenrecs are not closely related to the shrews and hedgehog.

Of special interest is, of course, the order *Primates*. Primates are divided into the *Strepsirrhini* ("wet-nosed primates", 139 species), which include the (Lemuriformes, Aye–Aye, etc.) and the *Lorisiformes* (lorises, pottos and galagos), and the *Haplorrhini* ("dry-nosed primates", 308 species). The latter are composed of the *Tarsiiformes* (tarsiers, 10 species) and the *Simiiformes* (monkeys and apes, 298 species), the latter of which are further divided into the *Plathyrrini* ("flat-nosed" primates) or New World monkeys (marmosets, tamarins, capuchins, squirrel monkeys, spider monkeys, etc., 139 species) and the *Catarrhini* ("downward-nosed primates", 159 species), i.e., Old World monkeys/apes). The latter include the *Cercopithecidae* or Old World monkeys, including macaques, baboons, langures, and the *Hominoidea*, which comprise the *Hylobatidae* (gibbons or lesser apes) and the *Hominidae* or greater apes (orangutan, gorilla, chimpanzee, bonobo, and *Homo sapiens*).

Primates of modern aspect (the "Euprimates") first appeared in the fossil record during the Paleocene-Eocene transition 60–55 mya in North America, Europe, and Asia. They are characterized by relatively large brains, enhanced vision brought about in part by optical convergence (forward-facing eyes, frontal vision), ability to leap, nails on at least the first toes, and grasping hands and feet. However, grasping abilities appear to have evolved prior to leaping and frontally oriented eyes; the latter two are not found in ancestors of modern primates (Bloch and Boyer 2002). They were arboreal graspers for terminal branch feeding rather than specialized leapers or visually directed predators. The evolution of grasping coincides with a major radiation of angiosperms in the Northern Hemisphere that resulted in an increased diversity of fruits, flowers, floral and leaf buds, gums and nectars during the late Paleocene, while leaping and convergent orbits probably developed later. The split between lemurs and lorisiforms occurred probably 75 mya; simiiforms appeared about 40 mya. The split between New World monkeys and Old World monkeys/apes occurred 35 mya or earlier.

9.5 What Does All This Tell Us?

This chapter was devoted to the second of the two large groups of animals, the deuterostomes, which, however, is much smaller than the protostomes or "invertebrates," but in many ways the "great alternative" to the latter. Nonetheless, modern developmental genetics show that despite the large differences in morphology, physiology, ecology, and behavior, both protostomes and deuterostomes appear to share the same fundamental developmental genes, including the general organization of an essentially tripartite brain. The deuterostomes are interesting for us, because in a descending taxonomic way, as members of the species *Homo sapiens*, we are deuterostomes, chordates, vertebrates, mammals, primates, and eventually hominins. Under a zoological perspective, we are ordinary members of the animal kingdom. This insight is by no means new, and is evidence that was accumulated in the eighteenth century. In order to attenuate this "humiliating" fact, at the same time philosophers as well as scientists started searching for traits that could preserve the status of humans as being "unique." This concerned, above all, mind, intellect, reason, etc. The decisive question now is whether such a view can be corroborated by looking at the brains of craniates-vertebrates in the hope of finding unique traits that could be correlated with the alleged uniqueness of the human mind. This will be the topic of the next few chapters.

Chapter 10
The Brains of Vertebrates

Keywords Origin vertebrate brain · Basic organization vertebrate brain · Urbilaterian brain · Medulla spinalis · Medulla oblongata · Cerebellum · Mesencephalon/Midbrain · Diencephalon · Telencephalon/Endbrain · Isocortex · Functional anatomy · Pallium birds · Mesonidopallium birds

10.1 The Basic Organization of the Vertebrate Brain

The origin of the craniate-vertebrate brain lies in the dark. A phylogenetic reconstruction is of little help, because the echinoderms, as apparently the most ancient form within the deuterostomes, do not possess a centralized, but a penta-partite radial nervous system plus nerve rings which are unlike anything else in the animal kingdom and faintly resemble that of jellyfish. The nervous system of the sister group of echinoderms, the hemichordates, has more resemblance to primitive nervous systems of protostomes than to that of craniates. Among the sister groups of craniates, the urochordates have an extremely simple nervous system, which may be secondarily simplified in the context of their sessile lifestyle. Only the nervous system of the cephalocordate *Branchiostoma* reveals similarities with the craniates' CNS and—most importantly—appears to possess most genes responsible for the ontogenetic development of the craniate brain, as described in Chap. 9.

Despite its apparently long history of at least 500 million years, the CNS of craniates, including vertebrates, reveals a highly uniform organization (for an overview, see Nieuwenhuys et al. 1998; Striedter 2005; Figs. 10.1, 10.2a–j). In its hypothetical ancestral form, it exhibits a tripartite organization, i.e., prosencephalon, mesencephalon, and rhombencephalon. As already mentioned, the question of whether such a tripartite organization, found in lophotrochozoans and ecdysozoans as well, is due to "deep homology" or to convergent evolution, is debated among experts. During ontogeny, this three-partite brain develops into a five-partite brain in the way that the rhombencephalon divides into a medulla oblongata and a metencephalon including a cerebellum, while the prosencephalon divides into a diencephalon (or "primary prosencephalon") and a telencephalon (or

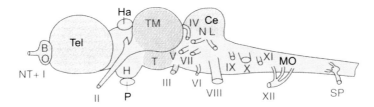

Fig. 10.1 Basic organization of the vertebrate brain. *BO* olfactory bulb, *Ce* cerebellum, *H* hypothalamus, *Ha* habenula, *MO* medulla oblongata, *NL* lateral nerves, *NT* terminal nerve, *P* hypophysis/pituitary, *SP* first spinal nerve, *T* tegmentum, *Tel* telencephalon, *TM* tectum mesencephali, *I-XII* cranial nerves. After Roth and Wullimann (1996/2000), modified

"secondary prosencephalon," endbrain). The mesencephalon with the isthmic region remains undivided. In birds and mammals, between mesencephalon and medulla oblongata and ventral to the cerebellum, a "pons" (Latin for "bridge") is formed. Medulla oblongata, isthmic region, pons, and mesencephalon together form the "brainstem." This basic organization of the brain is found in all vertebrates, as illustrated in Fig. 10.3, which shows representative cross sections through the brain of a frog as an example of that basic organization.

For more than 100 years, there has been a debate about the more detailed longitudinal and transverse organization of the vertebrate brain. Generally accepted is the existence of four longitudinal zones or plates on both sides, i.e., a floor plate, a basal plate, an alar plate, and a roof plate. The basal and alar plates are separated from the alar and roof plate by a "sulcus limitans," first described by the Swiss neuroanatomist Wilhelm His. These four longitudinal zones are clearly visible in the medulla oblongata, where the sulcus limitans separates the so-called somatosensory and viscerosensory zones, related to the processing of sensory information about the outer world and their own bodies, respectively, from the visceromotor and somatomotor zones controlling the intestines ("viscera") and the skeletal muscles, respectively. In the midbrain, the sulcus separates the dorsal sensory part with the tectum opticum and torus semicircularis from the premotor and limbic tegmentum, which in turn is divided into a dorsal and ventral zone (see below).

In addition to the existence of these longitudinal zones, it is now generally accepted that most parts of the brain, like the spinal cord, have a *segmental organization*, i.e., that the brain consists of "neuromeres" (Fig. 10.4). This view was first developed by the Swedish developmental neurobiologists Bergquist and Källén in the first half of the last century, and was confirmed more recently by the Spanish neuroanatomist Luis Puelles and the American neuroanatomist John Rubenstein. According to Puelles and Rubenstein (1993, 2003) and Pombal (2009), the rhombencephalon consists of seven *rhombomeres* (R1–7), which are marked (at least R3–R7) by the expression of genes of the Hox gene family. The mesencephalon is composed of an isthmic neuromere and a mesencephalic neuromere proper. Each neuromere exhibits the four plates already mentioned, i.e., a dorsal roof and alar plate and a ventral basal and floor plate.

10.1 The Basic Organization of the Vertebrate Brain

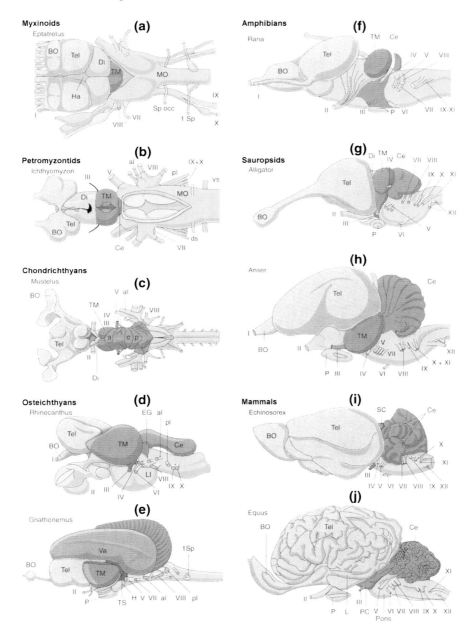

Leading neuroanatomists in the first half of the twentieth century, like C. Judson Herrick, believed that these four longitudinal zones, which are clearly recognizable in the medulla oblongata and midbrain, continue at least into the

◄ **Fig. 10.2** Brains of representatives of major groups of craniates. **a** hagfish, *dorsal view*, **b** lamprey, *dorsal view*, **c** spiny dog fish, *dorsal view*, **d** trigger fish, *dorsal view*, after, **e** elephantnose fish, *lateral view*, **f** frog, *lateral view*, **g** alligator *lateral view*, **h** goose *lateral view*, **i** moonrat, **j** horse. Abbreviations: *a* anterior cerebellar lobe, *al* anterior lateral nerve, *c* central cerebellar lobe, *BO* olfactory bulb, *Ce* cerebellum, *Di* diencephalon, *ds* dorsal spinal nerve, *EG* eminentia granularis, *Ha* habenula, *P* pituitary, *LI* inferior lobe, *MO* medulla oblongata, *p* posterior cerebellar lobe, *pl* posterior lateral nerve, *SC* superior culliculus, *Sp occ* spino-occipital nerve, *1Sp* first spinal nerve, *Tel* Telencephalon, *TM* tectum mesencephali, *TS* torus semicircularis, *Va* valvula cerebella, *vs* ventral spinal nerve, *I-XII* cranial nerves. After Roth and Wullimann (1996/2000), modified

diencephalon. Accordingly, the diencephalon was divided by them into four longitudinal zones, i.e., from dorsal to ventral in an epithalamus, a dorsal thalamus, a ventral thalamus, and a hypothalamus. The sulcus limitans of His would then separate the dorsal from the ventral thalamus. However, already Bergquist and Källén believed that the prosencephalon (i.e., diencephalon plus telencephalon), too, is composed of neuromeres. Puelles and Rubenstein accepted that view and called the diencephalic neuromeres "prosomeres P1, P2 and P3," together forming the "primary prosencephalon." Prosomere P1 corresponds to the pretectal or posterior thalamic region, P2 to the dorsal thalamus, and P3 to the ventral thalamus of the traditional nomenclature. In the opinion of Puelles and Rubenstein, the hypothalamus, together with the preoptic region, belongs to the telencephalon, rather than forming the ventral part of the diencephalon. As a consequence of the downward rotation of the diencephalon together with the telencephalon (as "secondary prosencephalon"), P1 still occupies a position rather orthogonal to the long axis of the hind brain, whereas P2 is tilted slighty forward, and P3 exhibits an oblique orientation (cf. Fig. 10.4). P6 now represents, with the optic chiasm and the eye stalks, the true anterior pole of the vertebrate brain, and not the olfactory bulbs, as is usually conceived.

While this segmentation of the diencephalon into three prosomeres is now well accepted, there are disputes about the segmentation of the telencephalon as a "secondary prosencephalon." In their revised model, Puelles and Rubenstein propose a "mixed" organization of the telencephalon in the sense that its posterior part is divided into the two prosomeres P4 and P5, which comprise the dorsal and ventral hypothalamus plus preoptic region in the classical sense and by rotation come to lie below the ventral thalamus (therefore classically called "*hypo*thalamus"). Prosomere P6, which includes the region of the optic chiasm and entrance of the optic nerves, lies even below P5 (Fig. 10.4). According to the two authors, the anterior, and in a strict sense dorsal part of the telencephalon, is *not* segmentally organized and consists of four *pallial* regions, i.e., a medial (MP), dorsal (DP), lateral (LP), and ventral pallium (VP), and 2–3 *subpallial* regions, i.e., the striatum (Str.), pallidum (Pa.), and possibly entopeduncular area. The caudal portions of each of these three subpallial regions form parts of the amygdala complex (Pombal 2009). In this model, the sulcus limitans of His follows the rotation of the prosencephalon and ends at the optic chiasm in P6, and does not—as often described—separate pallial

10.1 The Basic Organization of the Vertebrate Brain

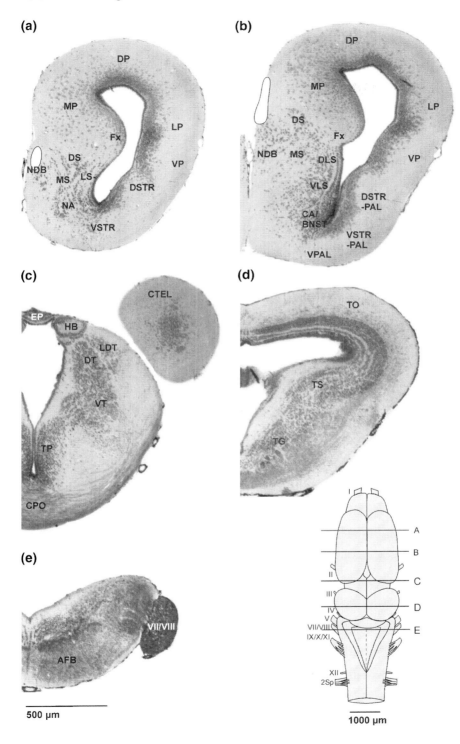

Fig. 10.3 Cross sections through the brain of the frog *Bombina orientalis*. Levels of cross sections **a–e** are indicated in the *dorsal view* of the brain at *lower right*. **a** rostral telencephalon at the level of the nucleus accumbens, **b** central telencephalon at the level of the dorsal and ventral striatum, **c** diencephalon at the level of the habenula and postoptic commissure, **d** midbrain with optic tectum and torus semicircularis, **e** rostral medulla oblongata at the level of entrance of 7th cranial nerve. Abbreviations: *AFB* descending fiber bundle, *CA-BNST* central amygdala-nucleus interstitialis of the stria terminalis, *CPO* commissura postoptica, *CTEL* caudal telencephalon, *DLS* dorsal lateral septum, *DS* dorsal septum, *DP* dorsal pallium, *DSTR* dorsal striatum, *DSTR-PAL* dorsal striatopallidum, *DT* dorsal thalamus, *EP* epiphysis/pineal organ, *Fx* fornix, *HB* habenula, *LP* lateral pallium, *LS* lateral septum, *LDT* lateral dorsal thalamus, *MP* medial pallium, *MS* medial septum, *NA* nucleus accumbens, *NDB* nucleus of diagonal band of Broca, *TG* tegmentum, *TO* optic tectum, *TP* tuberculum posterius, *TS* torus semicirularis, *VLS* ventral lateral septum, *VSTR* ventral striatum, *VSTR-PAL* ventral striatopallidum, *VP* ventral pallium, *VT* ventral thalamus, *VII/VII* 7th/8th cranial nerve, *2SP* 2nd spinal nerve. After Roth (2011)

from subpallial regions of the telencephalon. However, there is no last word on the true neuromeric organization of the telencephalon (for criticism, see Striedter 2005).

After this short description of the basic organization of the vertebrate central nervous system, I will briefly describe its major parts.

10.2 Medulla Spinalis and Oblongata

The medulla spinalis consists of an inner gray substance around the central canal, which is composed mostly of nerve cells, covered by white substance containing dendrites and ascending and descending nerve fibers. The gray substance is—according to the pattern mentioned above—divided into a dorsal somatosensory and viscerosensory region and a ventral visceromotor and somatomotor region. Nerve cells from the latter two regions innervate the various parts of the body via spinal nerves in a segmental fashion.

The *medulla oblongata* reveals the same dorsoventral organization as the medulla spinalis and contains from dorsal to ventral, somatosensory, viscerosensory, visceromotor and sensorimotor areas and nuclei of the cranial nerves V–X (or V–XII). These are the trigeminal nerve (5th cranial nerve, for facial sensations and motor functions), the abducens nerve (6th cranial nerve, an eye motor nerve), the facial nerve (7th cranial nerve, a sensory and motor nerve), the vestibulocochlear nerve (8th cranial nerve, also called acoustic-vestibular nerve), the glossopharyngeal nerve (9th cranial nerve, supplying the tongue and mouth region with taste), vagus nerve (10th cranial nerve, with control of the laryngeal and pharyngeal muscles and origin of the parasympathetic system). In tetrapod vertebrates, there are additional cranial nerves: the accessory nerve (11th cranial nerve, supplying, among others, the trapezius muscle of the neck, shoulder, and back), and the hypoglossal nerve (12th cranial nerve controlling the tongue). Hagfish have no eye muscles and nerves, but both are found in lampreys and all other vertebrates.

10.2 Medulla Spinalis and Oblongata

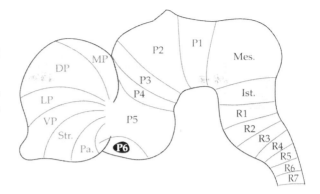

Fig. 10.4 Segmental organization of the craniate brain. *DP* dorsal pallium, *Ist* isthmic neuromere, *LP* lateral pallium, *Mes* mesencephalic mesomere, *MP* medial pallium, *P1-6* prosomeres, *Pa* pallidum, *R1-7* rhombomeres, *Str* Striatum, *VP* ventral pallium. After Striedter (2005), modified

The dorsal sensory roots of the cranial nerves carry ganglia containing the somata of sensory neurons, which with one "arm" of their axon extends into the muscles they supply, and with the other one into the dorsal regions of the medulla. The regions of sensory nuclei may undergo strong enlargement and complication, e.g., the gustatory vagal lobe in goldfish in combination with a highly evolved gustatory system in this teleost. A mechano- and electro-receptive system associated with cranial nerves is present in all fully aquatic anamniote vertebrates and was lost in some terrestrial amphibians and in all amniotes, i.e., in "reptiles," birds and mammals.

The *reticular formation system* is found inside the medulla oblongata and extends into the pons (present only in birds and mammals) and the tegmental midbrain. It is composed of a reticular (i.e., "net-like") structure in which important neuromodulator-producing centers (nuclei) such as the noradrenergic *locus coeruleus* and the serotonergic *raphe* nuclei are embedded (see Chap. 5). These nuclei send ascending fibers to almost all parts of the brain where noradrenaline (norepinephrine) and serotonin are released. The reticular formation controls centers for breathing and cardiovascular activity and gives rise to an ascending activation system for vigilance, awareness, and consciousness. The *pons* of mammals is situated in the rostral medulla oblongata and caudal tegmentum. It contains relay nuclei of fiber bundles that connect the cerebral cortex and the cerebellum. Similar pathways and nuclei are likewise found in birds, but most likely developed independently.

Despite its relatively conservative organization, the medulla oblongata undergoes several spectacular modifications in some groups of bony fishes (teleosts). The nucleus ambiguus is closely related to the glossopharyngeal and vagal nerve and innervates the mouth region and tongue. In most vertebrates, it has an inconspicuous structure. In cyprinid fishes (carps including the goldfish), however, it has grown enormously into the vagal lobe in parallel with the strong development of taste in these fishes. In cyprinids, it exhibits up to 15 cellular and fiber layers and can comprise up to 20 % of brain volume. Another conspicuous center for taste and touch in the medulla oblongata of bony fishes is the facial lobe, which is associated with the facial and, partially, the trigeminal nerve, and likewise supplies the mouth region and the lips. It is particularly large again in the cyprinids (in addition to the vagal lobe) and in catfish (siluriforms), which carry taste receptors all over their body surfaces, but also in the toxic Japanese puffer fish ("fugu"), belonging to the family *Tetraodontidae* (which means "fish with four teeth"). Likewise spectacular is another structure of the medulla oblongata, which is found in the weakly electric fish, i.e., the electrosensory lateral line lobe (ELL), which will be described in Chap. 11. Interestingly, electroreception apparently has evolved independently in the gymnotids and mormyrids, and the same holds true for their ELL. The ELL is closely connected with the enlargement of still another spectacular structure, the valvula cerebelli (see below and Chap. 11).

10.3 Cerebellum

The *cerebellum* is a formation of the dorsal rhombencephalon above the isthmic region. It is found in all vertebrates, but absent in hagfish, either primitively or due to secondary loss. This is somewhat surprising, because these animals are good swimmers, but the respective control functions are exerted by the spinal cord. In the ancestral state, the cerebellum is a processing center for information from the vestibular, mechanosensory, and electrosensory system including the lateral line systems, and its respective parts are called "vestibulocerebellum" and "spinocerebellum." These two parts are located in the lateral areas, the auricles, of the cerebellum of cartilaginous fishes, in the caudal lobe of the cerebellum of bony fishes, and the flocculo-nodulus of the mammalian cerebellum. A *corpus cerebelli* (called "vermis" in mammals) is found in cartilaginous and bony fishes and tetrapods. Across cartilaginous fishes, the cerebellum varies strongly in size and shape. Some groups of sharks (e.g., the squaliforms) and rays (e.g., the torpediniforms and rajiforms) have small cerebella with little or no folding of their surface, while other groups of sharks (e.g., the galeomorphs) and rays (the myliobatiforms) have cerebella that are voluminous, covering large parts of the rest of the brain, and with a heavily folded surface. These differences correspond well with the lifestyles of the respective groups: Animals with small and unfolded cerebella are slowly moving ground dwellers, while those with large and highly folded cerebella move actively inside the water column of the ocean (Lisney et al. 2008).

Teleosts generally possess a large cerebellum (cf. Fig. 10.2e, f), which consists of four lobes and a structure found only in teleosts, the valvula cerebelli (Wullimann and Vernier 2007; cf. Fig. 10.2f). The cortex of the corpus cerebelli exhibits a three-layered organization, i.e., a deep, small-celled granular layer, a large-celled layer of Purkinje cells, and a peripheral molecular layer. In weakly electric fish (gymnotids and mormyrids), the valvula cerebelli is the most complex part of the cerebellum. It is a rostral bulge of the corpus cerebelli and in mormyrids covers the entire dorsal part of the brain (Fig. 10.2f). Here, the tripartite lamination of the corpus is modified in the sense that the granular layer is not located below, but lateral to the molecular layer and that the granular cells extend their innumerable

10.3 Cerebellum

fibers directly parallel to the surface and not, as usual, after bifurcation in a T-shaped fashion. In the valvula, mechanoreceptive and electroreceptive information is processed (see Chap. 11).

Among amphibians, the cerebellum is small in frogs, strongly reduced in size in salamanders, and absent in caecilians. The cerebellum of sauropsids reveals the standard organization with a large corpus cerebelli and a small flocculus. It is small in limbless reptiles such as snakes, and largest in crocodiles and birds. Compared to body size, birds possess a large cerebellum (Fig. 10.2j), with a strongly folded medial part homologous to the mammalian vermis, and flat lateral parts, the auricles, corresponding to the flocculus and paraflocculus of mammals.

The mammalian, including human cerebellum (Fig. 10.2k, l), consists of three parts. The first part, the vestibulocerebellum, consists of the lobus flocculonodularis. It is tightly connected with the nuclei of the vestibular system of the inner ear and responsible for the control of balance. The second part is the spinocerebellum, comprising the medial part of the cerebellum, the vermis, and adjacent parts of the hemispheres. It is homologous to the corpus cerebelli of other vertebrates. It receives fibers from the spinocerebellar pathway that carry information about the state of activity of muscles and tendons and is responsible for posture. The third part, the cerebrocerebellum, also called neocerebellum or pontocerebellum, is a novelty of mammals and has evolved together with the neocortex (or isocortex, see below). It receives afferents from the pons carrying information from the motor and premotor cortex and is responsible for the fine-tuning of movements, ideas and words and their "smooth" execution. The cerebellar cortex of most mammals is strongly folded, while exhibiting the "standard" organization, i.e., a deep granular layer, with billions of very small neurons in large-brained mammals and relatively few larger "Golgi cells." The middle layer contains very large "Purkinje cells" (80,000 in the human brain), which have large, flat dendritic trees oriented parallel to each other like espalier trees. The superficial layer, i.e., molecular layer, contains stellar and basket cells, the dendritic trees of the Purkinje cells, and the ascending axons of the granular cells in the deep layer, which together form the parallel fiber system. Deep inside the cerebellum, the cerebellar nuclei are found which represent the output system of the cerebellum.

Among mammals, there are large differences in size and shape of the cerebellum. A relatively simple cerebellum with small hemispheres is found in the monotremes, while marsupials, insectivores, and hoofed mammals have insignificant hemispheres. These are larger in rodents and carnivores and largest in primates, elephants, and whales. The size of their cerebella corresponds well with the size of their neo(iso)cortex. This is not surprising, because the cerebellar hemispheres and the neocortex are closely connected via the pons.

Besides vestibular, somatosensory, and sensorimotor functions, the cerebellum of mammals, and perhaps of birds, is likewise involved in "higher" cognitive functions, such as thinking and action planning as well as language in humans (Ivry and Fiez 2000). To date it is difficult to find cognitive or motor functions that do not involve cerebellar activity. Most probably, the cerebellum has to do with the processing of information regarding the *temporal sequence* or *time differences* of events, whether motor responses, sensory signals, thoughts, or words.

10.4 Mesencephalon

The mesencephalon consists, from dorsal to ventral, of the *tectum* (in mammals called "colliculi superiores"), the *torus semicircularis* (in mammals called "colliculi inferiores") and the *tegmentum*. The latter exerts predominantly premotor functions. In its ventral part, we find the nuclei of the oculomotor and trochlear nerves (3rd and 4th cranial nerves, respectively) involved in eye movements. The dorsal tegmentum contains the fasciculus longitudinalis medialis and the dorsal tegmental nucleus, which have close connections with the tectum/colliculus superior and exert vestibular functions, especially with respect to head movements. There are other nuclei, such as the dorsal and ventral tegmental nuclei and the tegmental pedunculopontine nuclei, which are important relay stations between the limbic centers in the diencephalon and telencephalon (e.g., amygdala, cortical limbic areas) and limbic regions in the brainstem such as the periaqueductal gray, the reticular formation, and visceral regions in the medulla oblongata in the context of emotional and autonomic responses.

The tegmentum also contains the substantia nigra, which is considered part of the basal ganglia. Together with the adjacent ventral tegmental area (VTA), the substantia nigra is a major site for the production of the neurotransmitter/neuromodulator dopamine. Both areas are closely connected with the diencephalic and telencephalic limbic centers; of special importance is the projection of the substantia nigra to the dorsal corpus striatum in the context of control of voluntary actions, and the projection of the ventral tegmental area to the nucleus accumbens in the context of reward, reward expectation, and consequently motivation.

The *torus semicircularis* (colliculus inferior of mammals) is, in its plesiomorphic state, the midbrain relay station for auditory, mechano- and electrosensitive projections ascending to the diencephalon and telencephalon (cf. Fig. 10.3d). It is characterized by a combination of nuclear and laminar organization. In teleosts, its size and cytoarchitectural complexity corresponds well with the sensory equipment of groups. In bony fishes without electroreception, the torus processes auditory-vestibular and mechanoreceptive information. In the passively electric fishes such as the catfish (*Ictalurus*), processing of electrosensitive information is added. In the mormyrids and gymnotids, which use the electroreceptive system for "active" object location and communication, the dorsal torus is subdivided into many areas, which serve the processing of the respective signals. As shown in Fig. 10.5, these areas receive afferent from the electrosensory lateral lobe and the cerebellum and reveal a highly complex structure with a total of 12 layers and 48 different types of nerve cells. Here, the emitted echolocation and communication signals are compared with those received (cf. Chap. 11). There is additional visual input from the tectum.

10.4 Mesencephalon

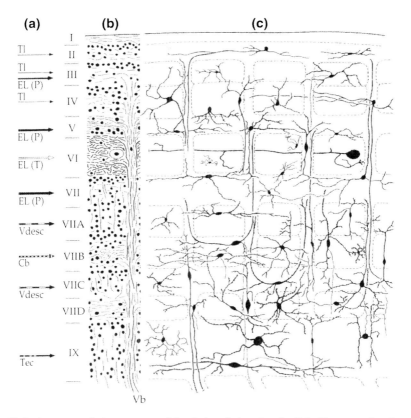

Fig. 10.5 Anatomy of the torus semicircularis of the electric fish *Eigenmannia virescens* exhibiting a spectacular laminar organization. **a** Afferents from different brain regions terminate in different layers of the torus, **b** laminar organization of the torus in Bodian staining, **c** Cytoarchitecture of the torus in Golgi staining. Abbreviations: *Cb* cerebellum, *EL(P)/EL(T)* electrosensory P and T type afferents, *Vdesc* nucleus descendens of the trigeminal nerve. *Tec* Tectum opticum, *Tl* Torus longitudinalis. After Nieuwenhuys et al. (1998), modified

In amphibians, the torus semicircularis is the main processing center for auditory, vibratory, and vestibular afferents as well as for afferents from the lateral line system, if present. The torus is relatively large in frogs, especially in those with well-developed auditory communication, and is divided into five different parts, organized partly in nuclei and partly in layers. In salamanders and caecilians, which do not produce sounds, the torus consists of a poorly developed periventricular layer.

In all anamniotes ("fish" and amphibians), all sauropsids and many mammals, but not in primates, the tectum mesencephali is the major integration center for somatosensory, visual, and auditory information. In most vertebrates, it reveals a laminar organization, i.e., a combination of cellular and fibrous layers, each of which has different input and/or output. Lampreys and cartilaginous fishes have a well-developed tectum, but its lamination is rather diffuse and not as precise as in frogs and teleosts. In the latter, we find 7–9 layers with up to 15 different cell types (Meek and Schellart 1998). The main input comes from the optic nerve and is therefore visual. In some bony fish, the optic nerve may contain nearly 1 million fibers. The second strongest input originates in the torus longitudinalis, which in actinopterygians extends along the midline of the tectum. This input is assumed to carry information about eye movement and from the mechano- and electroreceptive system (if present), which is then compared with the visual information to distinguish self-induced motion from passive motion.

In the frog tectum, we find eight alternating cell and fiber layers (cf. Fig. 10.3c), while the tectum of salamanders and caecilians is composed of a periventricular cell layer and a superficial fiber layer. A phylogenetic analysis suggests that this latter situation is not primitive, but the consequence of secondary simplification in the context of paedomorphosis (Roth et al. 1993). This means that—as a consequence of an enormous increase in genome size in salamanders and caecilians (cf. Chap. 3)—late developmental differentiation processes, like cell migration in the tectum, were abolished and that in a number of brain structures these organisms remain at a larval level (Roth et al. 1997; Dicke and Roth 2007). In the amphibian tectum as well as in that of other anamniotes, primary visual afferents from the retina invade the superficial fibers and cellular layers, while secondary visual afferents from the thalamus, pretectum, and nucleus isthmi terminate somewhat deeper, together with auditory, mechanosensory, and electrosensory afferents (if present). These deeper layers are the site of origin of descending projections to the tegmentum, medulla oblongata and spinal cord and of ascending projections to the diencephalon and telencephalon.

In sauropsids, including birds, the tectum reveals a similar degree of complexity, as in teleosts, in the context of a highly developed visual system. The bird tectum is composed of 14 layers, and the most superficial ones receive direct retinal input. These are integrated with visual afferents from thalamus, pretectum, hypothalamus, and basal optic nucleus. Nonvisual afferents come from the striatopallidum, the reticular formation, the midbrain tegmentum, and from the trigeminal system. There is a prominent ascending visual pathway from the tectum to the nucleus rotundus of the thalamus (see below).

In mammals, the midbrain roof is formed by the colliculi superiores and inferiores. The former are regarded homologous with the tectum, the latter homologous with the torus semicircularis of other vertebrates. The superior colliculi of mammals are involved in the control of visually or acoustically evoked orienting responses of gaze and head as well as goal-directed arm and hand movements and related spatial attention. The inferior colliculi are—like the torus semicircularis of other vertebrates—important centers of the auditory system (see Chap. 11). Compared to the midbrain roof of other vertebrates, that of mammals is relatively small, apparently as a consequence of the shift of important visual and auditory functions to the isocortex.

10.5 Diencephalon

In all vertebrates, the diencephalon is an important relay station for pathways ascending from the brainstem to the telencephalon and descending from there to the brainstem and spinal cord. It surrounds the third ventricle and is classically divided from dorsal to ventral into the epithalamus, thalamus, and hypothalamus (Fig. 10.3c). The *epithalamus* contains the *habenular nuclei,* which are important parts of the limbic system and present in all craniates. They project, via the fasciculus retroflexus, to the midbrain tegmentum. In many craniates, the epithalamus carries the pineal organ or "epiphysis," a small endocrine gland releasing the hormone melatonin, which affects wake-sleep patterns and seasonal functions.

The *thalamus* is classically divided into a dorsal and a ventral part and the posterior tuberculum (Fig. 10.3c). As we have already heard, this horizontal division deviates from the neuromeric model of Puelles and Rubenstein, which assumes that the dorsal thalamus is formed by the prosomere P2, the ventral thalamus by prosomere P3, and the posterior tuberculum by prosomere P1 (see above). Furthermore, the hypothalamus is conceived by Puelles and Rubenstein as prosomeres P4 and P5 of the telencephalon. Since most contemporary neuroanatomists still adhere to the classical description of Herrick, I will also use these terms.

The diencephalon of cartilaginous fish is inconspicuous compared to the other parts of their brain, but reveals all major functions typical of the vertebrate diencephalon. The mechanoreceptive lateral line system projects to the dorsal thalamus as well as to the lateral posterior tuberculum, while the electroreceptive system projects to the ventral thalamus, the lateral posterior tuberculum, and the hypothalamus. The somatosensory afferents terminate, like in all vertebrates, in the ventral and/or dorsal thalamus, and visual afferents from the retina terminate either directly or indirectly via the ventral thalamus in the anterior nucleus of the dorsal thalamus. The central posterior nucleus of the dorsal thalamus receives auditory information from the torus semicircularis. The dorsal thalamus sends sensory signals to the telencephalon, which, importantly, are not unimodal, but mixed visual, auditory, and somatosensory signals (Hofmann and Northcutt 2008).

The diencephalon of teleosts reveals some peculiarities. Here the central part of the pretectum, rather than the dorsal thalamus, receives the mass of visual afferents from the retina and is tightly connected with the cerebellum. The function of the central pretectum is the integration of visual and vestibular information. In the weakly electric gymnotids, the pretectum contains a nucleus electrosensorius, which receives massive afferents from the torus semicircularis and plays an important role in electrocommunication (cf. Chap. 11). The dorsal thalamus includes an anterior, a central, and a posterior region. The anterior region receives via the pretectal nucleus electrosensorius afferents from the electroreceptive system. It is unclear whether it receives direct visual afferents from the retina, as is typical of other vertebrates, and it does not project to the telencephalic pallium. In general, in teleosts the dorsal thalamus is not a relay station for ascending sensory projections to the telencephalon, as in other vertebrates. Instead, the strongest diencephalic projections arise in the preglomerulosus complex belonging to the region of the posterior tubercle (Wullimann and Vernier 2007). Ascending auditory, mechanoreceptive, and partially electroreceptive signals from the medulla oblongata and the torus semicircularis terminate in the lateral nucleus preglomerulosus, as do visual afferents from the optic tectum and gustatory afferents from the nucleus visceralis secundarius in the brainstem. The lateral nucleus preglomerulosus, in turn, projects primarily to the dorsal and less massively to the medial and lateral part of the dorsal pallium, and these pallial regions project back to the nucleus. In the weakly electric fish, there is another nuclear complex formed by the dorsal anterior pretectal and ventral thalamic nucleus, which together may correspond to the preglomerulosus complex of the other teleosts. This complex may have further developed in the context of a telencephalic control of electroreception and electrocommunication.

The dorsal thalamus of amphibians exhibits the standard division into an anterior, central, and posterior or pretectal nucleus (Fig. 10.3c), which occupy a position close to the ventricle. The ventral thalamus reveals instead some migrated nuclei in addition to periventricular nuclei, at least in frogs. These dorsal and ventral thalamic nuclei are important relay stations for sensory, motor, and limbic information between brainstem and telencephalon. Primary visual afferents do not reach directly the anterior dorsal, but indirectly via the ventral thalamus (Dicke and Roth 2007). In amphibians, the anterior dorsal thalamic nucleus is the only thalamic nucleus that projects to the medial and dorsal pallium via the *medial* forebrain bundle (cf. Fig. 10.6), and this projection contains, as in cartilaginous fish, mixed visual, somatosensory, and auditory information. The central dorsal thalamic nucleus receives afferents from the torus semicircularis and projects, via the *lateral* forebrain bundle, to the lateral amygdala and the nucleus accumbens, but not to the pallium. This twofold projection from the thalamus to the telencephalon, via the medial and lateral forebrain bundle, respectively, represents the starting point for the further development of thalamopallial and thalamocortical pathways in sauropsids and mammals, as described below.

In the epithalamus of sauropsids, i.e., "reptiles" and birds, we find the pineal organ, and in most lizards a parietal eye. The dorsal and ventral thalamus of "reptiles" is dominated by the nucleus rotundus, which occupies a central position and is the main receiver of visual afferents from the tectum (see above). This nucleus projects to a telencephalic structure called "anterior dorsal ventricular ridge—aDVR" situated in the lateral pallium, which will be described in more detail below. Ventromedial of the nucleus rotundus we find the nucleus reunions (in birds called nucleus ovoidalis), which receives auditory information from the torus semicircularis; somatosensory information from the brainstem terminates in the so-called nucleus-medialis complex. Both nuclei likewise project to the aDVR, although to separate sub-regions. The only unimodal visual projection to the dorsal pallium originates in the dorsolateral optic nucleus (DLON). The DLON is considered homologous with the mammalian lateral geniculate nucleus (see below). The dorsal thalamus of birds exhibits a similar organization. Here, too, the nucleus rotundus dominates; it receives visual information from the tectum and projects,

10.5 Diencephalon

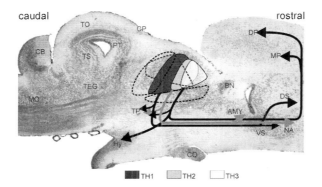

Fig. 10.6 Projections of nuclei in the dorsal and ventral thalamus of the fire-bellied toad *Bombina orientalis* identified by means of intracellular injection of the dye biocytin. The figure shows a longitudinal section through the brain from mid-telencephalon to the caudal cerebellum. Neurons in the anterior dorsal thalamus (*TH3*) project, via the medial forebrain bundle, to the medial (*MP*) and dorsal pallium (*DP*) as well as to the hypothalamus (*Hy*), while neurons in the central dorsal part (*TH2*) project to the amygdala (*AMY*) and the nucleus accumbens (*NA*). The caudal dorsal thalamus (*TH1*), projects, via the lateral forebrain bundle, to the dorsal (*DS*) and ventral striatum (*VS*). All three types of neurons likewise project to the posterior tuberculum (*TP*). Further abbreviations: *BN* bed nucleus of the stria terminalis, *CB* cerebellum, *CO* optic chiasm, *CP* commissura posterior, *MO* medulla oblongata, *PT* pretectum, *TEG* tegmentum, *TO* tectum opticum, *TS* torus semicircularis. From Roth et al. (2003), modified

via the lateral forebrain bundle, to the so-called nidopallium, which is considered to be homologous to the aDVR of "reptiles"—more precisely, to its rostral part, called entopallium (previously called "ectostriatum"). The DLON of birds receives direct retinal afferents, like in "reptiles," and projects via the medial forebrain bundle to the "wulst" as the visual part of the so-called hyperpallium. This means that in birds, there are *two* visual projections from the thalamus to the telencephalic pallium, i.e., a medial one to the wulst and a lateral one to the entopallium of the aDVR; wulst and aDVR—as we will see later—exert different functions. Somatosensory pathways from the brainstem terminate in the nucleus dorsalis intermedius ventralis anterior (DIVA), which, parallel to the DLON, projects to the rostral part of the wulst. Auditory afferents from the brainstem terminate in the nucleus ovoidalis, which in turn projects to "field L" of the nidopallium. As opposed to the situation found in amphibians and "reptiles," in birds unimodal somatosensory and auditory projections from the thalamus to the telencephalic pallium exist.

The diencephalon of mammals, when compared to that of the other vertebrates, exhibits an enormous differentiation. As a consequence of the strong enlargement of the cortex, it moved deep into the brain (cf. Fig. 10.7). The nuclei and nuclear regions of the large dorsal thalamus are divided into palliothalamic and truncothalamic nuclei. The projections of these nuclei are depicted in Fig. 10.8. The palliothalamic nuclei are relay stations of unimodal sensory afferents to restricted sensory areas of the cortex, from where they receive back projections, as well as

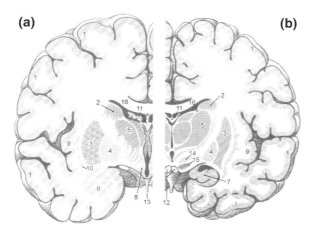

Fig. 10.7 Cross section through the human brain **a** at the level of hypothalamus, amygdala, and striatopallidum, **b** at the level of hippocampus and thalamus. Numbers: *1* cerebral cortex, *2* nucleus caudatus, *3* putamen, *4* globus pallidus, *5* thalamus, *6* amygdala, *7* hippocampus, *8* hypothalamus, *9* insular cortex, *10* claustrum, *11* fornix, *12* mammillary body, *13* infundibulum of pituitary, *14* nucleus subthalamicus, *15* substantia nigra, *16* corpus callosum. After Kahle (1976)

specific afferents to limbic cortical areas. This system of feedback loops is called the "thalamo-cortical system." The palliothalamic nuclei are further divided into an anterior, medial and lateral group, the pulvinar, the medial geniculate nucleus, and the lateral geniculate nucleus. The main nucleus of the anterior group is the nucleus anterior thalami. It is an important part of the limbic system, especially in the context of the control of emotional memories. The medial group, like the anterior group, is involved in emotional guidance and evaluation of behavior and consequently is part of the limbic system. The lateral group carries somatosensory information from the brainstem to the somatosensory cortex. The pulvinar is the largest thalamic nucleus and is involved in visual and auditory attention as well as in language and abstract-symbolic mental functions and projects to the posterior parietal cortex. The medial geniculate nucleus (or body) receives information from the auditory cochlear nuclei and projects to the primary auditory cortex, the so-called Heschl's gyri. Finally, the lateral geniculate nucleus (or body) receives direct visual input from the retina via the optic nerve and project to the visual cortex in the occipital lobe. We will learn more about these projections in Chap. 11.

The truncothalamic nuclei have limbic-emotional as well as modulatory functions. These include the so-called intralaminar and midline nuclei, which receive afferents from the reticular formation (see above) and project to the prefrontal and parietal cortex as well as to the striatopallidum. They are involved in the control of states of wakefulness, consciousness, and attention. The *ventral thalamus* and *subthalamus* (*zona incerta*) of mammals projects to telencephalic parts of the basal ganglia, i.e., corpus striatum and globus pallidus, and to the hippocampus.

10.5 Diencephalon

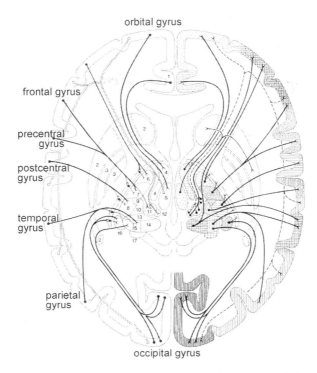

Fig. 10.8 Diagrammatic representation of the thalamocortical system. A horizontal section through the human brain is shown (*above* is the anterior-frontal, *below* the posterior pole of the brain). In the *middle* of the brain the numerous thalamic nuclei as well as other important subcortical centers, on the *outside* the various cortical areas with the major cortical gyri are shown. *Right* projections of thalamic nuclei toward cortical areas, *left* projections from cortical areas to thalamic nuclei. Abbreviations: *1* gyrus cinguli, *2* corpus striatum, *3* globus pallidus, *4* nucleus anterior thalami, *5* nucleus medialis thalami, *6* nucleus ventralis anterior, *7* nucleus ventralis lateralis, *8* nucleus ventralis posterior, *9* nucleus ventralis posterior, pars parvocellularis, *10* nucleus lateralis posterior, *11* nucleus centromedianus, *12* nucleus parafascicularis, *13* pulvinar, pars anterior, *14* pulvinar, pars medialis, *15* pulvinar, pars lateralis, *16* lateral geniculate body, *17* medial geniculate body. From Nieuwenhuys et al. (1988)

The *hypothalamus* and its appendage, the *pituitary* (*hypophysis*), are the main hormone-based control centers for basal homeostatic-autonomic functions. Cartilaginous and bony fishes exhibit a hypertrophy of the lateral hypothalamus (*lobus inferior hypothalami*), with unknown functions.

10.6 Telencephalon

The telencephalon of vertebrates and hagfish consists of a paired rostral part—the two cerebral hemispheres—and an unpaired part (the telencephalon impar), which is continuous with the diencephalon. In vertebrates, each hemisphere is divided

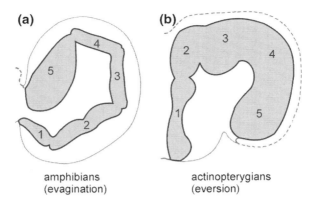

Fig. 10.9 Differences in ontogenetic development of the vertebrate telencephalon **a** Evaginated telencephalon as found in most vertebrates (here amphibians). **b** Everted telencephalon as found in actinopterygian bony fishes. *Numbers* indicate the major regions of the telencephalon: *1* ventromedial subpallium, *2* ventrolateral subpallium (striatopallidum), *3* lateral-ventral pallium, *4* dorsal pallium, *5* medial pallium. After Nieuwenhuys et al. (1998), modified

into a pallium ("cloak") around the dorsal, medial, and lateral parts of the ventricle and a subpallium surrounding its ventral parts (Fig. 10.3a, b). Accordingly, the pallium is composed of a medial, dorsal and lateral-ventral part, the latter including the olfactory amygdala (called cortical amygdala in mammals) and vomeronasal amygdala (called medial amygdala in mammals). In mammals, we find a basolateral amygdala in addition, probably originating from the lateral pallium. A ventral pallium is found only in terrestrial animals (cf. Fig. 10.3a, b). Rostral to the pallium and often separated by stalks, we find the olfactory bulbs. The subpallium consists of a medial septal region including the basal forebrain, the nucleus accumbens, and a lateral striatopallidum. In the caudal, unpaired part of the subpallium we find an autonomic or "central" amygdala as a continuation of the ventral striatopallidum.

The evolution of the telencephalon in craniates is debated. In all craniates, it receives olfactory information from the olfactory bulb as the only direct sensory input. Comparative neuroanatomists, therefore, previously believed that in its ancestral state, the telencephalon was a purely olfactory part of the brain. Later it was found that in all craniates, the telencephalon also receives information from other senses, e.g., visual, auditory, and mechanosensory, via pathways ascending from the diencephalon (see above). Accordingly, the telencephalon was regarded as "multimodal" in its ancestral state. Recent studies, however, revealed that in all craniates, except birds and mammals, these nonolfactory sensory afferents to the telencephalon are either multimodal or, if unimodal, do not form topographic representations. This would imply that the development of topographic representations of thalamic sensory afferents to pallial/cortical regions has happened independently in birds and mammals and would strengthen the "olfactory brain" interpretation.

10.6 Telencephalon

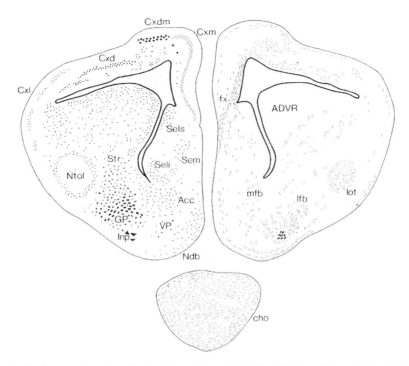

Fig. 10.10 Cross section through the telencephalon of the lizard *Tupinambis teguixin* at the level of the striatum. The anterior dorsal ventricular ridge (*ADVR*) bulges into the telencephalic ventricle. Abbreviations: *Acc* nucleus accumbens, *cho* chiasma opticum, *Cxd* cortex dorsalis, *Cxdm* cortex dorsomedialis, *Cxl* cortex lateralis, *Cxm* cortex medialis, *fx* fornix, *lfb* lateral forebrain bundle, *lot* lateral olfactory tract, *mfb* medial forebrain bundle, *Ndb* Nucleus of the diagonal band of Broca, *Ntol* nucleus of the tuberculum olfactorium, *Seli* inferior lateral septum, *Sels* superior lateral septum, *Sem* medial septum, *Str* striatum, *VP* ventral pallidum. From Nieuwenhuys et al. (1998), modified

The telencephalon of hagfish and lampreys is characterized by large olfactory bulbs, which in the latter are even larger than the rest of the telencephalon (Fig. 10.2a, b) and underscore the significance of the olfactory system for these animals. The telencephalic ventricles are almost invisible in hagfish and small in lampreys. Therefore, a precise delimitation of pallial and subpallial parts is difficult and disputed. The dorsal pallium of myxinoids exhibits a five-layered structure, which has developed independently of other lamination patterns found in craniates (see below).

Among cartilaginous fish, we find enormous differences in the size and structure of the telencephalon. Squalomorph sharks, chimaeras, and torpediniform and rajiform rays have telencephala of moderate size (Fig. 10.2c), which are of the evaginated type (see below). The pallium of these groups is divided into a medial pallium (MP) with strong cellular migration, but without recognizable lamination, a relatively large DP with three cellular laminae, and a lateral pallium (LP) giving rise to long stalks carrying massive olfactory bulbs. The telencephala of galeomorph sharks and myliobatiform rays are much larger relative to the rest of the brain and reach relative proportions found in mammals. The olfactory bulbs are not carried by long stalks, but are directly attached to the LP, where olfactory information is processed. The pallium is unlaminated, and its dorsomedial part receives numerous ascending pathways from the thalamus, posterior tuberculum, hypothalamus, and neuromodulatory (i.e., noradrenergic, dopaminergic, and serotonergic) nuclei in the isthmic tegmentum. In the center of the telencephalon we find a characteristic *central nucleus*, which apparently is formed by fusion of parts of the dorsal and medial pallia of both hemispheres. It is an important sensory convergence center, because it receives massive ascending tracts from the visual, lateral line, and probably auditory systems and gives rise to tracts that descend to the thalamus, optic tectum, and medulla oblongata.

A specialty of the shark telencephalon is a nuclear complex in its ventrocaudal part called area basalis. Studies by Hofmann and Northcutt (2008) demonstrated that this area receives massive olfactory input and in turn projects to the dorsomedial pallium. Such an olfactory pathway is found only in sharks and emphasizes the importance of olfaction for these animals. The dorsomedial pallium then projects—as is usual for vertebrates—to the hypothalamus. Electrophysiological recordings from pallial neurons in sharks reveal unimodal olfactory responses, but only mixed responses in the other senses. Accordingly, in sharks there are no unimodal, i.e., separate visual, auditory, and somatosensory pathways from the dorsal thalamus to the pallium.

The structural and functional organization of the telencephalon of the actinopterygian bony fishes and possible homologies of its pallial regions with those of the other vertebrates are a matter of debate. At first glance, the dorsal telencephalon of the actinopterygians does not reveal any similarity with that of other vertebrates. Rather, it seems to consist of two compact hemispheres without further cellular or fiber subdivisions. The hemispheres are covered by a thin neuroepithelial membrane. These striking differences are best explained by assuming that the telencephalon of the actinopterygians has an *everted* pallium, while the telencephala of the other vertebrate groups have an *evaginated* pallium (cf. Nieuwenhuys et al. 1998). This difference is illustrated in Fig. 10.9. In the *evagination* type, the walls of the embryonic telencephalon (1-5) bulge outward, encircling the lateral ventricle and divide into the mentioned pallial (3-5) and subpallial (1-2) parts. In the *eversion* type, the subpallial parts (1-2) remain in a medial position along the midline of the telencephalon, while the pallial parts (3-5) bend outward and then downward. As a consequence, the MP of this everted telencephalon now occupies a lateral and increasingly ventral position. Accordingly, there is a new "medial zone Dm," continued laterally by a "central zone Dc," and a "lateral zone Dl" and finally, in a caudal position, a "dorsoposterior zone Dp." According to Wullmann and Vernier (2007), this latter "dorsoposterior zone" receives olfactory input and therefore corresponds with the lateral, olfactory pallium of the other vertebrates. The dorsolateral zone receives predominantly visual afferents from the diencephalic nucleus preglomerulosus (see above), the lateral, central,

10.6 Telencephalon

and medial zone receive afferents from the mechanoreceptive system, and the dorsolateral and dorsomedial zone afferents from the auditory system via the torus semicircularis. The dorsomedial zone receives additional input from the gustatory system via the nucleus preglomerulosus.

Thus, all these zones appear to represent standard parts of the pallium. Unclear is the homology of the "central zone Dc" of the pallium. It gives rise to numerous pathways ascending to the olfactory bulb and descending to various diencephalic and mesencephalic nuclear regions, and in some teleosts, via both the medial and lateral forebrain bundle, to the torus semicircularis cerebellum and valvula. Thus, the central zone is the main efferent station of the actinopterygian pallium (Wullimann and Vernier 2007).

In contrast to the actinopterygians, the telencephala of lungfish and amphibians belong to the evaginated type and closely resemble each other (Nieuwenhuys et al. 1998; cf. Fig. 10.3). This is not surprising, because lungfish and amphibians are considered sister groups. The medial and dorsal pallium of amphibians receives multimodal, i.e., mixed visual, auditory, and somatosensory information via the anterior dorsal thalamic nucleus, as found in cartilaginous fishes. The lateral pallium is the site of the processing of signals from the main olfactory bulb, while the ventral pallium (VP) processes information from the vomeronasal (or "accessory") olfactory bulb. As shown in Fig. 10.3, the pallium of amphibians is generally unlaminated despite extensive cell migration in medial and dorsal parts, while the MP and DP of lungfishes (dipnoans) display some lamination.

In "reptiles," i.e., turtles, lizards, snakes and crocodiles, pallial divisions are called medial (Cxm), dorsomedial (Cxdm), dorsal (Cxd) and lateral cortex (Clx), and the already mentioned dorsal ventricular ridge (Nieuwenhuys et al. 1998; Fig. 10.10). The medial and dorsomedial cortex correspond to the MP of amphibians and probably the hippocampal formation of mammals, the lateral cortex to the lateral (olfactory) pallium of amphibians and the mammalian "piriform" cortex, and the dorsal cortex of "reptilians" is homologous to the DP of amphibians and possibly to the isocortex of mammals. The mentioned cortical regions exhibit three layers, which are clearly visible only in the medial and dorsomedial part and not completely continuous throughout the cortex. This lamination pattern has probably arisen independently of other cases of pallial or cortical lamination found in vertebrates.

The dorsal ventricular ridge is a structure unique to reptiles. It bulges in the medial direction into the ventricles and for a long time was considered a hypertrophic part of the corpus striatum as the major telencephalic component of the basal ganglia (Fig. 10.11). Accordingly, parts of the DVR of birds were called *ectostriatum, neostriatum*, and *hyperstriatum*. However, studies by the American neurobiologist Harvey Karten and colleagues in the 1970s of the last century demonstrated that the DVR is not homologous to the striatum, but is of pallial origin (cf. Karten 1969, 1991). First, it could be shown that the DVR of "reptiles" as well as the homologous "neostriatum" of birds (today called mesonidopallium—see below) receive visual, auditory, and somatosensory afferents terminating in different regions, which is not characteristic of the basal ganglia of other

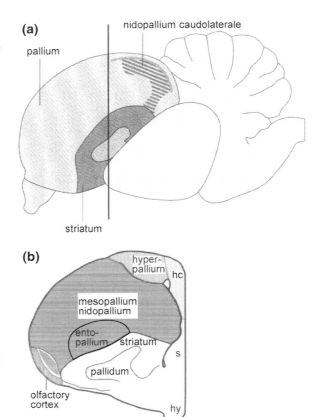

Fig. 10.11 Brain of a pigeon. **a** *Lateral view*. The telencephalon is composed of a pallium and a striatopallidum. A special pallial region is the nidopallium caudolaterale. After Güntürkün (2008) **b** Cross section through the telencephalon at the level indicated in **a**. Most of the pallium consists of the mesopallium and nidopallium; the entopallium (former ectostriatum) is the termination site of visual afferents from the nucleus rotundus. The hyperpallium is situated mediodorsally. Striatum and pallidum are located below the mesonidopallium. Abbreviations *Hc* Hippocampus, *S* Septum, *Hy* Hypothalamus. For further information see text. After Reiner et al. (2005), modified

vertebrates. Furthermore, histochemical staining showed that the larger dorsal part of the DVR and the ectostriatum/mesonidopallium do not contain cells containing the transmitter acetylcholine, which is typical of the basal ganglia. Such cholinergic cells are found only in the ventral parts of the reptilian and bird telencephalon below the DVR and ectostriatum/mesonidopallium. In the meantime, it has become generally accepted that the DVR and the mesonidopallium are of pallial and not subpallial origin.

The DVR of "reptiles" is composed of a larger anterior (aDVR) and a smaller posterior (pDVR) portion. The latter contains the amygdala complex and will not be discussed here. The aDVR receives, in its lateral part, visual afferents from the thalamic nucleus rotundus via the lateral forebrain bundle. In snakes with infrared sensing (such as the rattlesnake), afferents from parts of the tectum processing infrared signals terminate here, too. Somatosensory afferents terminate in the central, auditory afferents in the medial region of the aDVR.

In cross sections, the large dorsal telencephalon of birds does not reveal any similarities with either that of reptiles or of mammals (Fig. 10.11). Medially, we recognize a small, vertically oriented ventricle and a laterally adjacent mass of

10.6 Telencephalon

cells and fibers crossed by thin stripes of lower cell density. As already mentioned, this compact mass was considered to be part of the basal ganglia of birds, and this interpretation was easily adopted, because it was believed that the behavior of birds was guided primarily by "instinct" (assumed to reside in the basal ganglia) rather than by learned behavior (assumed to be located in the cortex). Accordingly, a ventral "paleostriatum augmentatum" was identified as a "primordial" form of the basal ganglia, as well as a newly formed part called "neostriatum," and above all a "hyperstriatum"—with unknown functions. Only the so-called "wulst" as a mediodorsal part of the hyperstriatum, with visual functions, was regarded as a "true cortex." As described above, this view has become obsolete, and a few years ago, a commission formed by bird brain experts decided to rename the problematic telencephalic regions such that the former neostriatum is now called *nidopallium*, and the former hyperstriatum ventrale is now called *mesopallium* (Reiner et al. 2005). The hyperstriatum accessorium containing the "wulst" in its rostral part is now called *hyperpallium*. In its rostral part, the *nidopallium* includes the *entopallium* (the former ectostriatum), and this is the site of the termination of visual afferents from the nucleus rotundus. The auditory afferents from the nucleus ovoidalis terminate in "field L" of the nidopallium.

The hyperpallium, situated dorsal to the nidopallium, receives both visual and somatosensory afferents and is considered to be homologous to the cortex of mammals. However, in its cytoarchitecture it does not resemble the mammalian cortex, but exhibits an apparently irregular arrangement of multipolar projection neurons and interneurons, which have no closer resemblance to the neurons found in the mammalian cortex (Tömböl et al. 1988). The same situation is found in the entire nidopallium, including the entopallium (the former ectostriatum). Here again, we find no lamination and rather uniformly looking multipolar neurons. The main type includes medium-sized projection neurons with dendrites that are moderately to heavily covered with spines, while some interneurons have rather smooth dendrites (cf. Fig. 17.1). Thalamic afferents with very thick diameters enter the entopallium ventromedially and quickly divide into secondary dendrites. These secondary processes then extend straight forward, divide again, and form a regular fiber network resembling the network of thalamic afferents to the mammalian cortex and making contact with projection neurons as well as with interneurons. There is another type of afferents with smaller diameters, which again run straight forward. This rather regularly arranged system of incoming fibers does not meet a regular, laminated arrangement of cells as in the mammalian cortex, but a seemingly irregular distribution of projection neurons and interneurons.

A special region of the nidopallium is the "nidopallium caudolaterale—NCL" (the former "hyperstriatum caudolaterale") situated—as its name indicates—in the caudolateral part (cf. Fig. 10.11). According to studies of the German biopsychologist Onur Güntürkün and his colleagues, functionally it strongly resembles the mammalian dorsolateral prefrontal cortex (dlPFC—see below), because it is involved in working memory, action planning, behavioral flexibility, and creativity—in essence, in "intelligence" (Güntürkün 2005). Like the mammalian dlPFC, it is a multimodal convergence center and receives a strong dopaminergic input from the VTA and nucleus accumbens. Despite these strong functional similarities, its homology with the mammalian dlPFC is unlikely on the basis of embryological evidence. The avian NCL, too, reveals no lamination.

The telencephalon of mammals is large to very large, relative to the rest of the brain, and in some groups, such as primates, elephants, whales, and dolphins, amounts to up to 80 % of entire brain volume. The mammalian telencephalic structures homologous to the pallia of other vertebrates are the iso- or neocortex, the hippocampus, and the limbic and olfactory cortex as well as the basolateral amygdala. The subcortical parts are homologous to those found in the other vertebrates, i.e., the striatopallidal complex, the subcortical parts of the amygdala, and the septal region including the basal forebrain. In the following, I will describe only the cortex and the closely related hippocampus (Figs. 10.7, 10.12).

The *hippocampus* consists of three layers representing a limbic or "allocortical" cortex and is situated close to the lower edge of the temporal lobe (Fig. 10.7b). It is reciprocally connected with all parts of the six-layered isocortex via the entorhinal cortex (another limbic cortex or allocortex), and in addition receives direct afferents from the basal forebrain, the amygdala, and other subcortical limbic centers. The hippocampus, together with the adjacent ento- and perirhinal cortex, are regarded as the *organizers* of the declarative memory, which in humans includes the conscious representation and reportability of events and knowledge.

The *orbitofrontal cortex* (Brodmann areas A11,12) is the only isocortical, i.e., six-layered, part of the limbic system (Roberts 2006; Barbas 2007). In humans, it is situated above the eye sockets or "orbits" and is adjacent to the ventromedial portion of the frontal cortex. In many mammals, it processes olfactory and gustatory information and is involved in the evaluation of quality and attractiveness of food. In primates, including humans, its posterior part processes the emotional and motivational aspects of control of behavior, particularly regarding the positive or negative consequences or aspects of past and planned actions. The gyrus cinguli (A23, 24) is part of the limbic cortex dorsally surrounding the corpus callosum. Two parts, an anterior and a posterior gyrus cinguli, are distinguished. In the view of many experts, the dorsal and posterior part of the gyrus cinguli has predominantly cognitive functions and is involved in the control of eye movement and visual attention. The anterior and ventral part, instead, is involved in the motivational control of attention, error recognition and error correction, sensation and evaluation of pain, and assessment of long-term gains and losses in decision making (Botvinick et al. 2004). The insular cortex (often called "insula") is found in a portion of the lateral cortex folded deep between the frontal, temporal, and parietal cortex and is, with the exception of a small opening ("operculum"), invisible from the outside. It is the site of awareness of body states, including pain sensation ("subjective pain"). In addition, in humans it is involved in recognition or imagination of painful events, including persons suffering from pain (Singer et al. 2004). In this context it is believed to be a center involved in empathy.

10.6 Telencephalon

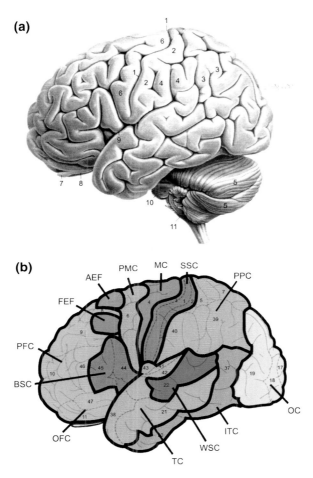

Fig. 10.12 a *Lateral view* of the human brain showing the cerebral cortex with its characteristic convolutions (gyrus/gyri) and fissures (sulcus/sulci) and the likewise strongly folded cerebellum. Abbreviations: *1* sulcus centralis, *2* gyrus postcentralis, *3* gyrus angularis, *4* gyrus supramarginalis, *5* cerebellar hemisphere, *6* gyrus precentralis, *7* olfactory bulb, *8* olfactory tract, *9* sulcus lateralis, *10* pons, *11* medulla oblongata. **b** Anatomic-functional organization of the lateral cortex. Numbers indicate Brodmann areas (see Fig. 11.14). Abbreviations: *AEF* anterior eye field, *BSC* Broca speech center, *FEF* frontal eye field, *ITC* inferotemporal cortex, MC motor cortex, OC occipital cortex, *PFC* prefrontal cortex, *PMC* dorsolateral premotor cortex, *PPC* posterior parietal cortex, *SSC* somatosensory cortex, *TC* temporal cortex, *WSC* Wernicke speech center (approximately). After Nieuwenhuys et al. (1988), modified

10.6.1 Functional Anatomy of the Isocortex

The isocortex of mammals is divided into four lobes, i.e., an occipital, temporal, parietal, and frontal lobe. It has six layers throughout and has a rather uniform appearance (Nieuwenhuys et al. 1988). The dominating cell type is the pyramidal cells, which in primates include about 80 % of all cortical neurons (Figs. 5.1 and 10.13). Pyramidal cells are exclusively excitatory and represent the projection neurons of the cortex. Their axons leave the immediate vicinity of the cell and enter the white substance, from where they either return to the gray matter or descend to subcortical regions of the brain. Pyramidal cells have a name-giving, pyramid-shaped soma. Their dendrites are covered with spines (in humans about 6,000 or more per cell), which are special sites of excitatory synaptic contacts. Often, one spine carries more than one presynaptic terminal (cf. Fig. 5.1), such that one neuron may receive excitatory input from about 20,000 other cortical cells. Inhibitory input comes from an average of 1,700 cells (mostly interneurons—see below) and does not terminate on the spines, but on the dendritic shaft, soma, or even axon. The dendrites of pyramidal cells located in the upper cortical layers reach the superficial molecular layer, where they bifurcate and form the horizontal fibers, which in primates extend for 100–200 μm, in large pyramidal cells up to 400 μm. The remaining cortical cell types are interneurons, i.e., cells that are involved in the local processing of information, with axons restricted to the immediate vicinity. These interneurons include stellate cells, basket cells, candelabra cells, and bipolar cells (Nieuwenhuys et al. 1988). Dendrites of stellate cells extend either radially or vertically. The surface of their dendrites is either smooth or carries only few spines. While the smooth stellate cells and the candelabra cells have inhibitory functions, the spiny stellate cells and the bipolar cells are excitatory.

Six cortical layers are distinguished (Fig. 10.13). The uppermost layer, I, is called the molecular layer and contains only few nerve cells, but primarily apical dendrites of pyramidal cells and the already mentioned horizontal fibers. Intracortical connection and input from "matrix" or M-type thalamic cells terminate there (Jones 2001) in contrast to "core" or C-type thalamic afferents that go to layer IV (see below). Layer II is called the "external granular layer" and contains small pyramidal neurons and numerous stellate neurons. Layer III is called the "external pyramidal layer" and contains mostly small and medium-sized pyramidal neurons as well as interneurons. This layer is the main origin of corticocortical efferents. Layer IV is called the "internal granular layer" and contains different types of pyramidal cells and many interneurons. It is the main cortical input layer for type-C afferents from the thalamus and of intrahemispheric connections. Layer V is the "internal pyramidal layer" and contains large pyramidal cells, such as the giant Betz cells, in the primary motor cortex. This layer is the main output layer of subcortical efferents, e.g., to the basal ganglia or to the brainstem and spinal cord constituting the pyramidal tract. Layer VI is the "polymorphic" or "multiform layer," also called "spindle-cell layer" and contains few large and numerous smaller spindle-shaped pyramidal cells and interneurons. It sends axons to the thalamus in a point-to-point fashion.

The mass of cortical afferents comes from the palliothalamic nuclei (Fig. 10.8), which end primarily in layer IV and lower layer III and strongly ramify there. They contact the small pyramidal cells as well as interneurons, and the latter make either excitatory or inhibitory connections (depending on the type of interneuron) with the pyramidal cells. Afferents from the truncothalamic and particularly intralaminar

Fig. 10.13 Cytoarchitecture of the six-layered mammalian isocortex. The *left side* of the figure shows the distribution of nerve cells, predominantly pyramidal cells, in a Golgi staining. In the *middle*, the distribution of cell bodies is shown in a Nissl staining. The *right side* shows the distribution of myelinated fibers in a Weigert staining. Roman numerals to the *left* indicate the gross lamination of the cortex, arabic numerals to the *right* indicate the sublamination based on a Nissl staining. After Vogt and Brodmann from Creutzfeldt (1983), modified

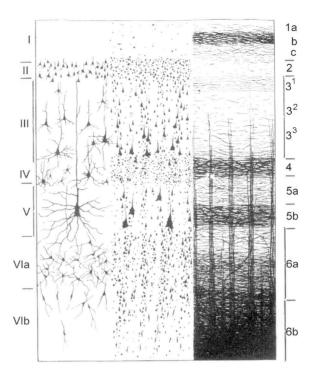

thalamic nuclei terminate in layers I and VI. Nonthalamic afferents come from the amygdala, the basal forebrain/septum (cholinergic afferents), the striatopallidum, the hypothalamus, anterior raphe nuclei (serotonergic afferents), locus coeruleus (noradrenergic afferents), and the mesolimbic system (VTA, nucleus accumbens with dopaminergic afferents) and enter the cortex in an oblique fashion.

In primates, cortical efferent fibers are five times more numerous than afferents. They originate predominantly from pyramidal cells in layers V and VI. Those in layer VI of a given cortical area project exactly to those thalamic nuclei, from where the area receives afferents, and this reentrant organization is part of the already mentioned thalamocortical system (Creutzfeldt 1983). Efferents from layer V run to the striatum and amygdala or constitute the pyramidal tract descending to the midbrain, pons, and premotor and motor centers of the medulla oblongata and spinal cord. The vast majority of cortical fibers, however, form intracortical projections called "association fibers," of which there are billions.

Isocortical areas differ in their precise cytoarchitecture, i.e., in the relative number of the different types of cortical cells present in the layers as well as in cell density, thickness of the single layers, and overall thickness. On the basis of such differences, the German neuroanatomist Korbinian Brodmann, at the beginning of the twentieth century, divided the entire cortex of humans and other mammals into anatomical areas, which in humans include 52 areas today called "Brodmann areas" (Brodmann 1909). They are labeled "A" for "area" or "BA" for "Brodmann area" (e.g., A1, A2, or BA1, BA2; cf. Fig. 10.14a, b). Although these divisions were based on purely morphological and cytological criteria, later studies revealed that they—*grosso modo*—also indicate functional areas. For the different functional systems (e.g., the visual, auditory, and motor areas) of the cortex, additional terms are used, which will be mentioned below. The following description relates mainly to the primate, including human, cortex (Fig. 10.12).

The *occipital* cortex contains exclusively primary, secondary, and associative visual areas, which are described in more details in Chap. 11. The *parietal* cortex includes the primary somatosensory cortex (A1–3), with information from the skin, muscles, tendons, and joints, and sends this information to the posterior parietal areas (A7), where they converge with visual, auditory, vestibular, and oculomotor informations. Together they are used for the construction of the body scheme, of a three-dimensional world with the localization of the sources of sensory stimuli, and of the position and movement of its own body inside this world. In addition, the posterior parietal cortex is important for the control of goal-directed head, arm, and hand movement. The parietal cortex reveals strong hemispheric specializations. The right parietal lobe is particularly involved in spatial localization and orientation and spatial attention, the realistic or mental construction of space with the possibility of changes in perspective. The left parietal lobe, including the gyrus angularis (A39) and gyrus supramarginalis (A40), is involved in processing of symbolic-analytic information in the context of reading, writing, and mathematics as well as "reading" maps and understanding the abstract meaning of pictures. Interestingly, the right gyrus angularis appears to be involved in the grasping of the metaphoric meaning of information (see "mirror neurons" in Chap. 13).

The *temporal* cortex includes, in its upper parts, the primary auditory area (Heschl's transverse gyri, A41), which processes simple auditory signals, such as pitch and timing of sounds, surrounded by the secondary auditory areas. In the latter, processing of complex auditory information takes place which are, among others, necessary for the understanding of vocal communication signals (cf. Chap. 14). In primates, the right lower and posterior temporal lobes are involved in the recognition of complex visual objects and situation. This includes, in the fusiform gyrus, the recognition of body parts such as hand, faces, eyes, and mouth, but also the distinction between living and nonliving objects. Recognition of the dynamics of faces (mimic) occurs in the superior temporal cortex (A22), while the fusiform gyrus or fusiform face area, located in the ventromedial temporal lobe (A20) is predominantly involved in the identification of faces.

The *frontal* association cortex includes the dorsal and ventral lateral prefrontal cortex, the already mentioned orbitofrontal and ventromedial cortex, the frontal eye field (A8), the supplementary eye field (A6), the supplementary motor areas (SMA, pre-SMA as medial part of A6), and the Broca speech-language center (A44, A45). The lateral prefrontal cortex (PFC) includes areas A9, A10, and A46. Many authors divide the primate PFC into a dorsolateral and ventrolateral part (dlPFC and vlPFC, respectively). These two parts of the PFC receive different input from other cortical regions. The dorsolateral PFC receives input mainly from

Fig. 10.14 Cortical areas after Korbinian Brodmann (1909). **a** *lateral view*, **b** *median view* of the brain. Numbers indicate the "Brodmann areas." From Roth (2003)

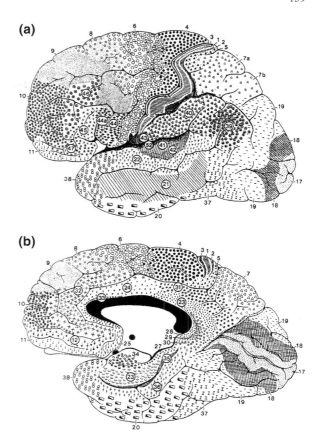

the posterior parietal cortex regarding position and movement of head, neck, face, and hands as well as information about spatial orientation and spatial aspects of action planning (Goldman-Rakic 1996; Fuster 2008). The latter input terminates primarily in the rostralmost, frontopolar region of the PFC (A10), which, according to some authors, is found only in humans (Wise 2008). The ventrolateral PFC, in contrast, receives input mainly from the temporal lobe carrying complex visual and auditory information, e.g., in the context of the meaning and relevance of objects and scenes, as well as language-related information from the superior and middle left temporal lobe. In the ventrolateral PFC, we find the Broca speech-language center (A44, 45), which is involved in grammatical and syntactical aspects of language (cf. Chap. 14).

The prefrontal-frontopolar cortex (A9, 10, 46) of primates is generally involved in the comprehension and processing of the temporal-spatial structure of sensory information and cognitive mental events such as thinking and imagining, predominantly in the context of action planning and action preparation, but also problem solving and decision making. It is also the site of working memory, as described in Chap. 2.

10.6.2 Are the Mammalian Cortex and the Mesonidopallium of Birds Homologous?

As we have just seen, only mammals possess a six-layered isocortex, and this isocortex is believed to be the site of complex cognitive functions, including consciousness. However, birds are capable of executing complex cognitive tasks similar to primates, as we will learn in Chap. 12; it is assumed that the mesonidopallium (MNP) of the bird telencephalon and especially the nidopallium caudolaterale are responsible for this capability. However, the anatomy and cytoarchitecture of the MNP has no resemblance to the mammalian isocortex. Thus, the question arises as to whether the MNP of birds and the isocortex of mammals are homologous or convergent structures. If they are homologous, then the great similarity in function would be of no surprise. If they are not, then we would be confronted with a striking example of convergent evolution of an "intelligence center" among vertebrates.

It is generally accepted that the MNP evolved from the "reptilian" aDVR, so that the homology question can be extended to the aDVR. For many years there has been a debate about the possible homology between the mammalian isocortex and the sauropsid aDVR/MNP. The competing interpretations are illustrated in Fig. 10.15. Harvey Karten, Anton Reiner, Onur Güntürkün and other colleagues argue that the aDVR and the MNP are homologous to the lateral temporal cortex (LC) of mammals and that phylogenetically as well as ontogentically both develop from tissue in the LP (cf. Reiner et al. 2005). This view is called "common origin hypothesis." The authors point to the fact that the MNP and DVR on the one hand, and the LC on the other, have very similar input and output connections. The mammalian LC indeed receives thalamic visual input from the pulvinar, which in turn receives input from the colliculus superior, as being homologous to the tectum of other vertebrates including birds (see above). In a similar way, the tectum of sauropsids projects to the thalamic nucleus rotundus, which in turn projects to aDVR and MNP. There is also an embryological argument of the authors: precursor cells in the embryonic mammalian brain migrate outward and then dorsally and later develop into the six-layered structure of an isocortex as being homologous with the dorsal pallium. In the opinion of the authors, such a migration and transformation of precursor cells does not occur in the bird, and instead of a lamination, the formation of nuclei takes place inside the region of the aDVR.

Other leading comparative neurobiologists, like Luis Puelles, L. Medina, and G.F. Striedter, reject such an interpretation (cf. Striedter 2005; Medina 2007) and contend that the MNP-DVR is not homologous with the dorsal pallium, but rather with the lateral (olfactory) and ventral (vomeronasal) pallium of other vertebrates. If this is true, then the LC and MNP-DVR originate from different phylogenetic-embryological material, i.e., dorsal and ventral pallium, respectively. This concept is called "de-novo hypothesis." The other authors likewise base their concept on embryological arguments and point to differences in the transcription factors responsible for the formation of the pallium. There are certain transcription factors that determine

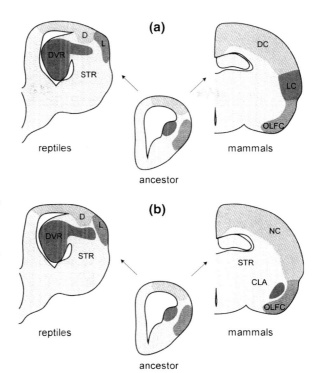

Fig. 10.15 Two hypotheses concerning the homology of the lateral mammalian cortex (*LC*) and the dorsal ventricular ridge (*DVR*) of "reptiles." **a** Hypothesis of "common origin" of the *DVR* and *LC* from the same embryonic material in the amniote ancestor. **b** Hypothesis of the *de novo* formation of *LC* and *DVR*. For further information, see text. Abbreviations: *CLA* claustrum, *D* dorsal cortex of reptiles, *L* lateral cortex of reptiles, *DC* dorsal cortex of mammals, *LC* lateral cortex of mammals, *NC* neocortex of mammals, *OLFC* olfactory cortex, *STR* striatum, From Striedter (2005), modified

the formation of the MP, DP and VP, which, however, are absent in the VP or at least parts of it. This appears to indicate that at least part of the VP is different from the rest of the pallium, which in sauropsids then develop into the DVR and MNP, while in mammals it gives rise to the cortical and basolateral amygdala and claustrum. Accordingly, MNP-DVR is functionally convergent, but not homologous to the LC of mammals, but rather homologous to parts of the amygdala and the claustrum.

Another important argument for the "de-novo hypothesis" is the fact that there are substantial differences between the visual cortex of mammals and the visual parts of the hyperpallium and nidopallium of birds. In the primary visual cortex of primates we find neurons that are either motion-sensitive or orientation-, contrast-, and color- (wavelength-) sensitive, as we will learn in more detail in Chap. 11. There are also disparity-sensitive cells as a basis for depth perception (stereopsis). In the pigeon, with a highly developed visual system, we find motion-sensitive neurons in the wulst as part of the hyperpallium, which most probably is homologous to parts of the mammalian isocortex, but no orientation-, contrast-, color-, and disparity-sensitive neurons. Instead, such cells are found in the nidopallium. Accordingly, in pigeons, lesions in the wulst lead to deficits in movement, but not in form and color perception, while the opposite is the case with lesions in the nidopallium. This speaks against a functional homologization of the bird MNP and the mammalian lateral (temporal) cortex. The latter does not have such visual functions, which instead are found in the occipital cortex (see above and Chap. 11).

10.7 What Does All This Tell Us?

While the vertebrates exhibit an impressive variability in ecology and lifestyle, their brains reveal high-level conservatism. Its pentapartite division into telencephalon, diencephalon, midbrain, cerebellum and medulla oblongata probably evolved 500 mya, perhaps from an ancestral tripartite brain present in the ancestor of all bilaterians. Until about 20 years ago, many comparative neurobiologists believed that certain nuclei and areae, like the noradrenergic locus coeruleus or the dopaminergic substantia nigra, were "not yet found" in so-called primitive vertebrates; but in the meantime, it became increasingly clear that even such structures and their respective functions are present in all vertebrates.

Dramatic differences instead occur with respect to the absolute and relative size of the brains and their various parts. This is true, above all, for the cerebellum, the roof of the midbrain, the dorsal thalamus, and the dorsal telencephalon, and these changes have occurred independently in most vertebrate classes. For example, some groups of cartilaginous and bony fish have small, and others very large to gigantic cerebella relative to overall brain size. Among amphibians, differences between frogs and salamanders-caecilians regarding cerebellum size are considerable, and the same is found among groups of mammals, e.g., with regard to the gigantic cerebella of elephants and cetaceans. The same holds true for the optic tectum and torus semicircularis in bony fish, amphibians, and birds, and finally for the dorsal pallium or cortex, which either is insignificant in size or became enormously large.

The simplest mechanism underlying a dramatic increase in relative (and absolute) size of a given brain structure is the multiplication of a certain modular organization, as has happened in the cerebella and cortices of mammals. This has led to an enormous increase, mostly combined with an infolding (gyrification) of the surface. More dramatic is an increase in structural and functional complexity, when—mostly in combination with a strong increase in volume—the number of layers (like in the tectum or torus) or number of nuclei (like in the thalamus) has increased. Such an increase normally goes along with an increase in number of types of neurons. In most cases, this is correlated with changes in lifestyle and sensory systems requiring an increase in multisensory information processing. In addition, there are cases of the formation of novel brain structures like the hypertrophic vagal and facial lobes in some teleosts in correlation with a highly evolved gustatory system—for example, in cyprinids. Other examples of such spectacular specializations are the enormous increase of the electrosensory lateral lobe and torus semicircularis in the weakly electric fish and of the optic tectum in teleosts and birds.

10.7 What Does All This Tell Us?

Of special interest are modifications and specializations inside the telencephalon, and here above all of the pallium, which always occurred together with changes in the thalamus. In actinopterygians, we find a morphological reorganization of the pallium in the form of an evagination instead of an eversion. Within the sauropsids, the ventral pallium evolved into the DVR of "reptiles" and the MNP of birds as the most important center for multimodal sensory integration, whereas mammals evolved the dorsal pallium into a six-layered isocortex, which then became enormously large in many groups. Thus, in amniotes, the inconspicuous pallium of amphibians underwent a dramatic evolution in two very different directions.

Together with the modifications of the telencephalic pallium, there were significant changes in the thalamo-telencephalic pathways. In the putatively ancestral state found in lampreys, cartilaginous, and bony fishes as well as in amphibians, the dorsal telencephalon is dominated by olfaction, while auditory, somatosensory, and visual information is mostly processed in the midbrain, cerebellum including valvula, and medulla oblongata. The pallium of these groups receives, via the anterior dorsal thalamus, primarily multimodal and only scarce unimodal afferents, which do not form topologically organized areas. Except for olfactory responses, only multimodal responses can be recorded in the pallia of these groups. Although there is still little understanding of the precise function of the MP and DP of anamniotes and "reptiles", one can assume that they have to do with multisensory integration, memory formation, emotion, and motivation, which, via the dorsal and ventral striatopallidum, influence centers that directly guide behavior.

In birds and mammals, the telencephalon has independently evolved into a site of unimodal sensory information processing, in addition to what happens in the medulla oblongata, cerebellum and dorsal midbrain, and this goes along with modifications of the dorsal thalamus. As to the visual system, in mammals, the lateral geniculate nucleus receives direct retinal input and projects to the occipital cortex, where numerous visual areas are found. In birds, the nucleus-geniculatus complex likewise receives direct retinal afferents and projects to the dorsomedial telencephalon called "visual wulst" as part of the hyperpallium. This visual pathway, however, is not the main one, which is the projection from the nucleus rotundus to the entopallium originating in the optic tectum.

Therefore, in birds and mammals, primary sensory pathways from the dorsal thalamus to the pallium have evolved independently, which in the case of birds take the course of the *lateral* forebrain bundle toward the entopallium (part of the MNP as a homologue of the "reptilian" aDVR), and in the case of mammals, the *medial* forebrain bundle toward the isocortex. In this way, the pallium of birds and the cortex of mammals evolved into the major site of processing of unimodal sensory information, in addition to the processing of limbic and multimodal-associative information. Apparently this was one of the decisive steps toward a strong increase in sensory and cognitive abilities in both groups of vertebrates.

Chapter 11
Sensory Systems: The Coupling between Brain and Environment

Keywords Sense organs—general function · Olfaction · Mechanical senses · Electroreception · Lateral line system · Auditory system · Visual system · Parallel processing visual system · Insect compound eye · Vertebrate eye · Retina

Under the aspect of evolution, comparative studies on the structure and function of sense organs and sensory processes are important, because sense organs and sensory processing mechanisms are commonly seen as the most telling evidence of the work of natural selection: nothing of an organism appears to be more closely related or "adapted" to survival and reproduction than the structures and functions of sense organs. Across the kingdom of animals, there seems to exist a bewildering diversity of them as well as a huge range in complexity, from sensory receptors in unicellular organisms to the eyes or ears of birds and mammals, including humans. However, at the same time, there is high uniformity and conservatism in the underlying principles, as we will see. This uniformity and conservatism partly results from basic physical and chemical constraints for the function of sensory receptors and sense organs, partly from phylogeny via "deep homologies" of developmental genes.

11.1 The General Function of Sense Organs

One major function of the nervous systems and brains is to generate a behavior that promotes survival and eventually successful reproduction (Barth 2012). In order to do so, they need relevant information about the environment and their own bodies. At the same time, brains, being composed of nerve cells, are *insensitive* to any event or "stimulus" in the environment, i.e., they cannot perceive these events directly. Therefore what they need are "mediators" or "transducers" between them and the environment in the form of *sensory cells* or *sensory receptors*, which are capable of interacting with both the environment and the nervous system. Thus, the impact of a stimulus on sensory receptors is transformed by *transduction* and *encoding* into the "language of neurons," i.e., into neurochemical or neuroelectrical signals eliciting action potentials (cf. Mausfeld 2013).

The process of *transduction* of the stimulus and the subsequent encoding into action potentials may take place either in one and the same cell, called *primary sensory cell*, or in two cells, a *secondary sensory cell* plus a neuron that is in close contact with that cell. The primary sensory cells are neurons which do not only transduce the impact of an environmental event (including their own body) into a graded potential, but produce action potentials and transmit them via an axon to neurons inside the central nervous system. This happens, for example, in the olfactory cells. In contrast, secondary sensory cells produce only graded potentials, and an action potential arises in the postsynaptic neuron which, as part of the central nervous system, sends an axon to the sensory periphery. This is found in the taste receptors, the hair cells in the inner ear or in photoreceptors.

The main principle of the function of sensory cells of both types is that chemical or physical events in the environment (odorant molecules, light, mechanical pressure, etc.) transmit energy onto the receptor site, i.e., specialized sensory molecules, inducing either directly or indirectly, i.e., via signaling cascades, changes in the membrane potential of the sensory cell by (mostly) opening sodium channels. This leads, either in specialized regions of the cell itself or in the postsynaptic neuron, to the generation of action potentials. The sensory molecules are often contained in specialized cellular structures such as hairlike cilia containing microtubuli or microvilli, also called stereovilli, i.e., protrusions of the membrane surface of sensory cells without microtubuli. They define the kind of sensory stimuli that are able to influence the sensory receptor cells in an "adequate" way, i.e., even at very low stimulus intensities and without harming or destroying them. One, therefore, speaks of the "adequate stimulus." Sensory receptors might also respond to stimuli other than the "adequate" ones, but at much higher intensity and often combined with damage. For example, the photoreceptors "adequately" respond to light particles, although they can be stimulated by strong mechanical force (e.g., when our eye is hit by a stroke). Then we will see something, e.g., "stars," but this kind of stimulation is not "adequate."

Besides sensing and transducing, sensory cells also *amplify* and *stabilize* the effect of sensory stimulation. The incoming physical and chemical stimuli are mostly very weak, but the stimulation effect is made robust and strongly amplified, for example, by summation of many single sensory effects and by the generation of the action potential itself.

Despite the importance of the function of sensory receptors, their working range is limited. First, a receptor responds "adequately" only to a particular form of stimuli, e.g., electromagnetic or acoustic waves, mechanical forces, etc., and within this *modality* it responds only to a tiny fraction, e.g., in the case of light sensitivity to wave lengths between 300 and 800 nm, while electromagnetic waves cover a total range of wave lengths more than 20 orders of magnitude.

11.1 The General Function of Sense Organs

Sensory stimuli differ in modality, quality (submodality), intensity, duration, temporal structure, and the location of impact on sensory surfaces. The *intensity* is encoded by the amplitude of a graded potential and the impulse frequency. This encoding starts as soon as a certain threshold of the receptor potential has been passed and usually follows in a logarithmic fashion, especially in sense organs with a very large operating range over several orders of magnitude, like eyes and ears, according to the Weber–Fechner law. According to this law, lower intensities are encoded at a higher resolution (i.e., in an expanded fashion) and higher intensities at lower resolution (and compressed fashion), until saturation is reached. The *duration* of a stimulus is encoded by the start and stop of discharge or in spontaneously active sensory cells by the start and stop of increased spontaneous discharge.

An important aspect of the graded potential and its encoding into action potentials is *adaptation* in the sense of a decline in the amplitude of the graded potential and the consequent decrease in discharge rate of the action potentials, while the strength (intensity) of the stimulus is maintained. There are sensory receptors that do not "adapt" and are called "tonic" receptors, while others adapt slowly and still others adapt rapidly, and both are called "phasic" receptors. The phenomenon of adaptation of sensory receptors is usually seen as preventing a "sensory overload," but also as a way of distiguishing between fast and slow changes in the environment as the basis of "habituation" (cf. Chap. 2).

The *site* of a stimulus in the environment is encoded by the site of excited and inhibited sensory cells within receptor surfaces, e.g., the retina or the skin. Usually, for exact spatial localization of the stimulus, additional mechanisms are necessary, e.g., the computation of differences between bilateral surfaces (e.g., the left and right retina or inner ear). The *modality* of a stimulus, i.e., whether it is visual, auditory, tactile, olfactory, etc., and its *quality* or *submodality*, e.g., whether within vision, light is perceived as color or as shape, is not encoded by the activity of sensory receptors or cells, but by the *site* or pathway of processing within the nervous system or brain connected with the sensory receptors or cells under consideration. This is called the principle of "labeled line." This processing occurs in a modality-specific way such that separate sensory pathways exist for each modality or quality including tracts, nuclei, and areas. Thus, what we experience as the most fundamental property of a sensory stimulus, i.e., its *modality*, is a construct of the brain on the basis of its topology, as Hermann von Helmholtz discovered in the nineteenth century.

In the following, I will deal with those sense organs and sensory systems that are dominant both in invertebrates and vertebrates, i.e., olfaction, touch-vibration-currents, audition and vision, and/or are interesting from an evolutionary perspective, like the mechanoreceptive and electroreceptive lateral line system of vertebrates. I will do that only in a highly abbreviated way. The focus lies on the attempt to understand the coupling between the evolution of sense organs and sensory systems on the one hand, and the evolution of nervous systems and brains on the other.

11.2 Olfaction

Together with the sense of touch, olfaction is the oldest sense. As we have seen in Chap. 6, both are already present in all unicellular organisms, while only some of them have a rudimentary visual system.

Olfaction relies upon the interaction of volatile or insoluble chemical (mostly organic) compounds ("odorants") with sensory receptors called chemoreceptors. Odorant molecules bind to G-protein-coupled receptors in the cell membrane of olfactory cells. The G-protein activates a downstream signaling cascade that causes increased levels of cAMP which binds to and opens sodium channels in the membrane. This depolarizes the cell and eventually leads to the generation of action potentials (Galizia and Lledo 2013).

The olfactory system is well developed in arthropods and in particular in insects. The antennae of insects as their olfactory organs carry up to 200,000 olfactory sensilla (mostly hairlike *sensilla trichodea*) and can detect the smallest quantities of odorants. The sensilla house several dendrites from 1–30 olfactory neurons (Fig. 11.1). The surface (cuticula) of the sensilla carries pores through which odorants move to the dendrites of the olfactory neurons, which carry specific odorant-binding proteins. The neuron is depolarized and eventually generates action potentials. The specificity of the sensilla is determined by 100–200 genes leading to a large variety of types of olfactory neurons. However, the different types usually have broadly and overlapping response profiles, and only in the

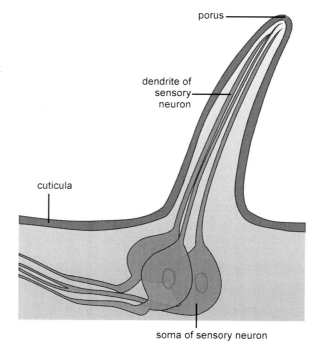

Fig. 11.1 Olfactory sensillum of an insect. The dendrites of the sensory cell reach into the sensillum, which carries a little opening (porus) at its tip. Modified from Hickman et al. (2008)

receptors that respond to pheromones, highly selective responses to sex odorants are found.

In insects, the activity of olfactory sensory neurons is transmitted via the antennal nerve to the paired antennal lobe in the deutocerebrum. This antennal lobe is composed of different numbers of "glomeruli," i.e., clusters of neurons; in the fruit fly *Drosophila*, we find 50, in the honeybee 160 of them. Each glomerulus receives input from several to many olfactory sensilla, each of which expresses the same gene for the receptor protein such that each glomerulus is characterized by the chemoprofile of one receptor type. Thus, inside the glomeruli there is a high degree of convergence of specific sensory information. The glomeruli represent the first stage of processing of olfactory information. Because the response profiles of the different receptors overlap, a complex odorant always activates a larger number of glomeruli in a parallel fashion. In this way, at the level of the antennal lobe, the olfactory environment is encoded in a *combinatorial* and *spatially overlapping* pattern of activation and inhibition. Every glomerulus contains 2–8 projection neurons which send their axons to the protocerebrum via the antennocerebral tract (ACT), terminating in the lateral horn and on the Kenyon cells (about 150,000 in the honeybee in each mushroom body) in the calyces of the mushroom bodies, where they are further processed and then combined with visual and mechanoreceptive input (cf. Chaps. 7 and 17). This input converges to output or projection neurons (about 400 in the honeybee) of the mushroom body that send their axons to the motor centers for walking, flying, and swimming located in the subesophageal ganglion and the thoracic ganglia.

A peculiar type of olfactory communication is based on the effect of pheromones, especially sexual attractants like those that are released by the female silk moth *Bombyx mori* or the tobacco hornworm *Manduca sexta*. The large and complex antennae of the males carry over 100,000 sensilla, half of which are pheromone-specific trichoid sensilla. A few molecules of the female sexual pheromones are sufficient to excite the sensilla. The impulses are transmitted to 1–2 particularly large glomeruli called "macroglomeruli" within the male antennal lobe. There are only 20 output or projection neurons per macroglomerulus, which means that there is an extreme convergence of about 2,000:1 or more, strongly increasing the sensitivity of the system. The activity runs to the mushroom bodies and the lateral horn of the protocerebrum, while the pheromone specificity is maintained.

The olfactory system of vertebrates is similar to that of insects. The olfactory epithelia of terrestrial vertebrates carry 10–20 million olfactory receptor cells. Most vertebrates possess ciliated olfactory receptor cells, but microvillar olfactory receptor cells are likewise widespread (Eisthen 1997). Teleosts and amphibians have both types, birds have receptors that carry both cilia and microvilli, and placental mammals have receptors with cilia and "brush cells," which is a type of microvillar receptor cells. Like in insects, the olfactory cells can respond to a huge variety of odorants, while one cilium carries only one type of receptor. There are about 1,000 different receptor types, the specificity of which is genetically determined—in mammals by about 1,000 genes, which is a substantial fraction of the entire number of genes (in humans 20–25,000). However, each receptor does not only respond to one odorant, but to many different odorants, although not with the same intensity (i.e., it is a "fuzzy" receptor). By this overlapping of response characteristics and combination of receptors, humans can distinguish between several thousand (some authors report 10,000) different odors, although their olfactory sense is regarded as poorly developed compared, for example, to mice or dogs.

The olfactory cells of vertebrates send their axons (the "fila olfactoria") via the olfactory nerve to the olfactory bulb, which has the same basic structure as the antennal lobe of insects, i.e., it likewise consists of glomeruli as well. In mammals, each of the approximately 2,000 glomeruli receives input from 1,000–6,000 fila olfactoria, and one fiber makes contact with only one glomerulus. Each glomerulus therefore receives input from many fibers of the same receptor type. This leads to an enormous convergence and amplification of the primary sensory effect.

The bewildering complexity of odors is again represented by a spatially distributed combination of activity of some to many of the 1,000 receptor types, which would be equivalent to an alphabet of 1,000 letters. The result is an encoding of a complex odorant by a map of distributed activity inside the olfactory bulb. This information is then sent to the secondary olfactory centers in the telencephalon, predominantly the lateral pallium and its derivatives, such as the cortical amygdala and piriform cortex of mammals, and is further processed there.

Like insects, many terrestrial vertebrates possess a pheromone-sensitive system called "vomeronasal system," in "reptiles" called "Jacobson's organ." Most often, the specialized receptor cells are located in a separate region of the nasal cavity. Fibers of these cells project, via the accessory olfactory nerve, to the accessory olfactory bulb, which likewise contains glomeruli. This bulb then projects to the vomeronasal amygdala (in mammals called "medial amygdala") and to the preoptic region of the hypothalamus, where the incoming information influences sexual and reproductive behavior. In humans, a separate vomeronasal organ has not been demonstrated, although humans respond well to pheromones. Probably, the pheromone-sensitive cells are intermingled with the "normal" olfactory cells.

11.3 The Mechanical Senses and Electroreception

The mechanical senses include a variety of separate senses including the sense of touch, balance, audition, vibration, air or water currents, and proprioception, as well as the mechanoreceptive lateral line system found in many aquatic vertebrates (cf. Albert and Göpfert 2013). The function of mechanoreceptive organs always involves the stretch or distortion of the membrane of mechanoreceptive cells, which in case of depolarization leads to the opening or closing of ion channels and a change in membrane potential, which eventually may lead to the generation of action potentials. In many instances, mechanoreception is based on the combination of a movable hair-like structure and a mechanosensory cell.

11.3.1 The Sense of Touch, Vibration, and Medium Currents

In arthropods, we find a large variety of mechanoreceptive structures corresponding to the large variety of mechanical stimuli in their environment that mostly respond to currents of air or water or strains in the cuticula. In spiders and a variety of insects (e.g., crickets), medium flow sensors called *trichobothria* or *filiform hairs* are found (Barth 1985). These fine cuticular hairs are of different lengths and originate from the bottom of a relatively wide and deep cup and are connected to the cuticula of the exoskeleton via a membrane of extreme flexibility. The least air movement is able to deflect them and to excite a single sensory cell or a group of sensory cells innervating them. The inner end of the hair shaft attaches to the dendrites of the sensory cell (or cells). Via the trichobothria, finest air currents can be perceived up to 600 Hz or even more. Spiders carry their trichobothria on the pedipals (i.e., appendages of the prosoma or "head") and on the legs, and a single leg may carry up to 100 trichobothria. These are able to respond to the airflow generated by the wingbeats of flies as possible prey at a distance of 70 cm.

Running prey is recognized by spiders by substrate or web vibrations. Many spiders do not live inside a web, but on solid substrates, like plants. A plant leaf transmits the vibration generated by prey in a wide frequency range of up to 5 kHz, and frequencies around 250 Hz or lower are of biological importance. The vibrations generated by a prey are perceived via *slit sense organs*, which are located on the metatarsi of the legs. Displacement of the tarsus of a leg by substrate vibration leads, via upward movement, to a corresponding compression of a stimulus-transmitting cuticular pad. This pad serves as a mechanical high-pass filter and eventually passes the forces to the adjacent metatarsal organ and compresses it, thereby activating the sensory cells innervating their slits (cf. Barth 2012). In this way, substrate vibration with amplitudes of only 0.1 nm (i.e., less than one millionth of a millimeter) is sufficient to activate a receptor. Two or more sensory slits closely arranged in parallel form the so-called *lyriform organs*, which have an enlarged working range with respect to stimulus magnitude and direction. Using temporal and intensity differences between the stimuli reaching the eight legs from the same source, spiders are able to locate prey objects precisely. The lyriform organs are also used for communication and courtship via self-generated vibrations. Information from the trichobothria and lyriform organs is processed in various parts of the subesophageal ganglion of the spider central nervous system.

A type of mechanosensilla of insects are the *campaniform sensilla*, which, like the slit organs, are activated by a strain of the cuticula. They are likewise used for the release of defense reactions, but also for the control of movement. In addition, there are *wind receptors* on the wings and head, which serve for measuring the animal's own flight velocity as well as the velocity and direction of air currents. An important wind sensor are the antennae, the bending of which is recorded by numerous campaniform sensilla.

Vibration sense organs of insects reveal the same mechanosensibility down to an amplitude of 0.1 nm. In this way, parasites or predators can be detected. The rear-most parts of the bodies of many insects, especially orthopterans such as crickets and locusts, carries cerci, i.e., long paired appendages that are covered with sensilla of different length responding to different frequency ranges of up to 2 kHz contained in the airflow. These sensilla are direction sensitive. The cerci of a cricket carry about 2,000 sensilla, which respond to air flows from all directions. Information provided by them is first processed in the last ganglion of the ventral cord, then transmitted to the thoracic ganglion and finally to the protocerebrum. Via efferent fibers, escape reactions (running or flying away) are released. Besides its enormous sensitivity, this system is characterized by extremely short response latencies of 50 ms or less, which is mediated by the thick axons of "giant" interneurons. In this way, a cricket can escape the tongue of a toad. However, some tongue-projecting salamanders (e.g., *Bolitoglossa* or *Hydromantes*) can make their tongues protrude within 6–10 ms and catch even springtails (collembolans) with an escape latency of 25 ms (Roth 1987).

For proprioception, i.e., information about the position of limbs and body appendages, insects use predominantly *chordotonal organs*, which serve as stretch receptors and, according to their function, are located mostly at the joints. They possess a special type of receptor called *scolopidium* which consists of a sensory cell with a mantled pin-like dendrite. Its tip extends into a movable cap that transmits mechanical stimuli to the dendrite. The membrane of the dendrite is depolarized by pressure from the cap. These receptors respond to minimal displacements of 0.1 nm. The hearing organs of insects, called *tympanal organs*, developed from these chordotonal organs (see below).

In vertebrates, mechanoreception is based, among others, on mechanosensitive hair cells. They carry tufts of stereovilli-microvilli plus one longer kinocilium on one side, which protrude from the apical membrane and are connected via tip links such that they move as a unit. Deflection of the hair bundle toward the kinocilium leads to a depolarization of the cell via K^+ influx, while a deflection in the opposite direction leads to a hyperpolarization. The hair cells respond to a displacement of 0.3 nm; the temporal resolution is likewise impressive and lies within the microsecond range.

The skin of vertebrates as the largest sense organ is densely packed with another type of mechanosensory receptors. The different layers of the skin, i.e., epidermis, dermis, and hypodermis, contain morphologically and functionally different types of cutaneous receptors. The *Meissner corpuscles* are egg-shaped receptors in the epidermis and in the transition zone to the dermis, especially of the fingers and toes. They are sensitive to light touch and adapt quickly. They are neighbored by the likewise egg-shaped *Merkel cells*, which, however, adapt slowly. Both are excellent detectors for static pressure and velocity of touch stimuli and enable vertebrates, including humans, to recognize objects by palpating their surface. In the dermis we find the slowly adapting *Ruffini corpuscles*, which, due to their dense distribution, have a high spatial resolution and respond well to fine changes in tension of the skin. Deeper in the dermis and in the hypodermis, *Vater-Pacini* (or *Pacinian*) corpuscles are located that adapt quickly and respond to pressure at a frequency range of between 20 and 1,000 Hz and therefore are sensible to vibration, but have a low spatial resolution due to their relatively low number. Finally, there are hair follicle receptors that surround the follicles of skin hair and adapt slowly.

Many vertebrates possess specialized touch receptors, for example, ducks and other birds that "dabble" with their beaks. At the rims of their beaks, but also at the tongues of woodpeckers and the follicles of contour feathers, we find numerous *Herbst corpuscles*, which are similar to the Vater-Pacini corpuscles. Spectacular are the vibrissae or whiskers found at the head, snout, forepaws, or belly of many mammals. These hairs are usually thicker and stiffer than other types of hair and are implanted in a special hair follicle that incorporates a blood capsule or blood sinus and are heavily innervated by sensory nerve fibers. The whiskers are often orderly arranged in grids or "barrels" and can be of different lengths. In mammals like mice, gerbils, hamsters, rats, guinea pigs, rabbits, and cats, each individual follicle is innervated by 100–200 primary afferent nerve fibers. The hairs are able to respond to displacements of 20 nm. Some mammals, e.g., rats or seals, can actively move their whiskers and palpate the surface of objects in the range of micrometers. This information is then processed in the primary somatosensory cortex (Brodmann areas A1–3).

11.3.2 The Mechanoreceptive and Electroreceptive Lateral Line System of Fish and Amphibians

Mechanoreceptive Lateral Line System

The mechanoreceptive lateral line system of fish and aquatic amphibians has great similarities with the vestibular system (not discussed here). Microvilli and kinocilia of hair cells are bent by water currents in different directions and to different degrees and can indicate the velocity and direction of the current as well as flows generated by other animals in the water. The microvilli and the kinocilium extend into a flexible and jelly-like cupula. Hair cells, cupula, and supporting cells together form a functional unit, the *neuromasts*.

In amphibians and lampreys, the neuromasts are located on top of the skin and are therefore called "epidermal" neuromasts. In bony and cartilaginous fish, in addition there are neuromasts located inside canals and therefore called "canal" neuromasts. The canals run below the dermis and are filled with fluid and connected with the surface via pores. With the epidermal neuromasts, the animals sense water movements, while with the canal neuromasts they sense pressure differences. The former include 10–12, the latter up to 1,000 hair cells (Fig. 11.2a, b). The lateral line system is extremely motion-sensitive: water movements of 0.1 μm or a displacement of the cupula by 2 nm are sufficient to activate the receptors. Epidermal neuromasts generally respond to lower frequencies of 10–60 Hz, canal neuromasts to higher frequencies beyond 50 Hz. The sensitivity of the lateral line system of a fish for slow

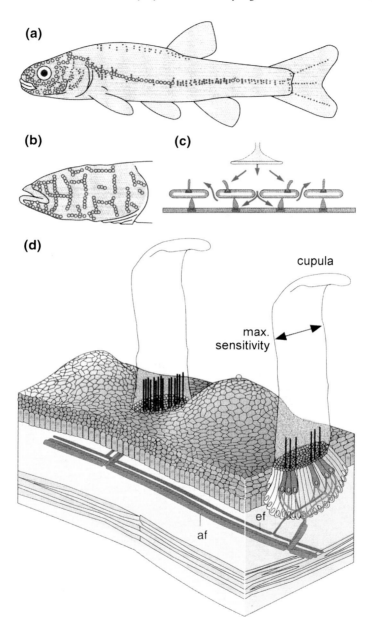

Fig. 11.2 Lateral line organ of bony fishes and amphibians. **a** Arrangement of lateral lines in the common minnow (*Phoxinus phoxinus*). *Circles* canal pores, *dots* epidermal neuromasts. **b** Head of a blind cave fish with horizontally and vertically oriented rows of neuromasts. **c** Schematic longitudinal section of a lateral line with epidermal and canal neuromasts, both with cupulae. **d** Epidermal neuromasts of the clawed toad *Xenopus laevis*. Afferent (*af*) and efferent (*ef*) nerve fibers supply the neuromasts. Modified from Dudel et al. (1996/2000)

or fast water movement is determined by a number of factors, such as the density and spatial distribution of neuromasts on the body surface and in the canals, the diameter of canals, the number, size and distribution of pores, the rigidity of the walls of the canals, and the size of the cupulae of the neuromasts. All these parameters closely correspond to the living conditions of the animals.

Fish with both canal and epidermal neuromasts and amphibians with epidermal neuromasts use their lateral line system for near orientation under conditions where visual orientation is difficult or impossible, e.g., in darkness, murky water or when blind, and for coordination in schooling behavior. Blind cave fish can generate, via movement of their fins, a water current field and use the reflections of the waves for localizing living as well as non-living objects (prey, obstacles etc.).

Signals from the neuromasts enter the medulla oblongata via the anterior and posterior lateralis nerve between the 5th (trigeminal) and 7th (facial) cranial nerves and terminate in the ipsi- and contralateral octavolateralis nucleus and the vestibulo-lateralis lobe of the cerebellum. Fibers run from the former nucleus to several mesencephalic nuclei and areas including the tectum and from there to posterior thalamic nuclei. Nuclei in the medulla oblongata are the origin of efferent fibers that leave the brain together with the facial nerve. In this way, the brain is able to fine tune the function of the neuromasts.

The Electroreceptive System

Electroreception (ER) is the ability to sense naturally occurring electric signals, mostly from living organisms (cf. Heiligenberg 1977; von der Emde 2013). It reveals a close structural relationship with the mechanoreceptive lateral line system, and experts believe that it evolved from it. However, electroreception exhibits a peculiar distribution among craniates/vertebrates. It is absent in hagfish, although these animals possess a mechanosensory lateral line system (MSLL), and it is unclear whether ER evolved only in vertebrates or was secondarily lost in the hagfish. Lampreys instead have both an MSLL and ER, as is the case in all cartilaginous fish. Within the bony fish, besides an MSLL, ER is found in "primitive" actinopterygian groups (chondrosteans) and in sarcopterygians (lungfish and *Latimeria*), but largely absent in neopterygians (holosteans and teleosts), with the exception of weakly electric fish, i.e., silurids (catfish), gymnotids, and mormyrids. It is generally believed that teleosts lost ER in the context of changes in body surface morphology, and that ER was reinvented independently in the three mentioned groups of teleosts (New 1997), most probably as modification of the existing MSLL.

Among mammals and sauropsids, electroreception is generally absent except in the two groups of monotremes, i.e., in the semiaquatic platypus (*Ornithorhynchus anatinus*) and in the four extant species of the terrestrial genus *Echidna* ("spiny anteater"). In both groups, electroreception has evolved independently as a specialization of the trigeminal system for detecting prey. Electrosensation is much more sensitive in the platypus than in *Echidna*. In the former, the 40,000 electroreceptors are located in rostrocaudal rows in the skin of the bill, whereas ordinary mechanoreceptors are uniformly distributed across the bill. In *Echidna*, there are between 400 and 2,000 electroreceptors on the tip of the snout.

Among electroreceptor cells of fish, the so-called *ampullar organs* are believed to be phylogenetically older than the other types. They are found in all fishes with ER and are sensitive in the lower range of frequencies, i.e., below 50 Hz. They are indented in an "ampullary" fashion into the skin of the animals and are in contact with the external water via jelly-filled canals. At the bottom of the ampullae there are between a few and few hundred receptor cells as highly sensitive voltage detectors. They are supplied by neuronal afferents generating a steady direct current, which is modulated by an alternating current from external sources. In cartilaginous fish living mostly in brine a negative, and in freshwater fish a positive polarity at the entrance of the ampullae leads to depolarization. The spatial distribution of the ampullary organs over the body enables the animals to detect and localize a low-frequency electrical field. This is called *passive* electrolocation. With their ampullary organs (here called "organs of Lorenzini"), sharks are able to sense even the orientation of the magnetic field of the earth via tiny induction currents generated by their own movement.

The weakly electric fish in addition possess receptors that are sensitive to high-frequency discharges up to 10 kHz. These types of receptors definitely evolved anew and independently in the mormyrids and gymnotids, most probably from MSLL receptors, but they lost the efferent innervation typical of MSLL neuromasts. They are used for *active* electrolocation. Besides location of objects, this mechanism is likewise used for electrocommunication. Gymnotids have *tuberous organs*, while mormyrids have *knollen organs* and *mormyromasts*. The tuberous organs of gymnotids are deeply embedded in the skin and covered with a plug of epithelial cells which capacitively couples the sensory receptor cells to the external environment. They respond only to high-frequency alternating currents. The knollen organs and mormyromasts are similar in morphology, but have evolved independently of the tuberous organs of gymnotids. Mormyromasts respond to *self-generated signals* and are used for electrolocation at a range of about half a body length. The knollen organs, in contrast, respond predominantly to *signals from external sources* and are used for communication at a distance of up to one meter. Important are the shape and discharge rate of the signals which are specific for sex, social status, and emotional state of the sender, e.g., in the context of aggression or courtship.

The electric discharges are generated by an electric organ located predominantly close to the tail. The single impulse has a characteristic waveform, and the sequence of pulses is highly variable. There are two major types of impulses: one up to 100 Hz, the other as a series of continuous, sinusoidal discharges up to 1,700 Hz. While mormyrids, e.g., the elephantfish *Gnathonemus petersi*, produce only the former type of pulses, gymnotids produce both types.

The afferent fibers of the different types of electroreceptors run in separate pathways to the electrosensory lateral lobe (ELL) in the medulla oblongata. There is a parallel processing of the different types of electrosensory signals. From the

ELL, signals run to the torus semicircularis for further elaborate processing (cf. Fig. 10.16), and from there either directly or via the tectum back to the ELL. In mormyrids, as already mentioned, a part of the cerebellum, the valvula cerebelli, has enormously increased in size and covers nearly the entire dorsal surface of the brain, which makes these brains the largest ones relative to body size among all animals (around 20 % of body volume) (Fig. 10.2e). Apparently, the valvula serves for comparison of temporal properties of electrical signals in electrolocation and communication.

While the independent re-evolution of ER in the weakly electric fish can be interpreted as an adaptation to life in muddy water, the loss of ER in all other neoptergygian fish remains unexplained (New 1997). Least convincing is the assumption that it was lost because it was "not needed anymore." This would not explain why cartilaginous fish, chondrostean, and crossopterygian fish as well as many amphibians retained ER.

11.3.3 The Auditory System

Hearing is generally based on the perception of sound waves, i.e., the propagation of periodic forward and backward movement (oscillation) of air or water molecules leading to zones of compression and rarefaction of molecules (cf. Ehret and Göpfert 2013). These oscillatory movements induce the same movements in the vicinity, resulting in the propagation of sound. *Sound pressure*, i.e., the strength of molecular movement, is perceived as loudness, while the *frequency* of the oscillations determines the pitch. In humans, hearing starts at oscillation frequencies of about 20 Hz (everything below is called "infrasound") and ends at about 20 kHz in young individuals (everything beyond is called "ultrasound"). Sound pressure propagates spherically and decreases with the square of distance. In addition, there is absorption of sound through the medium, with lower frequencies being less and higher frequencies more strongly absorbed. Air absorbs sound much stronger than water. The speed of sound propagation depends on the density of the medium; therefore in water, sound speed is 4–5 times higher than in air.

Most hearing organs are based on structures that can be activated by the pressure component of sound. In insects, specific hearing organs are the *tympanal organs*, which may be found on the thorax, the base of the wings, the abdomen and legs of the insect body. They developed from the already mentioned chordotonal organs in the way that a thin cuticular membrane sits on top of the scolopidium cap. The tympanal organ is surrounded by air sacs, which mechanically uncouple the organ from the body. Tympanal organs may contain up to 2,000 scolopidia. Local differences in thickness of the tympanal membrane lead to best frequencies. In some insects, such as crickets or locusts, the scolopidia are frequency analyzers, probably because of differences in sizes and in distribution. They respond best to a frequency of 30 kHh, with an upper range of 80 kHz or even higher.

Mosquitos and other insects like *Drosophila* hear with an organ called "Johnston's organ," which is found at the basis of the antennae of the animals (Fig. 11.3). It has a very thin flagellum that transmits the mechanical stimulus to just a few to hundreds to thousands of scolopidia, which are arranged around the flagellum. Each scolopidium contains a mechanosensory chordotonal neuron. The resonance frequency of males of the mosquito *Anopheles* is precisely tuned to the sound of 380 Hz produced by the wingbeats of females, but is insensitive to the higher frequency of their own wingbeats.

The hearing organ of vertebrates has evolved from the vestibular organ, which has hair cells with tufts of stereovilli and one kinocilium that are displaced by otoliths responding to gravity and linear acceleration. In terrestrial vertebrates, hearing plays an important role for orientation and communication via airborne waves. A special problem, however, is the difference in the mechanical impedance from the "soft" (i.e., compressible) air (i.e., gas) to the "stiff" (i.e., incompressible) lymph (i.e., fluid) of the inner ear constituting an impedance mismatch between the two media. This mismatch would lead to a nearly complete reflection of the sound waves at the boundary between air and inner ear lymph. This problem is partly solved by the middle ear, where a tympanum (eardrum) receives the sound wave and transmits it via a lever system to the lymph-filled inner ear, thereby amplifying the wave pressure. In amphibians and "reptiles" we find a lever system consisting of a *stapes* (stirrup) or *columella* and an *extracolumella* attached to the eardrum. Because the eardrum is much larger than the footplate of the columella sitting on the inner ear, there is an amplification of sound pressure that partially compensates the loss of amplitude between air and lymph. However, the transmission via extracolumella and columella restricts sound transmission to

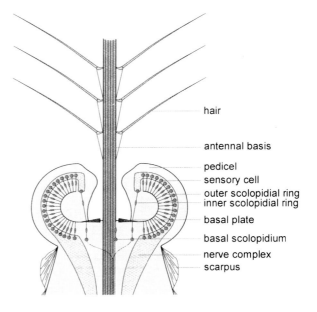

Fig. 11.3 Johnston's organ of a mosquito at the base of an antenna. The sense organ is composed of several rings of scolopidia with sensory cells which are stimulated by movements of the antennal basis in different directions and by vibration. Modified from Dudel et al. (1996/2000)

the inner ear to relatively low frequencies. Movement of the columella moves the lymph of the inner ear, and this deflects the hairs of the sensory cells of the basilar papilla.

In crocodiles and birds, the inner ear is strongly elongated and consists, like in mammals, of an upper *scala tympani* and a lower *scala vestibuli* and, in between, the *papilla basilaris*. As a consequence, the range of hearing frequencies increases up to 6–8 kHz in these animals, which, however, when compared with mammals, is still limited (see below). An exception is the barn owl, which can hear frequencies up to 10 kHz. The number of hair cells of the basilar papilla (or membrane) of different bird taxa ranges from 3,000 (canary) to 17,000 (emu). Hearing in birds probably functions predominantly according to the resonance principle and not, as in mammals, to that of traveling waves (see below), while the spatial representation of frequencies is the same, i.e., high frequencies are represented at the basis and high frequencies at the end of the papilla.

Among vertebrates and with the exception of owls, mammals have the most highly developed auditory system including substantial modifications of the middle and inner ear and as a novelty of an external ear with pinnae (or ear conches) and an external auditory canal. The pinnae serve to center the sound and help to localize the sound source. The middle ear consists of the tympanic cavity, housing the three ossicles: the *malleus* (hammer) attached to the eardrum, the *incus* (anvil), and the *stapes* (stirrup), the last of which is homologous to the columella of amphibians and "reptiles" and attaches to the oval window. Like the columella-extracolumella system, the three middle ear ossicles serve for impedance matching and, to a minor degree, for amplification. Most of the amplification of sound pressure results from size differences between the relatively large eardrum and the relatively small oval window, which amounts to 17:1 in humans, 35:1 in the cat, and 50:1 in some bats. In addition, the middle ear ossicles are able to transmit much higher frequencies than the columella-extracolumella system, i.e., up to 50 kHz and in some cases beyond 100 kHz.

The inner ear of mammals consists of the cochlea ("snail shell"), which evolved from the vestibular organ into a long tube that eventually became coiled up, thereby substantially increasing its length. Like in crocodiles and birds, in cross section it consists of three parts, the upper scala vestibuli, the lower scala tympani, and, in between the *scala media* with the organ of Corti (named after its first form, Alfonso Corti) (Fig. 11.4). The fluid (i.e., the sodium-rich and potassium-poor perilymph) of the scala tympani is connected with the oval window moved by the stapes, and the scala vestibuli by the round window, which serves for pressure balance, and both fluids are continuous via the helicotrema at the tip (cupula) of the cochlea. The scala media is filled with the sodium-poor and potassium-rich endolymph. The organ of Corti sits upon the basilar membrane and contains the inner and outer hair cells, supporting cells, and the tectorial membrane above the hair cells.

The inner hair cells are innervated by endings from bipolar sensory neurons located in the spiral ganglion located inside the cochlea. One arm innervates the inner hair cells and the other one contributes to the auditory nerve (8th nerve,

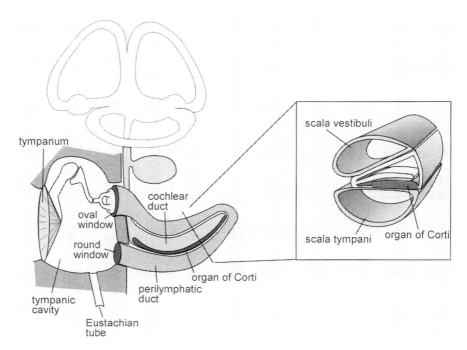

Fig. 11.4 Schematic diagram of the ear of a mammal. The middle ear with the three ossicles, the tympanum, the *oval* and *round* window as well as the cochlea with the organ of Corti are shown. To the *right*, a more detailed representation of the cochlea is given. Modified from Müller (2009)

N. stato-acusticus), which runs to the medulla oblongata of the brain. The neurons of the spiral ganglion also innervate to a minor degree the outer hair cells, while the majority of efferent fibers come either from the superior olive located in the brainstem or directly from the cochlear nucleus. Thus, the nerve fibers innervating the inner hair cells and then running to the brain are *afferent* fibers carrying nerve impulses to the brain, and those running from the superior olive to the outer hair cells are *efferent* fibers carrying nerve impulses from the brain to the inner ear. Accordingly, it is said that the outer hair cells are under efferent control by the brain, as is the case in the mechanoreceptive lateral line system.

In humans, the Corti organ has about 3,400 inner and 13,000 outer hair cells. However, while each outer hair cell is innervated by just one efferent fiber, each of the inner hair cells is contacted by 20 fibers from spiral ganglion neurons. This situation corresponds with the different function of inner and outer hair cells, as we will learn in a moment.

The process of auditory reception starts with the displacement of the eardrum by airborne sound waves, which are taken up by the three middle ear ossicles and then transmitted to the oval window. The vibrations of the oval window are propagated to the incompressible cochlear fluid and the elastic basilar membrane and are compensated by outward bulging movement of the round window. Thus, vibrations start at the oval window and generate a *traveling wave* that runs along the basilar membrane toward the helicotrema. Because the stiffness of the basilar membrane is high at the basis near the oval window and low near the helicotrema, the membrane can follow higher frequencies only near the basis, while lower frequencies are able to move the membrane up to the helicotrema. This effect is amplified by the fact that the basilar membrane is narrower near the basis and wider near the helicotrema. As a consequence, high-frequency traveling waves have their maximum near the oval window and low-frequency waves near the helicotrema. This leads to a two-dimensional spatial representation of the sound frequency spectrum along the basilar membrane. In the inner hair cells, the shearing of stereovilli causes an influx of mainly K^+ ions into the cell, which depolarizes it and activates voltage-gated Ca^{2+}-channels. This in turn leads to an increase in intracellular Ca^{2+} concentration and release of the transmitter exciting the afferent nerve. Finally, the activation of Ca^{2+}-sensitive K^+ channels leads to an efflux of K^+ ions, repolarizes the cell, and a new cycle begins.

The precise organization of sound frequency representation, or "tonotopy," of the basilar membrane differs among mammalian taxa in correlation with their lifestyle and the importance of low, medium, and high frequencies in prey capture, orientation, and communication. In rodents, bats, and cetaceans, ultrasound frequencies for communication and echolocation are of great importance and therefore "over-represented" (a "looking-glass" effect).

The traveling waves lead to a rather broad displacement of parts of the basilar membrane which would not allow a precise representation of the stimulus. Here, the outer hair cells, which are in contact with the tectorial membrane, come into play. They are likewise stimulated by the traveling waves and generate by themselves high-frequency vibrations by changing their length, and this influences the inner hair cells in two ways. First, the maximal displacement of the basilar membrane is amplified by about hundred times, and second, the flanks of the enveloping curve of maximal displacement are narrowed down to a much smaller area of spatial frequency representation, which strongly increases pitch selectivity.

Afferent fibers from the spiral ganglion in the inner ear take up the activity of inner hair cells and transmit it via the vestibulo-cochlear nerve to the ventral and dorsal cochlear nucleus in the medulla oblongata. Here, by a number of different types of auditory neurons, a first processing of auditory information takes place. From the ventral cochlear nucleus, a main auditory pathway runs to the ipsi- and contralateral superior olive (nucleus olivaris superior) and from there, via the lateral lemniscus, to the ipsi- and contralateral colliculus inferior of the midbrain roof (cf. Chap. 10). The superior olive contains neurons which send efferent fibers to the inner ear (see above). Auditory fibers run from the inferior colliculus to the medial geniculate body in the dorsal thalamus of the diencephalon, which in turn sends fibers to the primary auditory cortex in the dorsal temporal lobe. In humans, this includes areas A41, also called anterior and posterior Heschl's gyri or transverse temporal gyri. The primary auditory cortex reveals a tonotopic organization, i.e., spatially ordered representation of sound frequencies. The secondary auditory cortex surrounds the primary one in a horseshoe-like fashion and does not reveal a clear tonotopic organization. In contrast to subcortical auditory neurons, those in the auditory cortex mostly do not respond to pure tones, but rather to complex ones and mostly in a phasic way, i.e., they indicate changes in pitch and amplitude. Compared to the visual cortex of primates, the auditory cortex is much less understood.

11.4 The Visual System

Vision is an old sense. As described in Chap. 6, some prokaryotic organisms, like the Archaean *Halobacterium salinarum*, possess photoreceptors that absorb light in the orange part of the spectrum and enable them for phototaxis. The molecular structure of the light-sensitive pigment, the *bacteriorhodopsin*, is similar to the rhodopsin found in the retinae of both invertebrates and vertebrates and undergoes a conformation change upon absorption of a photon (cf. Kretzberg and Ernst 2013).

It is debated whether the eyes of all multicellular animals have a common origin at around 540 million years ago or have evolved many times independently. If the latter were true, it must have happened at least 40, and perhaps even 65 times (Fernald 1997). The shared anatomical features of eyes may be due either to direct homology, to "deep homology," or to convergent evolution and the work of physical constraints, in the sense that the laws of optics "forced" eyes to evolve in similar ways. Many authors assume the involvement of the *Pax6* gene as basis of "deep" homology. The *Pax6* gene plays an important role in eye development in distantly related animal taxa such as fruit flies and mice, but the resulting morphologies, mainly compound eyes and lens eyes, are dramatically different in many morphological and functional aspects. Furthermore, *Pax6* also plays a role in the development of other tissue inside and outside the brain, and homologous genes are found in animals like nematodes or sea urchins that have no eyes at all. So the question of whether the eyes found in multicellular organisms are homologous or the product of convergent evolution remains unanswered.

Virtually, photoreceptors of all animals are adapted to a very narrow band of electromagnetic waves between 300 (UV light) and 750 nm. One reason for this restriction is the fact that frequencies within this band are maximally visible in water at a depth of about one meter, where the ancestors of all animals probably lived, because electromagnetic radiation at almost all other wavelengths is strongly filtered by the water. This range remained unchanged when animals became terrestrial.

In both invertebrates and vertebrates, the photoreceptive substance, *rhodopsin*, or visual purple, consists of the protein moiety *opsin*, which itself is insensitive to light, and the chromophore (i.e., the photoactive part) *retinal*, which is located inside a pocket of the opsin molecule. Absorption of light quanta by retinal induces a conformational change (isomerization) from 11-*cis*-retinal into *all-trans*-retinal. Via the G-protein *transducin* in the opsin molecule, a second messenger cascade is triggered, which eventually closes Na^+ ion channels and thus leads to a stopping of the "dark current" due to continuous influx of sodium ions. This leads to a reduction or stopping of steady transmitter release at the synaptic sites of photoreceptors.

Thus, photoreceptors are not activated, but *inhibited by light*, which, however, is of no great importance, since this inhibition is turned into activation via inhibitory interneurons inside the retina. Rhodopsin, found in the rods of the vertebrate retina, most strongly absorbs green-blue light and, therefore, appears reddish-purple, which explains why it is also called "visual purple." It is responsible for *monochromatic* vision in the dark.

Animal "eyes" range from photoreceptive spots for phototaxis, as just mentioned, shallow eye cups, pinhole or camera eyes that lead to directional vision with relatively sharp images to lens eyes in many invertebrate taxa and all vertebrates, often together with a lens accommodation mechanism, and to compound eyes in many other invertebrate taxa, often in combination with lens eyes (ocelli, cf. Chap. 7).

11.4.1 The Compound Eye of Insects

Compound eyes are widely distributed across arthropods. However, the homology of these eyes in chelicerates, crustaceans, and insects is debated. Best known is the compound eye of insects (Fig. 11.5). It is composed of *ommatidia*, ranging in number from a few in some ants, about 800 in fruitflies, to 30,000 in dragonflies. The honeybee has 6,000 ommatidia per eye. The ommatidia consist of a part for light refraction, the *cornea* or *cornea lens*, and a *crystalline cone* composed of four cells. The crystalline cone is optically homogenous like the vitreous body of the vertebrate eye and influences the optical path only slightly. The light then enters the region of 5–12 *retinula cells* which are wedge-shaped and carry on their inner edge the narrow band of light-sensitive, i.e., rhodopsin-containing *rhabdomers*. The rhabdomers of each retinula cell are either very close to each other (Fig. 11.5b) or fused (Fig. 11.5a), forming the *rhabdom*. However, each single rhabdomer is still activated separately, because they are usually electrically insulated from each other. Each retinula cell sends one axon to the optic neuropils (see below). The ommatidium is optically shielded from its neighbor by pigment cells, which can expand and contract such that neighboring ommatidia can work either together or independently.

In the compound eye, there needs to be a compromise between visual acuity and light sensitivity as well as a compromise between precise object recognition and movement perception. The cornea lens of a single ommatidium has a diameter of 15–30 μm and generates a focal spot at the distal end of the rhabdom with a diameter of 2–4 μm, equivalent to the diameter of the rhabdom. This defines the lower limit of spatial resolution, which is inferior to that of the vertebrate lens eye. The quantity of light within this focal spot is proportional to the diameter of the cornea lens, which means that the wider this diameter, the more light reaches the

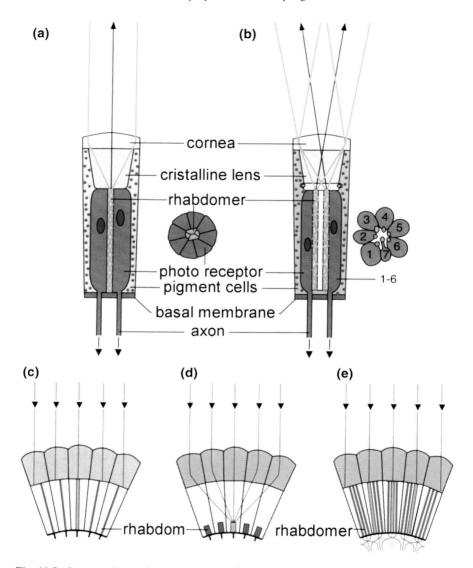

Fig. 11.5 Compound eye of an insect. **a** Longitudinal section through an ommatidium of the apposition eye of the honeybee. The rhabdomeres of the photoreceptors are in close contact with each other and together form the rhabdom, which has one visual axis. **b** Longitudinal section through the ommatidium of a superposition eye of a house fly with different visual axes. **c** Schematic diagram of an apposition eye as typical of diurnal insects. **d** Schematic diagram of an optical superposition eye as typical for nocturnal insects. **e** Schematic diagram of a neural superposition eye. For further explanation, see text. From Dudel et al. (1996/2000)

rhabdom, but the lower the spatial resolution. At the same time, spatial resolution of the compound eye likewise depends on the angle between the ommatidia—the larger the number of ommatidia and less curved the surface, the smaller the angle between them, and the higher the spatial resolution of the eye. Thus, at a given surface of the compound eye, a compromise between the diameter of the cornea lens and the number of ommatidia is necessary. One compromise may consist of in the fact that insect and crustacean eyes have a region of higher spatial resolution, corresponding to the fovea of the vertebrate eye, where the curvature of the eye is flatter.

Insects that are active during daylight (i.e., are *diurnal*) usually have *apposition eyes*, where each ommatidium has one focal point, which contributes to a mosaic image of high spatial resolution, albeit—because of small diameters of the cornea lens—of low light sensitivity. Insects that are active at dim light or during the night (i.e., are *nocturnal*), like moths, in contrast, often have *superposition eyes*, in which light rays from many (often 30) cornea lenses and crystalline cones are deflected to one rhabdom, which enormously increases light sensitivity at the expense of spatial resolution (Fig. 11.5d). However, in many insects, via migration of the pigment, apposition eyes can be turned into superposition eyes, e.g., during transition from daylight to darkness conditions. Finally, the so-called *neural superposition eye* is a compromise between the two types of compound eyes. It yields both high spatial resolution and high light sensitivity by neuronally connecting the rhabdomers of neighboring ommatidia with the same orientation of the optical axis (Fig. 11.5e). Many insects with apposition eyes or neural superposition eyes have a high temporal resolution, i.e., they can perceive 200–300 images per second, which is favorable for sharp vision during fast flight. This ability considerably exceeds that of the human eye, having a temporal resolution up to 60 images under optimal conditions.

In invertebrates and vertebrates, *color vision* is based on very similar mechanisms, in the way that the photoreceptors carry types of rhodopsins that differ in their spectral absorption properties. Often, three types of receptors are found for long, intermediate, and short wavelengths (e.g., red/yellow, green, and blue; see below). Many arthropod taxa, but also fish, amphibians, and "reptiles" have an additional receptor for ultraviolet light, i.e., at wavelengths between 300–400 nm, while many insects lack a receptor for long wavelengths (600–740 nm). However, there are diurnal butterflies with four or even five receptor types (tetra- or pentachromacy). Many hymenopterans that feed on flowers have trichromatic vision with a UV, blue and green receptors, but no red receptor, which means that they cannot perceive long waves as "reddish."

One spectacular ability of many insects is the perception of the plane of polarized light. Sunlight, when scattered within the atmosphere or reflected from shiny surfaces such as water, is partially polarized. Many insects make use of this fact during foraging flights for determining the position of the sun, when the latter is not directly visible. Since the degree of polarization is strongest in the range of short wavelengths, insects perceive the polarization pattern with their UV or blue receptors (Barth 2012).

Activity of the complex eye runs to the three to four optical integration structures, i.e., lamina, medulla, and lobula (in flies, lobula plus lobula plate). The lamina is responsible for contrast and temporal resolution, the medulla resembles the mammalian visual cortex in the sense that here, too, color, shape/contours, and motion are processed in parallel (see below). The lobula/lobula plate complex specializes in motion analysis, including direction sensitivity, local movement of objects in the visual field as well as optical flow of the entire visual field.

11.4.2 The Vertebrate Eye and Retina

Vertebrates generally have lens eyes. In terrestrial vertebrates, light refraction occurs predominantly through the cornea, because only its material has a sufficiently higher optical density than the surrounding air. The relatively small and flat lens serves primarily for focusing the light rays onto the layer of photoreceptors in the retina. In aquatic vertebrates, in contrast, light refraction occurs mostly via the large and spherical lens because there is no sufficient gradient in optical density between water and cornea. The light ray passes through the vitreous body to the retina on the inner side of the eye bulb opposite the cornea.

As shown in Fig. 11.6, the vertebrate retina consists of six layers, i.e., (as seen from the vitreous body) the layer of axons of retinal ganglion cells, followed by the somata of the retinal ganglion cells as output elements of the retina. This layer is followed by the inner plexiform layer containing connecting fibers and contacts between retinal ganglion cells and cells of the next layer, the inner granular layer containing the amacrine, bipolar, horizontal and interplexiform cells. Then follows the external plexiform layer consisting of processes of the mentioned cells and the "feet" of photoreceptors, i.e., rods and cones. The photoreceptor somata form the external granular layer. Then follows the layer of the outer segments of photoreceptors, which extend into the pigment layer.

These outer segments of the photoreceptors are the site of photoreception. Vertebrates, in contrast to many invertebrate taxa with *everse* eyes, generally possess *an inverse eye*, i.e., the photoreceptive structures—here the outer segments of rods and cones—"look away" from the light. It is unclear why this is the case (presumably for developmental reasons), but this fact is largely irrelevant for the quality of vision, because the light can pass almost unfiltered through the other layers of the retina. In addition, in the eyes of many vertebrates, there is a fovea region in which the photoreceptors are not covered by the other layers of the retina, and the light can reach them completely unimpaired.

The retina of vertebrates exhibits two basic types of photoreceptors: *rods* and *cones*. Rods have only one absorption maximum at 498 nm wavelength; therefore, the rod system is incapable of color vision. In contrast, they are very light-sensitive, and just one light quantum is sufficient to stimulate a rod. Rods are the basis of light-dark, or *scotopic* vision. Light sensitivity is further increased by the fact that the activity of many rods is summed up (convergence), albeit at the expense of spatial resolution. Rods are also more sensitive to motion than cones.

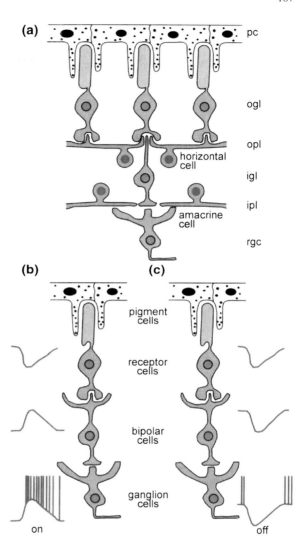

Fig. 11.6 Cytoarchitecture of the vertebrate retina.
a General anatomy of the retina with the layer of pigment cells (*pc*), the outer granular layer (*ogl*) containing the outer segments of photoreceptors and their somata, the outer plexiform layer (*opl*) the inner granular layer (*igl*) containing the horizontal, bipolar, and amacrine cells, the inner plexiform layer (*ipl*), and the layer of retinal ganglion cells (*rgc*). Photoreceptors, bipolar cells, and retinal ganglion cells constitute the *radial* flow, horizontal and amacrine cells the *transverse* flow of activity. **b** Flow of activity through an "on" bipolar and an "on" retinal ganglion cell. **c** Flow of activity through "off" bipolar and an "off" retinal ganglion cell. Modified from Dudel et al. (1996/2000)

There are up to four different types of cones—one for ultraviolet light, and three for the range of short, middle, and long waves, with absorption maxima around 420, 534, and 564 nm, which therefore are called blue, green, and yellow-red receptors according to human color perception. The outer segments of many fish and birds contain oil droplets, which function as color filters and can massively influence the spectral sensitivity of their retina.

Differences in the light absorption spectra of the types of cones are the basis of color vision. Subjective perception of different colors originates from neuronal integration of the relative contributions of the three types of cones within the actual range of the light spectrum. Light is absorbed over almost the entire visible spectrum by the blue, green, and yellow–red cones, although with different absorption maxima. For example, the perceived color "red" results from the dominance of activity of cones in the long wavelength range. Thus, wavelength and color are two different things: while the former is an "objective" property of light, the latter is a construction of our brain, based on the integration of differences in the relative contribution of the three cone types to the entire spectrum.

Cones are not only the basis of color vision, but also of high visual acuity. At the same time, as a consequence of their photopigments, they are much less light-sensitive than rods and therefore work only under daylight or *photopic* conditions (above about 0.25 lux equivalent to full moon light), while rods are active even in the dark, where subjectively no colors exist. I have already mentioned the "conflict" between spatial resolution (visual acuity) and light sensitivity in the visual system, and in vertebrates this conflict is at least partially solved by the spatial distribution of cones and rods within the retina. In diurnal animals and especially in those with a predatory lifestyle, we find either many cones distributed over the retina or densely packed within a central fovea as a retinal region of high visual acuity (in the human retina about 7 million). In contrast, nocturnal animals tend to have predominantly rods. Interestingly, prey animals which during the daytime live in open land, often have a horizontal band of densely packed cones running in parallel to the horizon, rather than a circular fovea, and this enables them to detect possible predators appearing on the horizon.

The primate ancestors of humans were nocturnal, and consequently their retinas consist mostly of rods, about 120 million in number, and of 7 million cones concentrated predominantly inside the fovea centralis. Thus, secondarily diurnal animals, such as humans, have color vision and high visual acuity only via their fovea. This seems to contradict our impression that the entire world around us and not just a small spot is colorful and can be seen sharply. The solution to this problem lies in the fact that our fovea is rapidly moving via involuntary and voluntary eye movements, over the visual field. Our brains integrate these kinds of information into a virtually stable, sharp, and colorful image—one of the most remarkable achievements of the brain.

In birds of prey, the density of very slender cones inside the fovea is twice as high compared to that of the human fovea and reaches the limits of physical optical resolution. Some animals, including birds, even possess two foveae, one for frontal and another one for lateral vision at high acuity.

11.4.3 Parallel Processing in the Visual System of Vertebrates

The visual system of vertebrates provides a good example of the basic principle of parallel-distributed processing of sensory information. This means that the different types of photoreceptors and their postsynaptic neurons respond only to certain

11.4 The Visual System

single properties of the normally complex visual stimuli like size, contrast, color, position inside the visual field, movement direction, velocity, and movement pattern. Inside the retina, and at later stages of the visual system inside the brain, these different kinds of information are processed at least partially separately and in parallel. Only at later stages is there "crosstalk" and convergence of visual information and eventually integration with nonvisual information. In this way, complex perceptual states are constructed. This is schematically shown in Fig. 11.7.

Inside the retina of mammals (e.g., cats and monkeys), we find two major types of retinal ganglion cells as output elements. One type has small somata and is driven primarily by cones with different spectral sensitivity and responds best to small, high-contrast, and colored stimuli. This type is called the X-type in non-primate mammals and P- (or "parvocellular") type in primates. It is the basis of color and contrast vision combined with high visual acuity as the basis of contour and object vision. The other type of retinal ganglion cells has larger somata and is driven primarily by rods. Consequently it is more sensitive to changes in illumination and movement (often causing slight changes in illumination). This type is called Y-type in non-primate mammals and M- (or "magnocellular") type in primates. In most vertebrates, we find a third type with large somata, which responds best to changes in ambient illumination (dimming) and those elicited by large objects, and in mammals are called W-type retinal ganglion cells.

The axons of retinal ganglion cells constitute the optic nerve that leaves the eye through the "blind spot" or optic disc. The optic nerve runs to the bottom of the diencephalon, where the two nerves from the left and right eye form the optic chiasm. In anamniotes and sauropsids, the fibers of the two nerves cross almost completely to the other side of the brain, while in mammals there is only partial crossing, i.e., up to half of the fibers cross, while the other half remains at the site of entrance. From there, fibers run to the optic tectum (superior colliculus in mammals) and to nuclei in the dorsal thalamus, in mammals the lateral geniculate nucleus/body (LGN/CGL). The optic tectum in turn projects to the dorsal thalamus, which projects to the medial and dorsal pallium in amphibians, to the occipital cortex in mammals and to the "visual wulst" and the entopallium (via nucleus rotundus) in birds, as described in Chap. 10.

Depending on the three different types of retinal ganglion cells, i.e., X-P, Y-M, and W, there are three major visual subsystems in all vertebrate brains, one responsible for the perception of shape/object and color, one for motion, and a third one for changes in ambient illumination. Within the well-studied visual system of mammals, information carried by X-P and the Y-M system is processed inside the visual cortex on the basis of segregated signals from the retina and the LGN/CGL. Inside the six-layered CGL of primates, four layers receive afferents from the P and two layers afferents from the M system from the left and right eye. Within the primary visual cortex of primates, we again find separate processing of P-mediated and M-mediated information as well as information coming from the left and right eye, and this kind of information is processed in various layers and sublayers of the primary visual cortex (V1, A18), mostly of layer 4, and eventually in the so-called blobs and interblobs in layers 2–3 (cf. Fig. 11.7). From the blobs

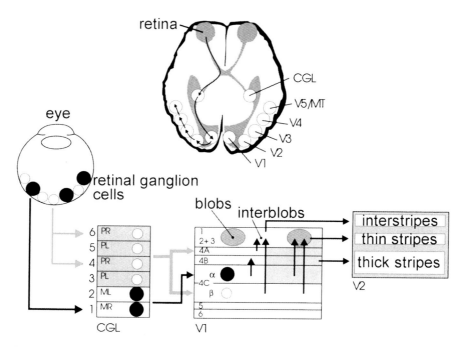

Fig. 11.7 Schematic diagram of the visual system of primates including humans. *Below*: P and M retinal ganglion cells (*white and black dots*, respectively) send their axons via optic nerve/tract to separate layers of the LGN/CGL inside the thalamus; axons from the left and right eye likewise terminate in separate layers of the CGL (PR/PL and MR/ML). From there, P and M cells project to the primary visual cortex (V1) terminating in different sublayers of layer 4: P cell axons terminate in layer 4A and 4Cβ. From there, cells project to the "blobs" and "interblobs" in layers 1–3, and cells from there to the "thin stripes" and "interstripes" in the secondary visual cortex (V2). M cells in the CGL project to layer 4Cα of V1, and from there axons run to layer 4B and from there to the "thick stripes" in V2. *Above*: From V2, two different "visual pathways" originate, one (the "dorsal" pathway) runs to areas V3 and MT (medial temporal) to the posterior parietal lobe and is involved in the perception of motion and space as well as in action preparation, the other one (the "ventral pathway") runs to area V4 and then to the inferior temporal lobe and has to do with the perception of objects and scenes. From Roth (2003)

and interblobs, separate projections run to the secondary visual cortex (V2, A19), where they again terminate in separate structures, i.e., the so-called thin and thick stripes as well as interstripes. From there, two major pathways originate: one is called the *dorsal path*, which runs via cortical areas V3/A19 and MT to the posterior parietal cortex and is involved in the perception of motion and space as well as spatial orientation of real and imaginated or intended body, arm, and hand movements. The other is called the *ventral path*, and runs via area V4 to the inferior temporal lobe and has to do with the perception of objects, faces, persons, and scenes and their meaning and also with color perception.

Thus, the different aspects of complex visual perception are processed and represented in different areas of the cortex, often merging with nonvisual sensory information. A much-discussed problem is how and where these separately processed and represented kinds of information "come together," because, seemingly paradoxically, in our subjective perception, the shape, contrast, movement, spatial position, and also meaning of objects form a unity and are not perceived as separate entities. However, there is no "highest" center for the full representation of objects inside the brain. This is called the "binding problem" (cf. Singer 1999; Koch 2004). Some experts believe that such a "highest" center actually exists, but has not yet been found, while others assume that there is no need for such an anatomical convergence site, but that binding occurs via physiological processes, e.g., patterns of oscillation and synchronization among neuronal networks that process similar information (cf. Singer and Gray 1995; Engel et al. 1991; Singer 1999). I will come back to that problem in later chapters of this book.

11.5 What Does All This Tell Us?

Biologists have always been impressed by the diversity of sense organs in form and function, as well as their close correspondence with the environmental conditions and lifestyles of their carriers. For many evolutionary biologists, sense organs are among the best examples of the effect of natural selection and their result, i.e., *adaptation*. However, textbooks are mostly filled with only few spectacular cases for the "fit" between organism and environment, while the far more numerous examples of organisms with modest sense organs remain undiscussed. Why did they not evolve more sophisticated sense organs, if these are so favorable? Why did many animals simplify or even lose sense organs, like eyes or electroreception, instead of making them more complex? I will come back to this crucial question of evolutionary biology in Chap. 16 of this book.

The evolution of sense organs and sensory systems is a fascinating mosaic of conservatism and innovation. The basic types of sensory receptors, transduction, and encoding evolved very early and are at least 1 billion years old. Likewise, highly sensitive and effective sense organs originated very early, e.g., the olfactory systems of nearly all animals, the insect compound eyes or the vertebrate lens eyes, and similar things can be said about mechanoreceptive organs including auditory and lateral line systems. Equally impressive is the number of seemingly independent origins of sense organs exhibiting very similar or even the same construction properties. Good examples are the olfactory system with respect to the formation of glomeruli and the spatially distributed mode of encoding olfactory information, the formation of lens eyes or compound eyes and the principles of parallel-distributed processing of the different aspects of object vision or color vision, and the parallel reinvention of electroreceptive systems in mormyrids and gymnotids—just to mention a few. Here, it is still unclear whether the great similarities are due to truly independent evolution under the strong functional pressure of physical and chemical working conditions, or due to the effect of "deep homologies," e.g., in the case of *Pax6* (see above).

There is a complicated relationship between the evolution of sense organs and nervous systems or brains. In some phyla, complex sense organs, such as the rhopalia of medusae or lens eyes in the scallop *Pecten* (cf. Chap. 7) are found without a counterpart of a complex nervous system. On the other hand, we often find an increase in sense organs to be paralleled by an (often enormous) increase in size and complexity of brain centers devoted to the processes of unimodal sensory information like the optic lobes in insects, the electric lateral lobe, the valvula cerebelli, and torus semicirularis in the weakly electric fish, the vagal and facial lobe in teleosts, together with a highly developed gustatory system, the optic tectum in teleosts and birds. In birds and mammals, in parallel, there is an "invasion" of *unimodal* sensory, i.e., visual, auditory, mechanosensory-tactile, and vestibular pathways into the pallium or cortex, which in anamniotes and most "reptiles" is dominated by olfaction and is only reached by multimodal afferents from the midbrain and diencephalon (cf. Chap. 10). In birds and mammals, this "invasion" enabled these systems to form extended unimodal topographic (i.e., retino-, tono-, and somatotopic) representations leading to a much more detailed representation of stimuli.

However, this impressive evolution of sense organs and sensory systems is limited not only by space, but also by the fact that the increasing amount of sensory information must be adequately processed in order to lead to successful behavior, survival, and reproduction. A maximum amount of sensory data does not automatically make sense. Therefore, secondary and tertiary centers and areas for processing of sensory information are formed. This comprises a comparison between information from the same types of receptors distributed over sensory surfaces as well as across different types, computation of time differences in the activation of the receptors leading to two- or three-dimensional representation of environmental stimuli, and integration of different modalities leading to multi-modal representation of complex events.

Finally, more abstract levels of information processing in the sense of abstraction and categorical learning evolved in many birds and mammals and in at least some invertebrates as well. Working memory systems evolved as a guide for attention and as a filter for long-term memories. At the same time, learning and memory are increasingly dominate over reactive behavior: the animals do not exclusively orient their behavior along actual sensory information, but on the basis of previous experience they form *predictions* about what is *most likely to happen* in the environment. Sensory information is then mostly used for recording *mismatches between expectation and factual events*. This leads to much faster and more economical adaptive behavior. In this sense, from a certain point in evolution brains "take over" and evolve more rapidly in size and processing modules than the sense organs.

Chapter 12
How Intelligent Are Vertebrates?

Keywords Cognition teleosts · Cognition amphibians · Mammals-birds: tool use · Tool fabrication · Quantity representation · Object permanence · Reasoning · Working memory · Social intelligence · Machiavellian intelligence · Gaze following · Imitation

One key publication in comparative intelligence research was the book *"Brain and intelligence in vertebrates"* by the British psychologist Euan MacPhail, which was published in 1982. Here, MacPhail proposed a thesis that shocked many of the experts in the field by stating that all vertebrates turn out to be equally intelligent, if their behavior is studied under "fair" conditions, i.e., under those who take into consideration their specific living conditions. At least—so the argument of the author goes—there are *only quantitative*, but *no qualitative* differences in cognitive abilities, i.e., those that are present in some, but completely absent in other groups. According to this view, birds and mammals are *not* generally more intelligent than teleosts, amphibians, or reptiles. The one exception is human beings, because only they have language and, consequently, consciousness.

Critics have argued that MacPhail's thesis is nothing but the trivial statement that all living vertebrates are equally intelligent, because they have survived successfully until now. However, one does injustice to MacPhail with such an argument, because in his book he presents much data on studies of animal behavior available at that time. Unfortunately, this data was mainly taken from laboratory experiments or controlled field experiments, and around 1980 such kinds of studies were not available in sufficient numbers for the behavior of non-mammalian or non-primate vertebrates. If one relies—as MacPhail did—mostly on experiments with classical (Pavlovian) and operant conditioning, one indeed gets the impression that—with the exception of humans—there is no substantial difference across vertebrates. Qualitative differences seem to be rare and difficult to prove, and if we refute circumstantial evidence, then we need rather sophisticated experiments. These, however, have been conceived and conducted since then in great numbers.

For comparing cognitive abilities of vertebrates, we usually first address basic learning and memory functions, which mostly are easy to study. "Higher" learning and memory functions are, as discussed previously, commonly subsumed under two different kinds of intelligence, i.e., *ecological* and *social intelligence*. Ecological intelligence in a narrow sense concerns feeding and prey capture habits, protection against predators, orientation, cognitive maps, and related memory functions. Within ecological intelligence, "physical intelligence" is often treated separately and deals predominantly with tool use and fabrication, but it is clear that both refer to *cognitive functions* in the traditional sense (Bates and Byrne 2010). Social intelligence involves aspects of social learning, imitation, cooperative behavior including altruism, and individual recognition, which may or may not involve "higher" cognitive abilities. Finally, there is the question of "mental" functions like mirror self-recognition, knowledge attribution, metacognition, "theory of mind," and, eventually, consciousness. This last question will be dealt with in the next chapter.

12.1 Cognition in Teleost Fishes

There are no detailed reports about the cognitive abilities of hagfish and lampreys, and also none of sharks and rays. This is a pity, because—as mentioned—there are great differences in relative brain size of about one order of magnitude and in brain complexity between chimaeras, squalomorph sharks and torpediniform and rajiform rays on the one hand, and galeomorph sharks and myliobatiform rays on the other, and it would be interesting to compare these differences with those in intelligence. Some details concerning the relationship between lifestyle and brain properties can be found in Lishey et al. (2008).

Teleosts, like perchs or carps, can be easily conditioned in the classical (Pavlovian) or operant way. Nevertheless, experts were surprised to learn from a review article by Bshary, Wickler, and Fricke about cognitive abilities of teleosts which—in the eyes of the authors—are comparable to those of primates (Bshary et al. 2002). The authors argued on the basis of the above-mentioned distinction between social and ecological intelligence. In their article, data from investigations of African cichlids plays a prominent role. Cichlids represent one of the behaviorally and ecologically most diverse groups of vertebrates; more than 1,600 species have been described, and new ones are constantly discovered. They are well known for having evolved rapidly and within large lakes, especially the African Great Lakes, and many species evolved only 100,000 years ago.

As to social intelligence in fish, Bshary and colleagues emphasize that in a number of teleosts, including many cichlid species, there is individual recognition on visual, but also on purely auditory cues in the context of brood care. Individual recognition, however, is observed only in stable groups and not in swarms, but may exist even there in the context of partner choice, foraging, and predator inspection. In some cichlid species with ranking orders, subordinate individuals often exhibit submissive or appeasing behavior toward high-ranking group members in order to reduce aggression, and such behavior is also shown during courtship. As in mammals, and particularly in primates, cheating is answered by exclusion and punishment of the cheater, which means that the individuals are capable of remembering their partners' behavior during past interactions. Also, teleosts (e.g., guppies) collect socially important information by observing the interactions between conspecifics. An observer will orient its behavior according to the observed mating or fighting success of other conspecifics. Cichlid males will actively intervene in female-female aggression in favor of an unfamiliar female, and this behavior increases the probability that the new female will settle in the group. There are many reports about social learning by observation in the context of site preferences, food sources, anti-predator behavior (e.g., mobbing an *Octopus*), and learning by young fish what to eat and what to avoid by observing adults. There are likewise many examples for cooperative hunting (e.g., in mackerels) and even for interspecific cooperative hunting, e.g., between giant moray eels and red sea coral groupers. The latter seemed to behave opportunistically (in the eyes of human observers) by soliciting the moray eels to hunt together with them.

Particularly impressive is the cleaning symbiosis of the so-called cleaner fish, which provide a service to other fish by removing dead skin and ectoparasites. Cleaning symbiosis apparently is a successful behavior, since cleaner fish can be found in many different fish families, and one cleaner, like the cleaner wrasse (*Labroides dimidiatus*), may serve clients belonging to more than 11 different species. The cleaner fish know their clients and their habits well and distinguish between resident species and "foreigners" by preferring the former. At the same time, they tend to cheat their clients by feeding on healthy tissue. In that case, clients often respond to cheating by cleaners with "punishment," i.e., with aggressive chasing and avoiding the cheaters, which in turn will evoke appeasing behavior toward the "angry" clients. In summary, in the eyes of Bshary and colleagues, teleosts clearly demonstrate social intelligence by exhibiting individual recognition, altruism, cheating and social punishment, appeasement behavior, and cooperation to an extent previously known only in mammals and particularly primates.

According to the authors, the same holds for "ecological intelligence" of teleosts. They adopt special techniques while feeding on sea urchins (e.g., by blowing water streams to turn them over), or manipulate their environment by removing obstacles to reach hidden prey. They use stones of various sizes for shelter-building (jawfish). Some teleosts, like the intertidal gobies, have excellent spatial memory. During low tide, these animals stay in tide pools and are able to jump from one pool to another without being able to see that pool, and they may even jump through a series of pools in order to escape to the sea. Experiments have demonstrated that these fish, while swimming over the tide pools at high tide, form an effective memory of the topography around their home pools. Long-term memory is relatively rare in the animal kingdom, but some teleosts, like the anemonefish and carps, exhibit memories lasting for at least several months.

Continuing this survey on fish intelligence, a team that included the German neurobiologist Hans A. Hofmann, published a report on the correlation between environmental complexity, social organization, and brain features in African cichlids (Pollen et al. 2007; cf. also Shumway 2008). The authors studied the behavior of the cichlids, the properties of the habitat (sandy, rocky, or intermediate, with increasing habitat complexity regarding depth, steepness, roughness of surface, and stone size), as well as social behavior regarding number of species, of individuals, and lifestyle (e.g., monogamous vs. polygamous). The brains of the cichlids were measured in detail, i.e., volumes of total brain, olfactory bulbs, telencephalon, hypothalamus, midbrain, midbrain tectum, cerebellum, and dorsal medulla. The study showed that sandy sites are inhabited by fewer species compared to intermediate or stony ones, and that habitat complexity is positively correlated with larger brains and larger cerebella, and negatively with the size of the olfactory bulbs and dorsal medulla. There was a tendency toward a positive correlation between habitat complexity and telencephalon size, but no correlation was found between habitat complexity and the size of the tectum and hypothalamus, which at least with respect to the tectum (the main sensorimotor coordination center in the teleost brain) is somewhat surprising. As for social organization, there was a significantly positive correlation between a monogamous lifestyle and the size of the telencephalon and a negative one with the size of the hypothalamus, while in polygamous species the hypothalamus was significantly larger compared to the monogamous ones.

Many teleost taxa stand out by their highly evolved sensory and communicative systems, e.g., the weakly electric fish (cf. Chap. 11), which at the same time have astonishingly large brains in relative terms. Unfortunately, no systematic studies exist on their cognitive abilities beyond sensory and communicative functions.

12.2 Learning and Cognitive Abilities in Amphibians

Traditionally, frogs and salamanders are regarded as being highly instinct-bound, if not simple "reflex machines" (cf. Tinbergen 1953). Extended behavioral and neurophysiological studies, however, demonstrated that these animals possess—like most other vertebrates—a complex system of sensory control of behavior, including some learning abilities (cf. Dicke and Roth 2007).

During the past 30 years, my colleagues and collaborators and I have investigated in detail the question of experience-dependence of prey preferences and feeding behavior in frogs and salamanders. These animals, like all vertebrates, possess some inborn object and movement pattern preferences including prey schemes, which, however, can be modulated substantially by early experience, i.e., by the type of food (e.g., flies, crickets, mealworms) they had been raised with (cf. Roth 1987). This means that adult salamanders and frogs will prefer that type of prey characterized mainly by size, shape, and movement pattern they grew up with.

During the last few years, together with a number of colleagues, we carried out numerous conditioning experiments with the fire-bellied toad *Bombina orientalis* using reward, omission of reward, and punishment in the context of feeding behavior (Jenkin and Laberge 2010; Dicke et al. 2011). As prey stimuli we used real-sized video images of naturally moving crickets (the familiar food) and mealworms (unfamiliar food), which were presented on a monitor in front of the toads. At the beginning of the experiments, spontaneous snapping toward the prey dummies was reinforced for two weeks by occasional feeding with one live cricket in front of the screen, until all animals snapped at 100 % of stimulus presentation toward the dummies. After having reached this criterion, one group was "overtrained" for an additional two weeks, whereas a second group received no further cricket reward. Now, in the first type of experiments, we tested the effect of the omission of reward in both groups. The results were somewhat surprising: The "overtrained" toads kept snapping at the dummies for weeks without any further reward inside the test box (they were fed outside the box), whereas the toads that had not been "overtrained" significantly reduced their snapping responses, and increasingly hesitated before snapping. When we presented video mealworms (unknown both as natural food and video prey dummies), then toads of both groups quickly decreased in their snapping responses at a similar rate. This indicates that the "overtrained" toads had formed a highly stable prey preference, which was not modified by omission of reward, while those with less or no further reward experience were sensitive to omission of reward.

In another series of experiments, another group of pre-trained toads were confronted with "punishment" in the form of a harmless electric shock applied to the feet after every snapping, while sitting on a platform with fine electric wires. At presentation of cricket dummies, this kind of "punishment" led only to a slight decrease in the rate of snapping responses, which was even slower than at omission of reward, while the presentation of mealworms (again unfamiliar to the animals) combined with "punishment" led to a significantly stronger decrease of snapping responses. The strongest decrease was observed, when—in addition to omission of reward—the presentation of mealworm dummies was combined with "irregular punishment," i.e., uncorrelated to the presentation of prey dummies. This type of punishment did not affect their natural feeding motivation, because the toads fed on natural crickets as readily as the "unpunished" ones.

These experiments demonstrate that in amphibians, too, feeding behavior of young as well as adult individuals can be modified by experience, but only if a specific type of prey or behavior is not strongly consolidated ("overtrained"). Interestingly, omission of reward turned out to be more effective than punishment. This may be explained by the fact that many amphibian species feed on defensive or unpalatable prey (bees, wasps, ants) such that "painful" prey items are part of their normal diet (Roth 1987).

Little is known about "higher" cognitive abilities of amphibians. As shown in a number of studies, human infants and monkeys select the larger of two numerosities in a spontaneous forced-choice discrimination task. The same method was adopted by Uller et al. (2003) in the red-backed salamander (*Plethodon cinereus*). Salamanders were able to select the larger of two numerosities when the paired numbers were 1 vs. 2 and 2 vs. 3, but not 3 vs. 4 and 4 vs. 6. Animals apparently recognized that 2 is more than 1 and 3 is more than 2. These experiments suggested that salamanders show a rudimentary ability to "go for more" that had previously only been shown in the primate lineage.

In a follow-up study by Krusche et al. (2010) in our laboratory, again with salamanders of the genus *Plethodon* (here *P. shermani*), animals were challenged with two different quantities (8 vs. 12 or 8 vs. 16) in a two-alternative-choice task using live crickets, videos of live crickets or images animated by a computer program. Salamanders reliably chose the larger one of two quantities when the ratio between the sets was 8 vs. 16 and stimuli were live crickets or videos thereof. However, magnitude discrimination was not successful when the ratio was 8 vs. 12, or when the ratio was 8 vs. 16, and when the prey stimuli were computer-animated and did not exhibit a natural movement pattern. This suggests that the salamanders used *movement* as a dominant feature for quantity discrimination. These results corroborate the view that salamanders as well as most vertebrates make use of two number systems, one system for small sets (≤ 4) that is precise but limited, as it works by keeping track of individual entities; and another one for larger sets that is independent of absolute set size, works on imprecise analogue magnitudes following Weber's Law, i.e., discrimination is a function of the ratio between the numbers in question.

Ursula Dicke and her collaborators from the University of Bremen recently demonstrated that salamanders of the species *P. shermani* possess an attentional system used in visual object selection resembling that found in mammals/primates. While in amphibians the optic tectum is the major site of visual information processing, ascending visual pathways relay in the dorsal thalamus and further extend to forebrain structures, which modulate the neuronal responses in the tectum via a feedback loop. The neuronal responses to the presence of two objects in the visual field are strongly inhibited toward the ignored object, and in this way favor the processing of the salient object. This inhibition effect is stronger, the "more attractive" the competitive stimulus (Schuelert and Dicke 2002, 2005). After lesion of the ascending pathway, salamanders stop turning toward one object and ignore another one, and therefore remain "undecided;" at the same time, the inhibitory effect in the tectum is no longer present (Ruhl and Dicke 2012). Although primates use the cortex for visual object selection rather than the tectum (here the colliculus superior), it is—like in salamanders—part of the dorsal thalamus, here the pulvinar that mediates the inhibitory effect onto visual processing mechanisms.

12.3 Cognitive Abilities and Intelligence in Mammals and Birds

During the past 20 years, studies on cognitive abilities and environmental-physical intelligence have been carried out in such a large number in mammals and in birds that it is impossible to give a comprehensive overview here. In the following, I will concentrate on the most frequently used paradigms for measuring environmental or physical intelligence, including tool use and tool fabrication, gaze following, imitation, intentional action and social learning, quantity representation, object permanence, reasoning, and working memory, as well as social intelligence, including imitation.

12.3.1 Tool Use and Tool Fabrication

There have always been anecdotal reports that animals use such objects as hammers, probes, anvils, weapons, sponges, or bait. Dolphins, for example, kill scorpion fish in order to use their stinging body to poke after a moray eel hidden in a crevice, and Galapagos finches (*Cactospiza pallida*) use cactus thorns to spike insects under the bark of a tree. Egyptian vultures (*Neophron percnopterus*) throw stones on ostrich eggs in order to break them up, and sea otter (*Enhydra lutris*) use different kinds of stones as anvils to break hard-shelled animals like snails or crabs. Their daughters specialize on those types of stones that are preferred by their mothers, which indicates a learned component of tool use in this species.

Tools can be divided into natural and artificial ones. The former include moss used as sponge or leaves for carrying water, a wooden stick for scratching a body part or removing ticks, and the latter comprise all natural products worked for a certain scope such as a stick that had been freed of twigs or bark by chewing them off (as happens in chimpanzees), a branch that is modified by elephants in order to use them as fly switches, but also the fabrication of a spit for nailing down insects as found in birds or of a spear for fishing, as happens in primates (Hart et al. 2008).

Tool use is commonly found among primates. Ring-tailed lemurs have recently been reported to successfully manipulate a puzzle feeder in the wild (Kendal et al. 2010), which is the only known case of lemur tool use in the wild. In captivity, however, manipulatory skills of lemurs with novel objects are roughly comparable to those of some New and Old World monkeys. The gray mouse lemur (*Microcebus*) mastered the opening of boxes in various ways, including the use of reversed images, and aye-ayes (*Daubentonia*) demonstrated a basic understanding of features of tools by solving a can-pulling task (cf. Fichtel and Kappeler 2010). Systematic tool use, including limited forms of tool making, is found in the capuchin monkey (Ottoni and Iza 2008; Visalberghi and Limongelli 1994; Visalberghi et al. 2009). Iriki and Sakura (2008) argue that in Japanese monkeys, latent cognitive abilities are made explicit by exposure to a proper environment. Furthermore, the successful training of tool use induced physiological, anatomical, and molecular-genetic changes in the brains of the animals.

Chimpanzees are known to fabricate and use a wide range of complex tools, and have been shown to vary in their tool use at many levels—for example, preparing twigs for ant and termite dipping (Goodall 1986; Boesch and Boesch 1990). In chimpanzee populations, tool kits consist of about 20 types of tools for various functions. Only chimpanzees appear to be able to use one type of raw material to make different kinds of tools, or make one kind of tool from different raw materials. They use tool sets in a sequential order, make use of composite tools, and combine tools to a single working unit (Sanz and Morgan 2009; McGrew 2010). In this context, chimpanzees as well as orangutans exhibit insightful problem solving (Mendes et al. 2007; Osvath and Osvath 2008). They are engaged in action planning, mentally pre-experiencing an upcoming event, and are able to select objects needed for a much-delayed future in tool use (cf. Mulcahy and Call 2006). I will come back to this topic in the next chapter within the context of consciousness.

Astonishing examples of tool use are likewise found in corvid birds. In nature as well as in captivity, corvids are shown to use natural objects as tools or to modify them until they have the right length or pass through an opening (Chappell and Kacelnik 2004; Weir et al. 2002). As demonstrated by Hunt and colleagues from the University of Auckland, New Caledonian crows (*Corvus moneduloides*) spontaneously make tools out of screw pine (*Pandanus*) leaves, in the way that they tear out stripes of leaf edges and use them as probes for removing insects from crevices. In that context, they fabricate strips of different size and shape for different purposes. This kind of tool use appears to have been invented once on New Caledonia (a group of islands east of Australia) and then spread from one individual to another.

In 2002, Chappell and Kacelnik and colleagues from the University of Oxford observed the New Caledonian crow Betty (which has died in the meantime) using a straight wire to remove a small bucket of food (i.e., pieces of pig heart) from a vertical pipe. After being unsuccessful, she spontaneously bent the wire into a hook and could now lift the bucket. In the following tests, Betty bent the wire into a hook nine times. According to Hunt, crows occasionally make hooks in the wild, too, and Betty had seen a hooked wire before, but this type of intentional tool-making is regarded as unique, and even chimpanzees are reported to have great difficulty with similar tasks.

A similarly impressive behavior of the New Caledonian crows is placing nuts in front of a vehicle on a street with heavy traffic and waiting until a car crushes the nut open. The birds wait at pedestrian lights with other pedestrians until they can retrieve the crushed nuts safely. Finally, Hunt et al. (2007) demonstrated that crows use short sticks to retrieve longer ones from a box, which is then used to retrieve food from a box (*metatool use*, i.e., use a tool to get another tool)—a behavior which otherwise has been observed in primates only.

The question of tool fabrication has been complicated by the finding that corvids, like rooks (*Corvus frugilegus*), which do not appear to use tools in the wild, are nevertheless capable of insightful problem solving related to tool use (Bird and Emery 2009). In a number of standard tests for intelligent tool use, including the use and modification of stones, sticks, fabrication of hooks, and finally "metatool" abilities, rooks turned out to be as able as the above-described New Caledonian crow, which regularly exhibit tool use and tool fabrication in the wild.

Here, too, the question arises of whether corvid (or other) birds exhibit a *causal understanding* of tool use, as is suggested by their ability to "purposeful" tool fabrication. Recent experiments by Taylor et al. (2009) cast some doubt on such an assumption. The authors conducted standard string-pulling experiments with New Caledonian crows, comparing experienced and naïve animals. The animals had to pull up strings with meat at one end. When they had full sight of the string and the meat while pulling, naïve ones solved the problem spontaneously in most cases. But when visual control of string-pulling was restricted, naïve animals could not solve the problem any more spontaneously and only after extended trial-and-error learning, and the performance drastically dropped even in the experienced ones. If they were offered visual feedback via a mirror, the performance rate increased. According to the authors, this demonstrates that in the string-pulling experiment, problem solving is based primarily on reinforcement learning, i.e., making the experience that pulling the string brings the meat closer, rather than on insight. The corvids did not appear to understand the causal relationship between strings and meat. Thus, the question of truly insightful problem solving in corvids remains open.

12.3.2 Quantity Representation

Lemurs are capable of controlling their impulsive gesture toward a larger option, when selection of a smaller quantity of food is rewarded with a larger one. They also learned to associate a graphic representation of the reward with the corresponding quantity, even though only one subject consistently selected the representation of the smaller quantity to be rewarded with the larger quantity of food (Genty and Roeder 2011). The fundaments of numerical abstraction appear to be present in prosimians. Nevertheless, numerical discrimination is superior in monkeys and apes. Capuchin monkeys are able to judge larger quantities of two sets, contrasting up to five items in food-choice experiments (Evans et al. 2009). Quantity-based judgments for two sets, with up to 10 items were tested in rhesus monkeys and great apes. Rhesus monkeys selected the larger of the two sequentially presented sets reliably when one set had fewer than four items and one set had more than four items (Beran 2007), whereas great apes did so, even when the quantities were larger (up to 10) and the numerical distance between them was small (down to one item) (Hanus and Call 2007). However, the performance decreased as a function of the numerical ratio between the sets (i.e., from 1:2 to 9:10). More recently, Byrne et al. (2009) reported an equal ability of Asian elephants to choose with considerable accuracy (70–80 %) larger quantities up to 12 items, even when the one quantity was only slightly larger than the other.

The Border Collie *Rico* revealed an excellent ability to select shown or named objects from a large collection of objects out of sight in a neighboring room (the record was 200 objects; cf. Kaminski et al. 2004), but at counting he did not go beyond five. Another "star" of cognitive performance was the Gray Parrot *Alex* (Pepperberg 2000), who could verbally express the number of shown objects up to four and verbally distinguish up to five colors and objects. According to his tutor, Irene Pepperberg, Alex had even understood the significance of "zero" and mastered verbally the concept "same" and "different." Nevertheless, even for him there was the magic limit of five, which appears to hold for all animals except apes, and even these animals go beyond only after prolonged training.

12.3.3 Object Permanence

Object permanence is divided into six major stages according to its gradual development in humans (Piaget 1954). At stage 4, human infants are able to mentally represent and retrieve an object hidden in a single hiding place. When the object is visibly placed into a new hiding place, infants continue to search the initial location. Stage 5 of object permanence characterizes the ability to find an object that has been hidden successively in multiple locations, whereas at stage 6, direct perception of an object is no longer required to infer an object's location. Lemurs successfully found objects (raisins) being visibly displaced and thus fulfill stage 5; they are capable of understanding and mentally representing visible displacements (Deppe et al. 2009). Lemurs as well as monkeys did not correctly locate objects during invisible displacements (stage 6). Monkeys accurately selected visibly displaced items, while invisible displacement was more difficult to track (Neiworth et al. 2003). The ability to locate an invisible moving object has consistently been reported only in great apes and humans (Barth and Call 2006; Collier-Baker et al. 2006).

12.3.4 Reasoning and Working Memory

For a long time it was believed that only humans possess reasoning and intelligence or the ability to think, and even today the question of whether animals can act rationally is debated among experts (cf. Povinelli 2000). The experiments described above on tool use in monkeys and birds suggest that these animals can use and even fabricate tools without a deeper understanding of the underlying physical mechanisms. However, Blaisdell and colleagues (Blaisdell et al. 2006) recently argued that causal reasoning in the form of predicting future events upon pure observation does not necessarily require such a "deeper understanding." They demonstrated that rats were able to change their behavior on purely observational learning—a result that cannot be explained on mere associative learning.

One common paradigm for testing the extent of causal reasoning in animals is the so-called "law of transitivity." One example of this law is: If A is larger than B and B larger than C, then it can be concluded that A is larger than C (cf. Chap. 8). The German–Argentinian psychologist Juan Delius carried out experiments with pigeons and initially came to the conclusion that these animals are capable of acting according to the law of transitivity (Delius et al. 2001). Later on, however, there were doubts as to whether or not their behavior was based on the higher pairing frequency of certain objects rather than on true logical reasoning. Experiments with primates, including apes, demonstrated that they, too, are poor at applying the law of transitivity genuinely, and only one chimpanzee gave the correct answer. Even human subjects demonstrated that when confronted with the same or similar problems, they often make use of simple heuristics ("rules of thumb") instead of true reasoning. Results are more clear-cut when the animals are confronted with cases of the law of transitivity of high biological or social relevance, for example, if animals have to guess if it is "reasonable" to fight against a certain competitor. When cichlids, crows, or mammals observe that a competitor against whom they had already lost, loses against a third animal, they mostly refrain from fighting against the winner. However, in that case again, a simple pairwise comparison is sufficient, as happens in bees (cf. Chap. 8), and animals would fail if they had to compare non-adjacent pairs like B and D. We can, therefore, conclude that causal reasoning is present in non-human animals only in restricted forms.

Another common task for testing cognitive-rational abilities is the understanding of serial orders, i.e., learning and memorizing a certain sequence of objects or actions. In a test that the American psychologist and behaviorist Herbert Terrace carried out some decades ago (Terrace 1987), pigeons had to peck on differently colored discs in a given order (red A, green B, yellow C, etc.). The pigeons were able to master that task only after 120 daily training sessions. They performed better if the three objects differed in shape as well.

One much-used task for testing working-memory abilities is the delayed-matching-to-sample task (DMTS) or the complementary form of delayed-non-matching-to-sample. In DMTS, the subject is shown a rewarded stimulus and has to keep it in mind for a variable period of time in which the stimulus was not visible anymore, and then identify it out of a pair of stimuli. Pigeons quickly master a delay of 5–10 s, but were successful at a delay of 1 min only after 17,000 training sessions. Macaques master a delay of 2–9 min again after long training. Dolphins reach a maximum of 4 min. In humans, the capacity of working memory is substantially increased by the "phonological loop" (cf. Chap. 14), but when human subjects are prevented from talking to themselves, then they are no better than dolphins and macaques.

More recently, the Japanese behaviorists Inoue and Matsuzawa (2007) reported the astonishing capacity for numerical recollection in three mother-offspring pairs of chimpanzees. In a numerical sequence task, the animals had to learn the sequence of Arabic numerals from 1 to 9, which appeared at different on-screen positions. Then the numerals were replaced by white squares. The subjects had to remember, which numeral had appeared at which location, and then touch the white squares in the correct sequence. All naïve animals mastered this task, but the performance of the three young chimpanzees was always better than that of the three mothers. Adult humans were slower than the three young chimpanzees. In another test (the "limited hold task"), the numerals appeared only for a very short and decreasing duration (610, 430, and 210 ms, respectively) and were then replaced by the white squares. It turned out that while in the best mother performer (*Ai*) as well as in human controls, the percentage of correct performances strongly decreased with decreasing duration, there was almost no decline in the best-performing young chimpanzee (*Aiumi*). It would be interesting to test young human children under the same conditions.

12.3.5 Social Intelligence

"Machiavellian" Intelligence

In the last decade of the last century, a wealth of data accumulated that suggested that in mammals, and particularly in primates, one dominant factor for the evolution of the large and complex brain was the necessity of developing complex social abilities involving deception, imitation, and gaze following. The most influential of this concept was, and still is, the "Machiavellian Intelligence" hypothesis of the British psychologists and behaviorists Richard Byrne and Andrew Whiten (Byrne and Whiten 1992; Byrne 1995), emphasizing the ability to act diplomatically, build coalitions, cooperate, establish dominance orders, deceive, recognize deception, and apply counter-deception. All this—according to the authors—is needed, if a group of animals reaches a certain level of social complexity.

Byrne argues that among primates, deception is much more common than previously assumed, and in the majority of species is learned by trial and error. Repeated deception provokes counter-deception. Apes, especially, appear to have a tendency to deceive, coupled with the ability for insight into one's own behavior. Deception is likewise found among birds, which often, however, is based on inborn behavioral patterns like distracting predators of ground-breeding birds (such as plovers) by displaying a broken wing behavior or "rodent run," i.e., pretending to be a small rodent. However, deception can likewise be learned in birds, and young birds deceive after being deceived for the first time (Emery and Clayton 2004).

Gaze Following

A special case of social intelligence is gaze following, which, except in dogs, is almost exclusively found in primates. Lemurs preferentially orient their eyes toward other lemurs and mirror the attentional state of others in their social group (Shepherd and Platt 2008). A study on gaze orientation and object-choice used a color photo of a conspecific as a model, oriented with eyes and head to a right- or left-sided reward (Ruiz et al. 2009). The response of the lemurs to the models' gaze significantly influenced their choice of behavior. The authors define this as *gaze priming*. In monkeys, a mentalistic understanding of the observing animal about the other's visual target remains an open question. In long-tailed macaques, gaze following was accompanied by frequent check-looks and was significantly more frequent in response to a signal of fear and submission than to a neutral facial expression (Goossens et al. 2008). Capuchin and spider monkeys spontaneously followed a human experimenter's gaze, and capuchin monkeys followed the gaze around barriers, but neither capuchin nor spider monkeys displayed any "looking back" behavior (Amici et al. 2009) and, thus, might lack perspective-taking.

Marmosets showed high proficiency in extrapolating gaze direction, but failed to show context-independent perspective-taking (Burkart and Heschl 2007).

According to the evidence we have now, monkeys appear to deal with a directed gaze without understanding visual perspective. Great apes are able to track gaze to hidden targets and look back to the human experimenter, when they do not find a target (Bräuer et al. 2005; Tomasello et al. 2007). However, great apes use both head and eye direction in gaze following, while human infants are much more attuned to the eyes.

Until recently, the only case of gaze following in non-primate animals was found in dogs. Dogs are believed to represent a special case of social intelligence because they have been trained over thousands of years to communicate with humans. Dogs—as every dog-owner knows—have excellent abilities to grasp the mood and intentions of a human familiar to them (particularly the dog owner), for example, whether one intends to go for a walk, and in this context they make use of emotional-communicative signals like body odor, body posture, gestures, and emotional valence of the voice. As shown by the Hungarian ethologists Miklósi et al. (2003) from Eötvös University in Budapest, dogs can understand human pointing and inform humans about hidden objects, look at the faces of humans and follow their gaze appears to identify. Since initially such behavior could not be found in wolves, it was concluded that it represented a specific adaptation of dogs to life with humans. Recently, however, Udell et al. (2011) demonstrated that wolves likewise are sensitive to human attentional states and that they are able to rapidly improve their perspective-taking abilities.

Imitation

Imitation was long considered an inferior kind of learning and typically called "aping" or "monkeying" in the sense of meaningless copying of a certain behavior. Only in recent years did it become clear that imitation is a higher order cognitive ability. However, to date there is no universally accepted definition of imitation, and some kinds of behavior previously seen as imitation are now interpreted differently. One of these imitation-like behaviors is *response facilitation* or *emulation*, found in a wide range of animals, which means that seeing an action "primes" the individual to do the same, and the individual, by trial and error, finds the same or a very similar solution to the problem. One famous example is potato-washing by macaques on the Japanese island of Koshima, apparently first observed in 1952 in a young female monkey named *Imo*. Fantastic stories came up reporting an unbelievably fast distribution of that habit once a critical number of monkeys (i.e., the "hundredth-monkey effect") was reached, but later this turned out to be completely fabricated. Actually, the habit spread very slowly over the macaque population of the island.

Of importance is the social component in emulation. For example, young baboons (*Papio*) quickly learn which kinds of fruit are edible, after one group member has tasted a fruit. Vervet monkeys (*Chlorocebus*), again Old World monkeys, learn this task more slowly, although they live in the same environment as baboons. The explanation for the difference may be that young baboons have a close social life and show great interest in each other, while this is not the case for the Vervet monkeys.

Another famous case of behavior misinterpreted as imitation is the breaking of foil caps of milk bottles at the doorstep of homes by the Great Tits (*Parus major*) in order to obtain the cream on top of the milk. This behavior was first noted in 1921 and then spread rapidly throughout England and parts of Scotland and Wales over the next two decades, and finally "jumped" to other bird species (Hawkins 1950; cf. Lefebvre 1995). Today, this behavior is viewed as a combination of *stimulus enhancement* and *reinforcement learning* (i.e., operant conditioning; cf. Byrne 1995): One bird discovers randomly that there is cream below the cap, and this type of behavior is reinforced by its success. Another bird watches the first one pecking at a certain object, and that attracts its attention and increases the probability that this other bird lands on the same bottle and pecks. By being rewarded, this bird is motivated to repeat this kind of behavior. The fact that cap pecking is not really imitated is demonstrated by the fact that birds apply different methods to open the cap (Byrne 1995).

In true imitation, an action is copied, which before was not part of the behavioral repertoire. Bates and Byrne (2010) distinguish between two types. In one type, called *action-level imitation*, all observed actions are copied exactly in detail. Apes, parrots, and dolphins, for example, often copy the behavior of other species in any detail in a meaningless fashion. Dolphins may copy exactly the swimming style and sleeping postures of sea lions. The other type is *"impersonation"* or *"program-level imitation."* In this type of imitation, a basic hierarchical structure of complex actions is adopted, but details of the execution are learned by trial and error. Mountain gorillas of a particular population feed on stems and leaves of certain plants covered with thorns and hooks. They use a special technique to get access to the edible parts of these plants. Other populations of mountain gorillas do not exhibit such behavior. From this follows that these techniques are acquired, and young animals learn them from their mothers and the alpha male (Bates and Byrne 2010).

Imitation of human behavior is frequently found among apes, e.g., in orangutans (*Pongo pygmaeus*). In the Indonesian Tanjung Puting National Park, orangutans were observed imitating everyday actions of the human park personnel (Pearce 1997). In some cases, the animals seemed to understand the sense of the copied action; in other cases, they carried out the actions "just for fun." This included pouring fuel from a barrel into a canister, sweeping trails, making fire, using a saw, mixing ingredients for a pancake, and washing dishes.

Also, imitation occurs when a social signal is conveyed; this type of social mimicry may depend on the action copied and the motivation behind the copying. Furthermore, *contextual imitation*, which is found in monkeys and apes, includes learning to employ an action already in the repertoire. *Production imitation* stands for learning a new motor skill by observation, and further comprises *program level* and *rational imitation*. In the former, fine detail is unimportant as long as the right result is obtained, while in the latter an understanding of the logic of how actions achieve their ends is present.

Chimpanzees and other great apes show imitative abilities beyond those of other primates. The recent view is that great apes display program-level imitation and explicit recognition of imitation, rational imitation, that they are capable of mentalizing about others and have some understanding of intentionality and causality (see next chapter). It is presently unclear whether copying an expert's use of a rule rather than just copying a certain motor behavior found in macaques (Subiaul et al. 2004) evidences contextual or production imitation at the monkey level. In macaques, posterior parietal and frontal areas, including the much discussed "mirror neurons" in frontal area F5, are dedicated to the execution and recognition of meaningful hand reaching and grasping as well as facial movements (Rizzolatti and Craighero 2004), but their significance for imitation remains unclear (see below). Chimpanzees are able to distinguish between an experimenter who is either unwilling or unable to give them food. Hence, they do not simply perceive the behavior of others, but also interpret it (Call et al. 2004). Recently, capuchin monkeys were shown to distinguish between intentional agents and unintentional objects (Phillips et al. 2009).

Learning from others' mistakes is as important as copying others' actions. Apes and children differ in the social learning mechanisms they use in problem solving. The tool use of a human demonstrator to retrieve an invisible reward from a puzzle-box was reproduced by chimpanzees imitating the overall structure of the task. In the visible condition, chimpanzees ignored the irrelevant actions in favor of a more efficient, emulative technique, while children employed imitation to solve the task in both conditions, at the expense of efficiency (Horner and Whiten 2005). Capuchin monkeys were unable to spontaneously compensate failures of a human demonstrator, who showed the monkeys an action to open or fail to open a baited box. However, when a conspecific was watched and failed to open the box, the other monkey successfully opened it (Kuroshima et al. 2008). Monkeys were able to refer to the outcome of the others' action as well as to the others' action per se, which suggests that monkeys, like humans and great apes, may understand the meaning of others' actions in social learning.

In the next chapter, we will discuss cognitive functions which traditionally are linked to "higher mental abilities," including consciousness, and then draw conclusions from the two chapters.

Chapter 13
Do Animals Have Consciousness?

Keywords Consciousness · Conscious attention · Mirror self-recognition · Metacognition · Theory of mind · Dolphin intelligence · Elephant intelligence

In the preceding chapter, I have asked how intelligent vertebrates and, in particular, birds and primates are. There is a continuing debate as to what degree the different kinds of intelligence discussed are accompanied by consciousness (cf. Bates and Byrne 2010). Certainly, it is conceivable that some kinds of intelligent behavior, including acts of "social intelligence," need not be accompanied by consciousness, but may be a case of fast implicit learning.

In humans, consciousness includes very different phenomena which only have in common the fact that we have subjective awareness of them. These include, among others, (1) *wakefulness* or *vigilance*, (2) *conscious perception*, (3) *attention* as a state of increased and focused consciousness, (4) *conscious mental activities* such as thinking, remembering, imagining, and planning, (5) *identity awareness*, (6) *autobiographic consciousness*, and, finally, *self-awareness*, i.e., the ability of self-recognition and self-reflection (Roth 2000).

In Chap. 2, I addressed the central question of whether animals have at least some forms of consciousness, and how one can test that question. The starting point is that we humans can accomplish certain cognitive tasks, such as mirror self-recognition, metacognition, i.e., the ability to know what one knows, theory of mind, and focused attention only while being conscious, and that it is unlikely that animals, but not humans, can exert those tasks without consciousness (in a "zombie-like" fashion, cf. Chap. 2).

13.1 Mirror Self-Recognition

The ability of mirror self-recognition is usually taken as evidence of "higher" mental states of consciousness eventually leading to self-consciousness. However, the proof of this ability turned out to be rather complicated. The question of whether at least some animals, like humans, are able to recognize themselves in a mirror had already interested Charles Darwin. While visiting a zoo, he held a mirror up to an orangutan and carefully observed the ape's reaction, which made a series of facial expressions. Darwin found the results of this simple test ambiguous because it remained unclear whether the ape made these expressions toward himself or toward a conspecific—or was playing with a new toy.

The American psychologist Gordon Gallup was the first to develop a method to (relatively) reliably test mirror recognition abilities in animals and infants using the "mark" or "rouge test" (Gallup 1970). He demonstrated that—besides humans—at least some chimpanzees and orangutans are capable of recognizing themselves in a mirror. These experiments proceed as follows: First one has to test how animals behave in front of a mirror (which needs to be unfamiliar to them), i.e., whether or not they show threatening gestures or other social reactions or look behind the mirror, as small children do at the beginning. After the animals get acquainted with the mirror, some of them start using it to investigate their bodies. Finally, a mark of paint or cream is applied under anesthesia or when distracted, to the front of the animal (mostly in case of primates) or of a body region that cannot be inspected without the use of a mirror, and the animals are confronted again with the mirror. They then test whether the animals spontaneously touch the mark of the mirror image or their own bodies. From age 18 months, a human child will immediately touch its own front, and this is taken as evidence that they recognize themselves in the mirror. The same happened in Gallup's experiments with chimpanzees and orangutans, but only in less than half of animals tested and not reliably in those that passed the test. It is mostly the young animals that display mirror self-recognition, and even they rapidly lost interest in such experiments. Later, it was shown, even with great difficulty, that gorillas, too (here *Koko*), are capable of mirror self-recognition.

In the following years, much effort was invested in tests for mirror self-recognition in other animals believed to be intelligent and highly social, but mostly with negative or equivocal results. Finally, the two American behaviorists, Reiss and Marino (Reiss and Marino 2001), succeeded in demonstrating that captive-born bottleneck dolphins (*Tursiops truncatus*) are capable of mirror self-recognition. At the beginning, the dolphins showed great interest in the marks attached to their bodies, which were invisible to them without the help of the mirror, but like the chimpanzees and unlike young (as well as older) humans, they rapidly lost interest in the procedure. As the last of the large-brained mammals, the elephant remained to be tested for mirror self-recognition. After a number of failures, some years ago Plotnik and colleagues demonstrated mirror self-recognition in at least one out of three Indian elephants, *Elephas maximus* (Plotnik et al. 2006). But here again, the successful elephant lost interest quickly.

13.1 Mirror Self-Recognition

From all that, one can draw the conclusion that mirror self-recognition is a highly cognitive ability which must have evolved several times independently in highly social animals with large to very large brains, since apes, elephants, and dolphins are only very distantly related. Remarkably, these intelligent and social animals exhibit no great interest in their own mirror image, and this may be the reason why tests for mirror self-recognition are so cumbersome. High intelligence and sociality appear not to be strictly connected to this ability, because neither the smart parrot *Alex* nor the equally smart Border Collie *Rico* passed the test.

The story seemed to have come to an end, but a team of biopsychologists from Bochum University in Germany, led by Onur Güntürkün, demonstrated that at least the Common Magpie (*Pica pica*), a corvid, passes the mark test (Prior et al. 2008). When the plumage of the magpies was marked below the beak, i.e., at a site that under normal circumstances was out of sight, the animals started cleaning themselves and trying to touch the spot after they had discovered it. They did not respond to pictures, padded or alive, marked or unmarked magpies behind a glass pane, i.e., they did not confound their own mirror image with that of conspecifics. Magpies are likewise highly social animals and exhibit an unusual ability to relocate cached objects (food, but also glittering objects). They are capable of recognizing conspecifics and other animals individually. This adds another case to the unusual cognitive abilities of corvid birds.

However, it remains unclear *why* other highly intelligent and highly social animals, like parrots, dogs, or baboons did not evolve the ability for mirror self-recognition, and it is still hotly debated, mostly among philosophers, whether mirror self-recognition really can be regarded as a pre-stage of self-reflection and the formation of an ego. At least it seems difficult to explain this ability without referring to conscious perception.

13.2 Metacognition

Until recently, the question of whether animals possess *metacognition*, i.e., the ability to know what they know and what they do not know, was considered irrelevant, because it seemed impossible to test for metacognition in animals because they cannot communicate verbally. However, in a recent overview, Smith (2009) showed that tests for metacognition are possible and have been carried out. The principle of such experiments is that suitable animal subjects like monkeys, apes, or dolphins are confronted with tasks in which they have to discriminate between two tones of different pitch or two pictures showing grains of different size. The differences between the two tones or pictures are now reduced stepwise, such that they become increasingly difficult to distinguish. Correct answers are rewarded; for incorrect answers there is a "time out." However, in addition to the decision between the two stimuli, there is the possibility to carry out an *uncertainty response* (*UR*) if animals have great difficulties choosing the correct answer. This allows them to immediately perform the next trial. In the experiments with macaques, chimpanzees, and dolphins, such URs occurred exactly at moments when human subjects likewise had problems with distinguishing the pattern, and started disappearing when it became increasingly easy for the human observer to distinguish the patterns. Usually, the URs were preceded by hesitation. Likewise, URs occurred more frequently, the longer the distance between stimulus presentation and decision—a typical working memory task. Smith emphasizes that it was impossible to explain the behavior of the animals on the basis of simple conditioning. Remarkably, pigeons as well as capuchin monkeys failed in these experiments. This data makes the presence of mental representations and conscious access to one's own knowledge very likely at least in some primates and in dolphins: the animals have difficulties with recognition of a pattern, and they are *aware* of this difficulty.

13.3 Theory of Mind: Understanding the Others

Under the topic "theory of mind—ToM," experts discuss a variety of related functions, e.g., individual recognition of the others, understanding the intention of others and knowledge or "false belief" attribution.

Individual recognition of conspecifics has often been regarded as being closely related to or a necessary prerequisite for ToM. Individual recognition has been documented in a wide variety of animals including insects, fish (see preceding chapter), bullfrogs, rodents, horses, sheep, dogs, dolphins, a number of birds including corvids, and, particularly, primates (Seyfarth and Cheney 2008; Bates and Byrne 2010). The contexts in which the ability for individual recognition of conspecifics occurs varies greatly, e.g., discriminating neighbors from nonterritorials, maintenance of dominance hierarchies, mating interactions, and cooperativity. Individual recognition of conspecifics may be based on just one cue, e.g., voice (as in the case of the bullfrog) or odor, or on more complex stimulus arrangements, like faces or body movement. Some taxa, like sheep, dogs or primates are even able to recognize individuals from other taxa, e.g., humans. However, all this does not necessarily include a deeper knowledge of the way of thinking or feeling of other individuals.

The question of whether or not nonhuman animals possess a ToM, i.e., the ability to understand another individual's mental-emotional state, is hotly debated, as is the related ability to ascribe to a conspecific a certain knowledge or false knowledge (or false belief) and to take both into account in the planning of one's own behavior. Knowledge attribution may occur in a two-step fashion, i.e., "I know that he knows," or a three-step fashion, such as "I know that he knows that I know." About 30 years ago, the two American primatologists Premack and Woodruff published the groundbreaking article "Does the chimpanzee have a theory of mind?" (Premack and Woodruff 1978). In the 1980s and 1990s, the US-American anthropologist Daniel Povinelli and others continued this kind of research. At first it seemed that at least chimpanzees possessed the ability to attribute certain knowledge to other chimpanzees or humans and to take that knowledge into account. Chimpanzees were confronted with the task of identifying, out of several cups, the one that was baited with food. Two people pointed toward one of the containers, trying to "help" the chimpanzees, but only one of the helpers could know the "truth," i.e., the one who had baited the container, while the other one had been out of the room or had been prevented from seeing the baiting. After long training, at least some of the chimpanzees gave correct answers (Povinelli et al. 1990, 1993). Rhesus monkeys completely failed at this task.

Later, however, Povinelli became skeptical and could find no convincing evidence for the existence of ToM and knowledge attribution in chimpanzees or other animals, but argued that his own findings could be better explained as the result of operant conditioning (Povinelli and Vonk 2003). Other primatologists strongly disagree and point to substantial drawbacks in the method applied by Povinelli and colleagues. Among them is the primatologist Tomasello, working at the Max-Planck Institute for Evolutionary Anthropology in Leipzig, Germany. Initially, Tomasello was skeptical, but—like other primatologists such as Richard Dunbar—now believes that chimpanzees possess at least some aspects of a ToM and knowledge attribution comparable to that of human children aged 3–4 (Tomasello et al. 2003).

O'Connell and Dunbar (2003) compared chimpanzees with a group of autistic children (assumed to lack ToM) and children at ages 3–6. "False belief" was tested using nonverbal tests. The chimpanzees performed better than autistic and 3-year-old normal children; they were equal to 4–5-year-old and inferior to 6-year-old children. This would corroborate the view that chimpanzees exhibit at least some aspects of ToM. At present, the existence and degree of ToM in nonhuman primates remains controversial. Call and Tomasello (2008) report that chimpanzees understand the goals and intentions as well as the perception and knowledge of others, but found no evidence for understanding false beliefs, while Penn and Povinelli (2007) argue that there is no evidence that nonhuman animals possess anything remotely resembling ToM.

Most relevant in this context are the experiments by Call and Tomasello on the behavior of chimpanzees in the so-called "Ultimatum Game—UG," which were undertaken to study the extent of cooperativity and fairness in humans and apes (see Jensen et al. 2007). In the UG, two individuals are assigned the roles of proposer and responder. The proposer is offered a sum of money and can decide whether to divide this sum with the responder at any ratio from 0–100 %. The crucial point of the UG is that the responder can accept or reject the proposer's offer. If the responder accepts it, both players receive the proposed division; if the responder rejects it, both get nothing. On the basis of the classical "rational choice theory," one would expect that the proposer will offer the smallest possible share and that the responder will accept any non-zero offer. However, this is not what happens with human players. While results vary across cultures and settings, proposers typically make offers of 40–50 % and responders routinely reject offers under 20 % (cf. Sanfey et al. 2003; Camerer 2003). This finding is usually interpreted in the sense that humans are sensitive to fair offers and punish those that make unfair offers by rejecting those offers even at a cost to themselves. This kind of behavior has gained some fame under the title of "altruistic punishment" (de Quervain et al. 2004).

Tomasello and colleagues tested chimpanzees in a "mini-ultimatum game," where the animals had to cooperate in order to get food, which the proposer could then share with the responder at ratios varying from 0:10 (i.e., the responder gets all) to 10:0 (i.e., the proposer gets all), and the responder could reject the offer. In contrast to human players, the chimpanzee responders accepted offers lower than 20 % and even zero offers, while—again unlike human responders—not showing tantrums or other signs of arousal. This finding is difficult to interpret. It could be that proposers and responders offer and accept, respectively, the zero offer because of lack of empathy or ToM (they are simply disinterested in the mind or emotions of conspecifics), or they act egoistically despite their ability for ToM. It has consistently been found that at least among primates, "true" altruistic behavior, i.e., helping others even if there is no immediate gain, is found only in humans and already occurs at an early age (Harbaugh et al. 2007).

The ability for "mind-reading" may not be exclusively found in primates. Recently, Bugnyar, from the Lorenz Forschungsstelle in Gruenau, Austria, reported that in a food caching experiment, a raven was able to take into account what another raven has seen or not seen, while a human experimenter was hiding food in a cache. Ravens were quicker at pilfering the human-made caches when facing a fully informed rave competitor compared to a partially or noninformed one (Bugnyar 2010). The author interprets his finding as a "precursor step to a human-like understanding of the others' mind."

The ability of humans for empathy, ToM, and imitation is often implicated with the existence of so-called *mirror neurons*. A group of Italian neurophysiologists at the University of Parma, among them Vittorio Gallese and the group leader Giacomo Rizzolatti, discovered this type of neurons in the premotor area F5 of the cortex in macaques (Gallese and Goldman 1998; Rizzolatti et al. 1996). These mirror neurons respond to goal-directed movements executed by the subjects themselves or observed in others, particularly grasping, manipulation, and placing of objects. The function and meaning of the activity of these neurons is still unclear. Initially it was believed that they were "imitation neurons," enabling the animal to copy certain hand movements and object manipulations. Such a view, however, has to struggle with the restricted presence of imitation in monkeys and their apparent disinterest in the intentions of conspecifics (cf. Corballis 2010). A possible interpretation is that the mirror neurons help macaques to understand meaningful and goal-directed actions of conspecifics, while in humans and apes they may indeed support imitation. However, in humans empathy involves completely different brain regions such as the anterior cingulate, medial prefrontal and orbitofrontal cortex, and, above all, the insular cortex (T. Singer et al. 2004). At present, many authors believe that empathy, the sense of fairness and justice and "true" altruism have evolved during the evolution of humans.

The fact that in humans, "imitation" or "empathy" neurons are found in the vicinity of Broca's speech center gave rise to interesting speculations about the question of how gestures and their observation had something to do with the evolution of human language. A recent review article by Lotto and colleagues (Lotto et al. 2009), however, came to the conclusion that there is no direct link between the mirror neurons in monkeys and the evolution of human language. I will come back to that topic in Chap. 15.

13.4 Conscious Attention

In Chap. 1 of this book, I addressed the question whether or not animals have consciousness and how to prove it. For sake of simplicity, in the following I will concentrate on consciousness in the form of focused attention, which is one prominent state of consciousness in humans. I will not discuss whether or not consciousness and attention can be regarded as two different states, because they are at least closely interlinked (cf. Koch and Tsuchiya 2007). In order to test for focused attention in animals, we begin with the idea that humans need focused attention to solve certain cognitive tasks, e.g., follow a sequence of variable items (words, letters, objects) in the presence of distractors. Consequently, we can confront animals with such or similar tasks. In addition, we can study whether or not the same brain regions are active in these animals while performing such tasks that are known to be involved in humans. The latter, of course, makes sense only in those animals (primarily primates) that have brains that are similar in anatomy and physiology to our brains.

The German psychologist Wolfgang Köhler (1887–1967) was the first to experimentally test tool use and fabrication abilities of chimpanzees on Tenerife during the First World War. Köhler was followed by my teacher Bernhard Rensch (1900–1990) from the University of Münster, Germany, and Rensch's favorite subject was *Julia*. As shown in Fig. 13.1, in a typical experiment Julia had to draw an iron ring out of a wooden maze covered with a glass plate by using a magnet. She could choose between two starting points, one at the left and one at the right side, of which only one led out of the maze, and she had only one move. Rensch and his collaborator Döhl (Rensch and Döhl 1967) started with simple mazes, but eventually confronted Julia with rather complex ones, which we humans can master only after carefully "wandering" with our gaze through the maze. Julia did exactly that and in most cases (86 %) chose the right path. Rensch (1968a, b) interpreted these findings as clear evidence that at least chimpanzees possess conscious awareness and can solve problems mentally.

My colleague Andreas Kreiter and his collaborators from the University of Bremen made extensive studies on the neural basis of attention in macaques (cf. Taylor et al. 2005). In a typical experiment, the animal is sitting in front of a screen where two series of objects are shown, one to the left and one to the right which undergo stepwise changes in shape ("morphing"). The monkey has to concentrate on one of the two series (the other serves as a distractor) and touch a lever, as soon as in that series of morphing objects a certain shape reappears that had been taught

Fig. 13.1 Chimpanzee Julia solving a maze task. At the starting point, Julia had to decide to which side to draw a metal disk using a magnet in order to move it out of a simple (*above*) or complex (*below*) maze. The maze was covered with a sheet of *acrylic glass*. Before making the move (mostly correct), she looked at the maze intensely for some time. From Rensch 1968a, b

to be the target. This is a difficult task even for human subjects, because one is easily distracted by the other series of morphing objects and requires full concentration. However, after relatively long training, the monkey masters this task perfectly. Based on the behavior of the animals and their performance, there can be no doubt that during this task they have conscious experience in the form of focused attention apparently equivalent to that of humans.

At the same time, the neuronal activity within small regions in the visual cortex of the monkeys, here area V4 (belonging to the associative visual cortex involved in the recognition of shape and color) is being recorded using a multielectrode set. What Andreas Kreiter and his colleagues found is that in the moment of recognition of the reappearing target stimulus, there is synchronous and oscillatory activity among the recorded neurons within a range of 30–70 Hz (the gamma band), and this kind of activity disappears as soon as the monkey does not concentrate on the target stimulus (Taylor et al. 2005). Such synchronous, oscillatory activity of neurons has long been assumed to be involved in focused, conscious attention (cf. Engel et al. 1991). There are numerous additional results from other laboratories showing that during experiments requiring visual attention, the activity of neurons significantly increases in amplitude (Treue and Mounsell 1996; Kastner and Ungerleider 2000). It can be assumed that both an increase in amplitude and synchronous oscillatory activity are two neuronal states closely linked to the state of focused attention that lead to an increase in the signal-to-noise ratio and/or to other kinds of improvement of information processing inside the specific visual area.

Critics may object that such experiments merely demonstrate that macaques and chimpanzees are able to master cognitive tasks which in human subjects are connected to focused attention, and that there are similarities in the neuronal activities occurring in comparable regions of the visual system. Whether the animals have the same subjective experience as we humans do, must remain uncertain, but experiments in primates on "blindsight" makes this more likely. In humans, lesions in the primary visual cortex (Brodmann area A17) result in the inability to consciously perceive objects or situations in front of them—the patients "see nothing" (Weiskrantz 1986). However, if they are urged to reach out for the "invisible" object in front of them (e.g., a coffeepot), they do this correctly, even if they consider that what they are doing is absurd—why should they reach out into an empty space!

A possible explanation for the phenomenon of blindsight is that the so-called ventral pathway of the visual system involved in the conscious perception of objects and colors (cf. Chap. 11) had been lesioned before, while the so-called dorsal pathway, where information for spatial orientation and the guidance of arm and hand movements is processed, had remained intact. This interpretation is corroborated by the fact that patients can recognize objects when these are heavily moved. An alternative explanation is that spatial visual attention and related working memory functions are disturbed.

In 1991, Cowey and Stoerig carried out experiments with macaques with a unilateral lesion of the primary visual cortex (V1). These animals revealed a unilateral "blindsight," i.e., they behaved as if they saw "nothing" in the visual hemifield contralateral to the lesion, while their visual recognition abilities were unimpaired in the other hemifield. This strongly suggests that the monkeys have the same disturbance of subjective conscious perception as the human patients.

Another test for conscious experience is "binocular rivalry." Here, to the left and right eye of a human or nonhuman primate subject, two different pictures are shown simultaneously that cannot be fused into one (e.g., three-dimensional) picture—for example, one picture with horizontal and another one with vertical stripes. Since fusion is impossible, the human observer perceives that the two pictures alternate, i.e., in one moment they see the horizontal and a few seconds later the vertical lines, but never a picture with both types of line.

In carrying out these kinds of experiments with nonhuman primates, the neurobiologists David Leopold and Nikos Logothetis (1996) trained monkeys to touch a lever, whenever they perceive one of two types of pictures (e.g., the one with horizontal lines) and not the other. As expected, the monkeys pressed the lever in roughly the same rhythm as humans do, which can only be interpreted in the way that the monkeys have bistable conscious visual perception in the same or very similar manner as the human subjects do. Additionally, similar changes, as in the human EEG, occurred in the neuronal activity of visual area V4. All this makes the assumption reasonable that not only apes, but also macaques and other monkeys have a conscious perception of visual stimuli in a way similar to that of humans, and that circumscribed lesions of the cortical areas involved lead to predictable disturbances.

With much greater efforts one can conduct similar experiments with other mammals with sufficiently large and well-studied brains, e.g., cats and dogs, and it is very likely that they, too, possess at least some of the states of consciousness found in humans. In smaller mammals (e.g., rats), this is even more difficult, and in birds, reptiles, amphibians, and fish, comparisons of brain activity are difficult due to the lack of a cortex, and we would mostly rely on behavioral data. In large-brained birds, experiments on "binocular rivalry" should not be difficult to carry out. The way a salamander fixates on unfamiliar prey object, e.g., a mealworm, looking at one end and then at the other end of the worm, moves its head forward and backward, before snapping (or not snapping), gives the strong impression that this animal, too, possesses some kind of focused attention. As described above, neuronal mechanisms have been identified that supposedly underlie these states.

There remains, of course, the question of whether or not animals possess states of "higher consciousness" like self-reflection or "ego-identity." This question is difficult to prove, but it is not unlikely that at least in the great apes we find some aspects even of these states. As we will learn in Chap. 14, it is reasonable to compare the cognitive and mental abilities of these highly intelligent animals to those of young children aged 3–5.

13.5 How Intelligent Are Dolphins and Elephants?

The question about intelligent mammals with very large brains, such as dolphins, whales, and elephants, really are has interested behaviorists for a long time, and, more recently neurobiologists as well. Since antiquity there have been many myths and speculations about the mental powers, especially of dolphins. Even today there are popular statements that these animals have an intelligence that is "far beyond" that of humans.

Dolphins (*Delphinidae*), with about 40 species, are the largest family of the suborder of toothed whales (*Odontoceti*) and of the entire order *Cetacea*. They are characterized by splendid and often acrobatic motor skills, are playful, friendly, highly social, and possess a highly evolved sound-producing and sound-perceiving system using frequency-modulated whistles (mostly for intraspecific communication) and clicks (mostly for echolocation). The bottlenose dolphin (*Tursiops truncatus*) is able to recognize individual conspecifics. Besides their excellent auditory system occupying large parts of their cortex, dolphins have a well-developed visual system. They can be trained to invent novel or "creative" behaviors and often produce water bubble rings not only for foraging on fish, but also "just for fun" and to play with the bubbles. Particularly regarding their large to very large brains reaching 8–10 kg in the "False Killer Whale" (*Pseudorca crassidens*), which in fact is a dolphin, one is inclined to expect equally developed cognitive-intellectual abilities.

Controlled experiments on the alleged super-intelligence of the dolphins yield somewhat mixed and often disappointing results. Their problem-solving abilities are by no means outstanding. The German-Argentine behaviorist Lorenzo von Fersen, who had studied the behavior of dolphins over decades in the zoo of Nürnberg, Germany, and Onur Güntürkün from Bochum University, regard the intelligence of dolphins to be equal of that of pigeons and rats (Güntürkün and von Fersen 1998). Although dolphins are able to distinguish between objects differing in shape, they are incapable of categorization, for example, of distinguishing between "round" and "triangular" objects and assigning an unfamiliar round or triangular object to one of these two categories—something that pigeons, crows, parrots, dogs, any kind of primates, and even bees are capable of. There is anecdotal evidence for tool use. As mentioned above, Reiss and Marino demonstrated that bottlenose dolphins exhibit mirror self-recognition (2001), although—as in similar cases—the interpretation of the results of this experiment is debated.

Elephants are the largest living terrestrial animals, and males regularly reach weights up to 6 tons or more. They form two genera, *Loxodonta* (*L. africana*, African bush elephant, and *L. cyclotis*, African forest elephant) and *Elephas* (*E. maximus*, Asian or Indian elephant). Besides their body mass, elephants stand out by their high sociality. Females spend their entire lives in tightly organized family groups formed by mothers, daughters, sisters, and aunts, and the group is led by the eldest female, called "matriarch." Adult males, in contrast, live mostly solitary lives.

Elephants are able to produce and use sound ranging from 5 to 9,000 Hz. While the high frequencies occur in high-frequency barks and trumpets, the former often occur in the infrasonic range up to two octaves below the lower limit of human hearing. They use these infrasounds as substrate-bound contact calls, which may travel over 19 km, in order to localize group members, but they are also able to hear distant events (rain showers, thunderstorms) over much larger distances, and this may help them find freshwater sources. Like dolphins, at least the Asian elephant exhibits a high teachability, which in Asia has been exploited by humans for thousands of years. In addition, they possess a magnificent spatial orientation that enables them to head for water ponds as distant as 60 km. Likewise, they can recognize human individuals after decades, as Bernhard Rensch and Rudolf Altevogt have demonstrated (Rensch and Altevogt 1955).

All this stands in sharp contrast to their rather unimpressive cognitive abilities (Hart and Hart 2007; Bates and Byrne 2010). As one example for tool use (already mentioned by Darwin), elephants use sticks for scratching their bodies and removing ticks, use bushes for fly-swatting, which they modify until they are long and effective enough, they throw mud or stones at rodents or humans. As to learning abilities, Rensch und Altevogt (1955) report their cumbersome work of carrying out a simple operant conditioning experiment teaching an elephant from the zoo of Münster, Germany, to distinguish between black and white or small and large objects. The ability for mirror self-recognition has been mentioned above, but as in dolphins, the results are somewhat equivocal. Also, there is no positive evidence that elephants possess a ToM in the above sense. This stands in some contrast to the ability to recognize a large number of conspecifics and humans using a variety of auditory, visual, and chemical cues and to exhibit over a hundred different gestures (Bates et al. 2007).

Many authors have speculated why dolphins and elephants are not as intelligent as expected. In dolphins, the lack of hands, and accordingly, hand use could have played a highly restrictive role, but the marine environment may not have been stimulating enough, as was the savanna for human ancestors (cf. Chap. 15). Elephants have a prehensile trunk, but it is not nearly as useful as the primate hand and cannot exert the "precision grip." I will comment on the importance of the hand in Chap. 15. In the next chapter, I will turn back to this question under the aspect of the large brains of dolphins and elephants.

13.6 What Does All This Tell Us?

We are now in a position to reject the hypothesis by MacPhail that was cited at the beginning of the previous chapter, saying that within vertebrates and with the exception of humans, there are no differences in intelligence. Comparative studies demonstrate that there are, of course, clear differences among vertebrate taxa at all taxonomic levels, from classes and orders to families, genera, and even among species and individuals. These differences, however, do not represent any sort of

13.6 What Does All This Tell Us?

"scala naturae," which means that there is no linear tendency "from fish to man." Rather, high intelligence has evolved many times independently. Nevertheless, there are evolutionary tendencies at a gross level.

Among craniate-vertebrates, the lowest levels of intelligence are found in hagfish and lamprey, followed by amphibians, which, however, may exhibit signs of "higher cognitive abilities," such as focused attention in visual object selection. Likewise, "reptiles" are not famous for superior intelligence, although detailed investigations are scarce. However, in some teleost groups, such as cichlids, some authors identify signs of "primate-like" intelligence, and the communicative abilities of intraspecific communication of weakly electric fish are spectacular. Unfortunately, there are no studies on the cognitive abilities of cartilaginous fishes, despite the fact that some groups have extraordinarily large brains compared to body size.

Many groups of birds and mammals show high levels of intelligence. Among birds, parrots and corvids stand out (cf. Lefebvre et al. 2004), and among mammals this holds true for cetaceans and particularly toothed whales including the dolphins, as well as for elephants and primates. However, many other groups of mammals, like dogs, bears, and even rats likewise reveal considerable intelligence. Within primates, prosimians exhibit manipulatory, perceptual, and cognitive capacities, although often only as a basic ability. New World monkeys possess moderate to well-developed capacities in various cognitive domains that partially overlap with those of Old World monkeys. The behavior of the latter shares characteristics with apes, although great apes clearly outperform the other nonhuman primate taxa in most respects. Clearest differences exist between prosimians (lemurs), monkeys, and apes, including humans. Whether there are substantial differences among the great apes, i.e., orangutans, gorillas, and chimpanzees, is unclear—many authors, e.g., Byrne, would consider the two chimpanzee species to be more intelligent than gorillas and orangutans. Taken together, this survey of recent data suggests that intelligent behaviour is distributed across vertebrates in a much more overlapping as well as mosaic manner than previously thought. The same "gradualist" view can be applied to "higher" mental abilities like theory of mind, knowledge attribution, metacognition and consciousness, which some decades ago had been believed to be restricted to humans. One important function has not yet been addressed—language. I will do this in Chap. 15.

In summary, levels of high intelligence based on systematic observations and experiments include the following abilities:

1. To adopt the perspective of conspecifics, e.g., in the context of deception and counter-deception. This is found in primates and a number of other mammals and birds (e.g., at food caching).
2. To anticipate future events, e.g., in the context of tool fabrication (including modification of natural objects) for future use. This is found predominantly in apes, but also in corvid birds.
3. To understand underlying mechanisms of processes, e.g., at fabrication and use of tools. This is clearly present in apes, while data from birds are equivocal.

4. Self-recognition in a mirror, which has been demonstrated unequivocally and systematically only in the great apes, and has been found in a few individuals of dolphins, elephants, and corvid birds.
5. Knowledge attribution and a theory of mind, which is found—albeit in preliminary form—only in the great apes.
6. Metacognition, i.e., knowledge about one's own knowledge. This seems to exist in dolphins, monkeys (at least in macaques), and apes.
7. Consciousness. Presumably, many if not most, vertebrates, have some form of consciousness, e.g., in the form of focused attention, while "higher" forms of mental activity, like self-awareness and reasoning, may be restricted to (some) birds and mammals or even only to apes.

While humans turn out to be superior in all these abilities, there appear to be only quantitative, but no qualitative differences between humans and nonhuman animals. What is still lacking in this list is language. The question about whether or not human language is unique or has some predecessors among nonhuman animals will be discussed in Chap. 15

Chapter 14
Comparing Vertebrate Brains

Keywords Brain-body relationship · Absolute brain size · Relative brain size · Corrected relative brain size · Encephalization quotient · Extra neurons · Cortex information processing capacity · Cortex modularity · Number of cortical neurons · Cytoarchitecture cortex · Specialties cortex

In the preceding chapters, we have tried to figure out how vertebrates differ in intelligence including mental functions, and we arrived at a certain ranking order. In this chapter we will ask to what degree these differences in intelligence can be correlated with brain traits. In the past, there have been many such attempts. The first trait that comes to our mind is *absolute brain size* (grams/kilograms or cubic centimeters), because many experts, including Rensch, were convinced that "bigger is better," i.e., bigger brains mean higher intelligence. Another much-discussed trait is *relative size*, i.e., percent of body size, of the entire brain, or of alleged "seats" of intelligence like the mesonidopallium in birds or the cerebral cortex in mammals. Since it becomes clear that much of brain size is determined by body size, experts have tried to determine the degree of *encephalization*, i.e., brain size beyond the mass related to body size, e.g., Jerison's encephalization quotient or corrected relative brain size. One could also look for neurobiologically more meaningful traits like the number of neurons in the entire brain or in the "intelligence centers," the degree of connectivity, etc., relevant for "information processing capacity." Finally, one could look for "unique" properties that could best explain the observed differences in intelligence. Let us first study the significance of absolute brain size.

14.1 Brain Size and Body Size

Animals vary enormously in body size (volume or weight, which is directly convertible). Some invertebrates, like nematodes and mites, are so tiny that they cannot be detected with the naked eye, while the largest invertebrates, the giant squids (*Architeuthidae*), may reach a body-tentacle length of 15 m. The smallest

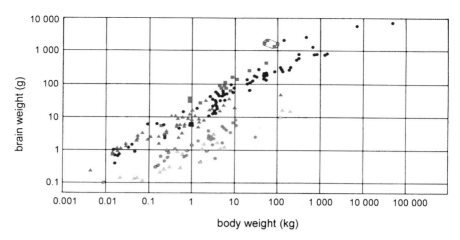

Fig. 14.1 The relationship between brain weight (ordinate, grams) and body weight (abscissa, kilograms) in 200 vertebrate taxa in double-logarithmic presentation. *Purple circles* bony fishes; *yellow triangles* "reptiles" *red triangles* birds; *blue circles* mammals except primates; *green squares* primates; *and encircled green squares* Homo sapiens. Further explanations in the text. From Jerison 1973, modified

vertebrates are found among teleosts and amphibians, with a body length well below 1 cm. The smallest mammal is the Etruscan (or pygmy) shrew *Suncus etruscus* with a body weight of 2 g, and the largest mammal and animal of all times is the blue whale (*Balaenoptera musculus*), with a length of 33 m and a body weight up to 200 t. The largest living terrestrial animal is the African elephant *Loxodonta africana* with a body weight of up to 7.5 t. Thus, among vertebrates (from the smallest fish to the blue whale) there is a range in body size or weight of about 11 and in mammals of 8 orders of magnitude.

The volumes or weights of the nervous systems and brains likewise vary enormously. As mentioned, for the study of the nervous system of a mite, we need an electron microscope. The smallest vertebrates (teleosts and amphibians) have brains that are less than 1 mm long and have a weight of less than 1 mg. The smallest mammalian brain, that of the bat *Tylonycteris pachypus*, weighs 74 mg in the adult animal, and the largest brains of all animals are found in the sperm whale and the "killer whale" (Orca) with up to 10 kg. Elephant brains have weights up to 6 kg. This is again an enormous range, here roughly of eight orders of magnitude in vertebrates and of five in mammals.

However, there are basic differences in average brain size across the different classes of craniates-vertebrates (cf. Figs. 14.1 and 14.2). Hagfish and lampreys generally have small to very small brains weighing between 16 and 50 mg, and even relative to body size, these brains are small. On average, hagfish and lampreys have brains that are ten times smaller than those of bony fish. Within the cartilaginous fish, chimaeras and squalomorph sharks have relatively small brains, while the brains of galeomorph sharks and myliobatiform rays are about ten times larger in relative terms. Teleost, in contrast, have small brains, both in absolute and relative

14.1 Brain Size and Body Size

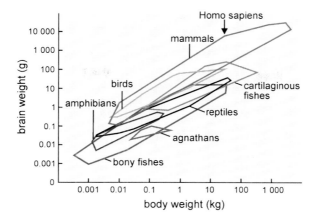

Fig. 14.2 The relationship between brain weight (ordinate, gram) and body weight (abscissa, kilogram) in the vertebrate classes in a double-logarithmic presentation using the polygon method developed by Jerison (see text). Mammals and birds generally have larger relative brain weights or volumina than "agnathans" (i.e., myxinoids and petromyzontids), bony fishes, amphibians, and "reptiles." The brains of cartilaginous fishes lie in between. The weight/volume of the human brain is on top of the distribution, when corrected for body size. After Jerison 1973, modified

terms, with the exception of weakly electric fish (mormyrids and gymnotids), the brains of which may occupy up to 20 % of the body mass. This is mostly due to a hypertrophy of their cerebellum, including their valvula cerebelli (cf. Chap. 8).

Amphibian brains are found in the upper range of teleosts both in absolute and relative terms, as are the brains of "reptiles" (i.e., lizards, turtles, snakes, tuatara, crodiles). Of interest are brain sizes of the extinct dinosaurs. Since brains do not fossilize, their sizes can be determined only through so-called cranial endocasts, i.e., the measurement of the cranial cavity (while taking into account the fact that in reptiles the brain does not fully fill the cavity). Jerison, a pioneer in such a procedure, assumes that the brain size of dinosaurs was within the range of extant "reptiles" (Jerison 1973), sometimes on the upper and sometimes on the lower limit. The "terrible" *Tyrannosaurus rex*, one of the largest land carnivores of all times, had a body weight of 7.7 t (more than a living elephant), but a brain weight of only 400 g, which is equal to the brain weight of a cow, and the giant *Brachiosaurus brancai*, with an estimated body weighing 90 t, had a tiny 300 g brain. This is 1/10–1/15 of the brain of a mammal with the same body size. These very small brain sizes may be surprising, given the great biological success of the dinosaurs, which dominated the animal kingdom throughout the Mesozoic, i.e., for about 200 million years. A similar situation is found in the bony fishes with absolutely and relatively small brains, as we have already seen, which are the most successful vertebrates in the number of species and variability of behavior.

Birds are surviving dinosaurs and closely related to crocodiles, but have brains that are 6–10 times larger than those of all "reptiles." Among birds, hummingbirds have the smallest brains in absolute terms (the smallest brain has a weight of 170 mg), but relative to body size, their brains are in the upper range.

Table 14.1 Brain weight, encephalization quotient, and number of cortical neurons in selected mammals

Animal taxa	Brain weight (in g)[a]	Encephalization quotient[b,c]	Number of cortical neurons (in millions)[d]
Whales	2,600–10,000	1.8	10,500
False killer whale	7,650		
African elephant	4,200–6,000	1.3	11,000
Homo sapiens	1,250–1,450[a]	7.4–7.8	15,000
Bottlenose dolphin	1,350	5.3	5,800
Walrus	1,130	1.2	
Camel	762	1.2	
Ox	490	0.5	
Horse	510	0.9	1,200
Gorilla	430[e]–570	1.5–1.8	4,300
Chimpanzee	330–430[e]	2.2–2.5	6,200
Lion	260	0.6	
Sheep	140	0.8	
Old World monkeys	36–122	1.7–2.7	840
Rhesus monkey	88	2.1	
Gibbon	88–105	1.9–2.7	
Capuchin monkeys	26–80	2.4–4.8	720
White-fronted capuchin	57	4.8	
Dog	64	1.2	160
Fox	53	1.6	
Cat	25	1.0	300
Squirrel monkey	23	2.3	450
Rabbit	11	0.4	
Marmoset	7	1.7	
Opossum	7.6	0.2	27
Squirrel	7	1.1	
Hedgehog	3.3	0.3	24
Rat	2	0.4	15
Mouse	0.3	0.5	4

[a] Data from Haug (1987), Jerison (1973) and Russell (1979). [b] Indicates the deviation of the brain size of a species from brain size expected on the basis of a "standard" species of the same taxon, in this case of the cat; [c] Data after Jerison (1973) and Russell (1979). [d] Calculated using data from Haug (1987). [e] Basis for calculation of neuron number

The absolutely largest bird brain is found in the ostrich with an average of 42 g, which, however, is in the lower range of relative brain sizes. Galliform birds (chicken, turkey, etc.) and columbid birds (pigeons, doves) have absolutely as well as relatively small brains, while song birds (Passeriformes), including corvids (crows, ravens, jays, magpies, nutcrackers, etc.) and parrots (Psittaciformes) have large to very large brains compared to the avian average. The much-studied New Caledonian crows (*Corvus moneduloides*, five specimens studied; Cnotka et al. 2008a) have a mean body weight of 277 g and a mean brain weight of 7.56 g. However, some parrot species, like *Ara arauna* and *A. chloroptera* as well as

Anodorhynchus hyacinthus, and also the black woodpecker *Dryocopus martius*, have absolute brain weights above 20 g and relative brain weights above corvids, although the differences are not significant (Iwaniuk and Hurd 2005).

Mammals, like birds, generally have brains that are about ten times larger than those of bony fishes, amphibians, and reptiles of the same body size (cf. Table 14.1). Within mammals, primates, with the exception of prosimians, generally have larger brains than the other orders with the same body size. In primates, brain size ranges from 1.67 g in the prosimian mouse lemur *Microcebus* to 1,350 g in *Homo sapiens*. Generally, prosimians and tarsiers have relatively small brains with a range of 1.67–12.9 g (average 6.7 g), followed by New World monkeys with a range of 9.5–118 g (average 45 g) and Old World monkeys with a range of 36–222 g (average 115 g), with the largest brains found in baboons. Among apes, gibbons have brain sizes (88–105 g) that lie within the range of Old World monkeys, while the great apes, i.e., orangutans, gorillas, and chimpanzees, have brain weights between 330-570 g (males).

Thus, in extant primates, we recognize five non-overlapping or only slightly overlapping groups with respect to brain size: (1) prosimians and tarsiers, (2) New World monkeys, (3) Old World monkeys and hylobatids, (4) the great apes, and (5) extant humans. The gap between non-human apes and humans is filled by brains of extinct australopithecines (e.g., *Australopithecus afarensis*, *A. africanus*) with reconstructed brain sizes of 343–550 g., *H. habilis* with brains of 550–780 g, and *H. erectus* with brains of 909–1149 g. (Jerison 1973). The largest hominine brain, that of *H. neanderthalensis*, had a mean weight of 1,487 g. (Falk 2007).

In summary, we learn that brain size varies enormously both within and across phyla, classes, and families. At the same time, we recognize that the basic assumption that "bigger is better" does not hold for intelligence. First, there are many animals, like corvid and psittacid birds, which have much smaller brains compared to other members of their taxa, and yet are at least equally or more intelligent. Also, within mammals, monkeys have much smaller brains than ungulates, and humans much smaller brains than whales and elephants and without a doubt are more intelligent. However, there are groups like primates, where "bigger is better" appears hold true, as we will learn further below.

14.2 The Significance of Relative Brain Size and of "Encephalization"

When comparative neurobiologists became aware that humans do not have the largest brain of all creatures (as is still often stated in non-neurobiological textbooks), they started looking at other brain characteristics, where humans could excel, and believed to have found that they possess the largest brain relative to body size (this is even more often stated in non-neurobiological textbooks). However, as we will learn in this chapter, this likewise is not correct.

What is the general relationship between an increase in body size and in brain size? One could assume that an increase in body size (volume or weight) is accompanied by a *proportional* increase in brain size, because the brain is involved in the control of the body and a larger body may require more brain mass. When such a proportional increase occurs, we speak of *isometric* growth, which means that the proportions between body and brain volume (or weight) remain the same. However, as we have already heard, among vertebrates this often is not the case: with respect to body size, we find an increase by 11 orders of magnitude, while brain size increases "only" by 8 orders of magnitude and among mammals a relationship of 8:5. This means that an increase in brain size dramatically "lags behind" an increase in body size. But the opposite may also happen in the sense that brains or parts of them, like the cortex, increase *faster* in volume or weight than the body. In both cases, we speak of *allometric* brain growth; in the former case, of *negative*, and in the latter of *positive* allometric brain growth.

When we compare the overall relationship between body size and brain size across all vertebrate classes, as is illustrated in Figs. 14.1 and 14.2, we easily see that this overall relationship is *negatively allometric*. In Fig. 14.1, the body-brain relationship (BBR) for 200 vertebrates is shown including data from teleosts (purple circles), reptiles (yellow triangles), birds (red triangles), mammals (blue circles), and primates (green squares), including man (4 measurements, encircled four green squares). The figures show the data in a double-logarithmic representation, which makes a nonlinear function, here a power function, linear.

The general power function for the BBR in vertebrates is $E = kP^\alpha$, in which E and P are brain and body weights or volumes, respectively, and k and α are constants; "k" is a proportionality factor, the meaning of which will become clear in a moment, and α is the allometric (or scaling) exponent, which indicates how strong the brain grows compared to body growth. With $\alpha = 1$ we would have an isometric growth, whereas $a > 1$ would indicate a positive and $a < 1$ negative allometry. In double-logarithmic transformation, we obtain the linear equation log $E = \log k - \alpha \log P$, where k is the intercept with the y-axis and α is the slope of the line.

The exact value of α is still a matter of debate. For vertebrates in general, von Bonin (1937) found a value of 2/3, which was confirmed by Jerison in his famous book *Evolution of Brain and Intelligence* from 1973, and for Jerison this relates to the fact that with an increase in volume, the body surface increases by 2/3. Here, he followed Snell (1881) by arguing that the most important factors of a brain are the sensory surfaces of the body and the processing of the information coming from them. However, already in his 1973 book, Jerison mentioned that there are differences in α across the different vertebrate classes. Later measurements confirmed such differences: for "reptiles," one finds a lower value of 0.53, and for birds and mammals, a higher value between 0.68 and 0.74. In primates, $\alpha = 1$ was found, which would indicate an *isometric* growth of brain size (Herculano-Houzel 2009, 2012; Herculano-Houzel et al. 2006). Finally, in extinct hominins plus living *Homo sapiens*, α amounts to 1.73 (Pilbeam and Gould 1974), which is the steepest increase in size during the entire brain evolution. However, so far there is no convincing explanation for the differences in α.

In Fig. 14.1 we recognize that teleosts, "reptiles," birds, mammals, and primates differ not only in α, but also in the proportionality factor k. The latter has the consequence that their scatter plots overlap only partially. This is better seen in Fig. 14.2, where the "minimum convex polygon" method developed by Jerison (1973) is applied. We arrive at such "minimum polygons," when we draw the shortest line around the data points from a group of vertebrates (classes, families, etc.) like a thread drawn around an area filled with needles. The figure shows polygons for "agnathans" (i.e., hagfish and lampreys), bony fish, amphibians, reptiles, cartilaginous fish, birds, and mammals. The message we have already obtained from Fig. 14.1 becomes clearer now: the single polygons, while having roughly the same slope of their long axis, are in part displaced with respect to each other. Those for hagfish and lampreys ("agnathans"), bony fish, amphibians, and reptiles considerably overlap, and the same is true for birds and mammals, but between these two larger groups there is no overlap. However, an overlap exists in form of the polygon for the cartilaginous fish, which results from the already mentioned fact that some groups of cartilaginous fish have brains similar in size to those of bony fish, whereas other groups have relative brain sizes reaching bird and mammalian levels.

All this tells us that groups of vertebrates may differ *both* in "k" and "α", which means that first there are groups of vertebrates (birds, mammals) that *generally* have larger brains, often by one order of magnitude, than others (bony fish, amphibians, reptiles), and second that the increase in brain size with increasing body size is *faster* in birds and mammals (and fastest in hominins) than in the other vertebrate classes. Overall, an increase in body size is "followed" by an increase in brain size with an allometric or scaling exponent α around 2/3 (Jerison 1973). As a consequence, we can arrive at the following three fundamental statements regarding BBR: (1) small animals have small brains and large animals large brains in absolute terms, (2) small animals have larger brains and large animals smaller brains relative to body size, and (3) up to 90 % of an increase in brain size, depending on the taxa under consideration, can be explained by increase in body size. Thus, animals mostly get absolutely large brains by becoming large, while their brains become relatively smaller!

This may sound trivial, but it is not, because it tells us two important things. First, since increase in brain size is to a major degree a consequence of increase in body size, an increase in brain size often is not the primary target of selection. Selective advantages for becoming large are quite numerous: the ratio between body size and body surface decreases, because body volume increases by a cubic function, while surface increases by a square function. A decrease in relative surface is favorable for thermoregulation and for nutrition (large animals need less food per unit of body weight). Furthermore, large animals tend to have fewer predators and can move faster. However, while many groups of animals actually have become larger over millions of years, others have done just the opposite and become small to very small, and again there are many advantages to becoming very small. Many mites have become so small that they cannot be detected with the naked eye, and are one of the most successful animal groups; the same is true for many nematodes. We see that in evolution there is no universal recipe for best survival. At any rate, it appears that in many cases, brains just became large without specific selective pressures by following the described negative brain allometry.

In this context, those cases are particularly interesting, where increases in brain size *positively* deviate from this standard rule. As already mentioned, there was a tenfold increase in the galeomorph sharks and myliobatiform rays among cartilaginous fishes, again a tenfold increase in birds with respect to ancestral sauropsids, and a roughly sixfold increase in corvids and parrots within birds, and a sixfold to more than tenfold increase in brain size in simian primates with respect to other mammals. Finally, as we will see, there was a dramatic increase in relative brain size in the lineage leading to *Homo sapiens* and *Homo neanderthalensis*.

In Fig. 14.3 we get a closer look at the situation in mammals, again in a double-logarithmic representation. The long axis (i.e., the regression line) drawn through the data polygon has a slope of 0.74, which is typical of the BBR in mammals. We recognize that the values for some shrew and mice species, for dogs, horses, and the African elephant lie more or less exactly on the line and therefore represent the average of mammals. The values for other mice species, for chimpanzees, humans, but also for dolphins, lie above the line and accordingly represent BBRs above average, while those for some other shrew species, bats, hedgehog, pig, hippopotamus, blue whale, and sperm whale are found below the line and accordingly

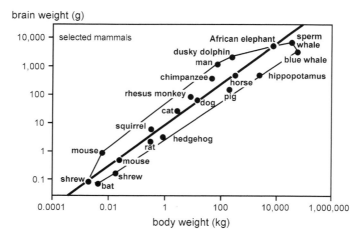

Fig. 14.3 The relationship between brain weight and body weight in mammals in double-logarithmic presentation. Some species of shrews, mouse, dog, horse, and African elephant have "average" brain weights; accordingly, their data points lie exactly on the regression line. Chimpanzees, humans, but also other species of mice and dolphins have brain weights above average, while some species of bats and shrews, hedgehog, pig, hippopotamus, blue whale, and sperm whale brain weights below average. After Nieuwenhuys et al. 1998, modified

14.2 The Significance of Relative Brain Size

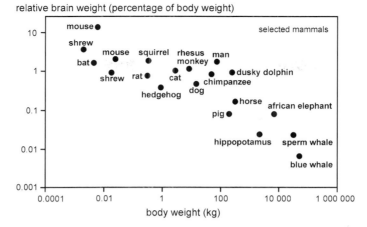

Fig. 14.4 The figure shows brain weight as a percentage of body weight for the same 20 mammalian species as in Fig. 14.3, again plotted in log-log coordinates. As can be seen, small mammals such as mice and shrews have much larger brains in relative terms (10 % or more of body weight) than cetaceans (less than 0.01 %). Humans, with a brain occupying 2 % of body weight, have a much higher relative brain size than expected. From van Dongen in Nieuwenhuys et al. 1998, modified

represent BBRs below average. The value for humans lies farthest above the regression line, which means that we humans have a brain size that is the largest one compared to mammalian average. I will return to this point later.

In Fig. 14.4, again in double logarithmic representation, we find the data from the same 20 groups of mammals as in Fig. 14.3, but now the y-axis does not indicate the absolute, but the *relative* brain weight as expressed in percent of body weight. We clearly recognize the fact that with increasing body weight, the relative brain weight decreases from more than 10 % in very small mammals to less than 0.005 in the blue whale. The human brain again ranks relatively high with roughly 2 % of body weight, but is close to that of apes and dolphins.

Thus, according to negative brain allometry, small animals tend to have relatively larger, and large animals relatively smaller brains. If this had any impact on intelligence, then the tiniest animals should be the smartest ones, which apparently is not the case in any vertebrate class or order. What we find, however, is more complicated: small animals with absolutely small brains can be surprisingly intelligent and large animals with absolutely large brains less intelligent than expected, as described in the previous chapters.

Harry Jerison was among the first who discovered that within classes (birds, mammals) or orders (e.g., primates), animals of the same body size may exhibit large differences in absolute as well as relative brain size. If we find a reference value for that class or order, e.g., *average brain-body ratio*, then we can assess to what degree brain size of a given species lies above or below that reference value, i.e., is extraordinarily large or small. He therefore tried to illustrate the observed deviations of brain sizes from average by calculating what he called the "encephalization quotient," EQ, according to the formula EQ = E_a/E_e. This quotient indicates the extent to which the relative brain size of a given species E_a deviates from the expected or average relative brain size E_e of the larger taxon (genus, family, order, etc.) under consideration. Accordingly, among mammals an EQ of 1 tells us that the mammal under consideration has a brain with (more or less) average relative size regarding all mammals investigated, while an EQ larger than 1 indicates that a brain is larger, and an EQ below 1 that a brain is smaller than expected, given a certain body size.

Jerison's findings are given in Table 14.1. This table tells us that rabbits, for example, have a brain size considerably below average, closely followed by mice and rats. The latter finding is rather surprising, because in view of the alleged intelligence of rats, we would expect a relative brain size above average. The cat has an average relative brain size and according an EQ of 1, while dogs, but also camels, fox, gorillas and whales have EQs slightly of above average. The value for the gorilla is surprisingly low, given the undoubtedly high intelligence of that species. Interesting is how the relatively low EQ of whales correlates with their only moderate degree of intelligence (cf. Chapter 12). Among primates, Old World monkeys have higher EQs on average than New World monkeys, with the exception of the capuchin monkey, and the highest EQs are found in dolphins and finally humans, the latter with an EQ of 7.4–7.8, meaning that the human brain is roughly eight times larger than an average mammal of the same body size. The high EQs of humans is of no surprise, and the likewise high EQ of the capuchin monkey is not unexpected in view of its high cognitive abilities and deserves further explanation.

Partly for these inconsistencies, Jerison (1973) tried to make his calculations again more realistic by distinguishing between brain parts necessary for the maintenance and control of the body (E_v) and those associated with improved cognitive capacities (E_c), in mammals mostly the cortex, which Jerison called "extra neurons" (N_c). The idea behind it is that the neuronal "expenses" for the control of a large body are not nearly as high as the "expenses" for the processing of complex sensory data and related cognitive functions. Therefore, we expect an increase in behavioral intelligence to be paralleled by an increase in "extra neurons." Calculating the number of such "extra neurons" removes some striking inconsistencies in the EQ list. For example, while the New World monkeys *Cebus albifrons* and *C. apella* have unusually high EQs compared even with large-brained apes, their N_c is much lower than that of the latter, and even lower than that of the Old World monkeys, which is consistent with their lower levels of intelligence. As with EQ, in extra neurons there is a huge gap between the great apes (3.2) and humans (8.8 for male humans) which, however, can be filled by an average N_c of 3.9 in australopithecines (Jerison 1973).

More recently, experts in brain allometry adopted a slightly different method to correct the relative brain size for body size, i.e., to remove that portion of observed brain size that is simply due to negative brain allometry (cf. Lefebvre et al. 2004; Lefebvre and Sol 2008; Lefebvre 2012). These authors start with a linear regression of brain size against body size in a given taxon (e.g., birds), and then measure the deviation of data points from the regression line, the "residuals." However, this method likewise does not solve the main problems of taking relative brain size even after correction for body size. For example, among birds, corvids and psittacids (parrots) have roughly equal values for corrected relative brain size that lie far above average, and both groups are considered to be comparably intelligent. However, parrots, as mentioned above, on average have larger to much larger brains than corvids, in absolute terms. Dolphins exhibit much larger corrected relative brain sizes than gorillas and even chimpanzees, but are not nearly as intelligent as the former. Thus, the relationship between brain size and body size turns out to be much more complicated than previously thought. It could be that it is not absolute or relative size of the total brain that counts for intelligence, but only certain parts, in mammals, above all the cortex, as the part of the brain that is believed to be most closely related to intelligence.

14.3 The Fate of the Cortex as the "Seat" of Intelligence and the Mind

14.3.1 Information Processing Properties of the Cortex

Various authors have tried to determine to what degree different parts of the vertebrate or mammalian brain changed in size relative to the size of the entire brain. Studies by the Canadian neurobiologist G. Baron (cf. Baron 2007) give evidence that while there is an increase in overall brain size in mammals, the olfactory bulbs and the medulla oblongata decrease relatively. In contrast, the cerebellum increases relatively, but is exceeded by the isocortex, which is on top of growth dynamics.

With increasing brain size in mammals, cortices increase in surface as well as in volume. The smallest mammals, shrews, have a total cortical surface of 0.8 cm^2 or less; in the rat we find 6 cm^2, in the cat 83 cm^2, in humans about 2,400 cm^2, in the elephant 6,300 cm^2, and in the "false killer whale" (*Pseudorca crassidens*) a maximum of 7,400 cm^2. Thus, from shrews to false killer whales we find a nearly 10,000-fold increase in the cortical surface, following exactly the increase in brain volume at an exponent of 2/3, as expected.

This dramatic increase in brain surface contrasts with a very moderate increase in cortical thickness, i.e., from 0.4 mm in very small shrews and mice to 3–5 mm in humans and the great apes. The large-brained whales and dolphins have surprisingly thin cortices, between 1.2 and 1.6 mm, and even the elephant, again with a very large brain, has an average cortical thickness of "only" 1.9 mm. If we compare cortical volume (surface times thickness) across mammals and ask for its relationship to brain size, then we recognize that the cortex grows faster than the rest of the brain, i.e., in a *positive allometric fashion* with an average exponent "*a*" of 1.13 (Changizi 2001). This exponent is slightly larger in primates and slightly smaller, but still greater than 1 in hoofed animals, whereas in whales and sea cows, as well as in the elephant, it is below 1. This means that in these animals, cortical volume, while increasing in absolute volume, decreases in relative volume in a *negative allometric fashion*.

However, while looking for anatomical correlations with intelligence, one might argue that the overall mass of the cortex is not as important as the volume of the *associative* cortex in the sense of Jerison's concept of "extra neurons." Of special interest in this context is the size of the *frontal* or *prefrontal* cortex, which is assumed to be the "seat" of working memory, action planning, and intelligence. Therefore, the question is whether supposedly very intelligent animals like primates, elephants, and whales-dolphins have a particularly large frontal-prefrontal cortex. Hence the much-cited statement of Deacon (1990) that humans have a prefrontal cortex that is three times larger in relative terms than that of the other apes.

Studies by Semendeferi et al. (2002), and Teffer and Semendeferi (2012) using structural MRI confirmed, on the one hand, that among primates, humans have the largest frontal, including prefrontal cortex (gray plus white matter!). The relative size (i.e., percent of total brain volume) of the human frontal-prefrontal cortex amounted to 38 %, and the same value was found in the orangutan. The gorilla had 37, the chimpanzee 35, the gibbon 30, and the monkeys 31 %. This means that in general, the size of the frontal-prefrontal cortex increases slightly positively with respect to the total brain mass, with an exponent of 1.14, but given the fact that humans have a brain that is more than two times larger than that of a gorilla and three times larger than that of a chimpanzee, the human frontal-prefrontal cortex is even smaller than expected—it should have a relative size of more than 40 %. According to Semendeferi et al. (2011), inside the human frontal cortex, what has increased in size is mostly the dorsal part and particularly the frontopolar area (A 10), which appears to have twice the size of what one would expect. The ventral parts, i.e., the orbitofrontal and ventromedial cortex, have become relatively (although not absolutely) smaller.

In searching for a more direct neurobiological basis of intelligence, the *number of neurons*, particularly of *cortical* neurons as well as the *effectiveness of their wiring* and *processing speed*, comes to mind quite naturally. Brains and cortices of the same volume may contain very different numbers of neurons depending on packing density (NPD), which—among others—depends on the size of the neurons, including their dendritic trees. Processing speed largely depends on interneuronal distance (IND) and axonal conduction velocity, which in turn largely depends on the degree of myelination. Fortunately, at least for mammals, there is sufficient data to make a rough comparison.

The mammalian cortex consists, as already mentioned, of roughly 80 % of pyramidal cells; the rest are different kinds of excitatory and inhibitory interneurons (Creutzfeldt 1983). However, the size or volume of pyramidal cells (measured in cubic micrometers) varies greatly among mammals and roughly increases with an increase in brain size, i.e., larger brains and cortices tend to have larger pyramidal cells (Changizi 2001). The average size in mammals is 2,300 μm^3. Accordingly, cetaceans and elephants with large to very large brains have large to

14.3 The Fate of the Cortex as the "Seat" of Intelligence and the Mind

very large pyramidal cells. The bottlenose dolphin *Tursiops truncates*, with 5,400 μm^3, has "giant" pyramidal cells, followed by the elephant with 4,100 μm^3. Primates generally have small pyramidal cells. Accordingly, very small volumes slightly below or above 1,000 μm^3 are found in macaques, chimpanzees, and humans (Haug 1987).

An increase in the volume of pyramidal cells is accompanied by a decrease in packing density with a negative exponent of −1/3 (Changizi 2001). This is the consequence of several factors. On the one hand, larger neurons have larger dendritic trees, and the arborization of local axon collaterals is wider. This enlarges the entire space occupied by a neuron and its appendages. Additionally, the number of glial cells and blood vessels tends to increase, albeit with large deviations, with increasing neuron size. Glial cells play an important role in nutrition, and the supply of oxygen and sugar as well as other substances increases with increasing cell volume.

According to measurements of the late German neuroanatomist Haug, neuronal packing density (NPD) is high to very high in primates (Haug 1987). Here, the prosimian mouse lemur and the New World marmoset have the highest NPD with about 75,000 neurons/mm^3, followed by the New World squirrel monkey and baboons with about 60,000 neurons/mm^3. Macaques, talapoins (*Miopithecus*), and chimpanzees have about 40,000, spider monkeys, woolly monkeys, gorillas, and humans 25–30,000 neurons/mm^3. In contrast, the cortices of whales and elephants have a very low NPD with 6–7,000 neurons per mm^3. These results contradict the much-cited statement by Rockel et al. (1980) that in all mammals a cortical column with a given cross-sectional area, e.g., one square millimeter, contains the same number of neurons independent of the size of the cortex. Instead, while in monkeys such a cortical column may contain 190,000 and in humans an average of 50,000 neurons (ranging between 30,000 and 100,000, depending on cortical areas), in cetaceans and whales we find only 19,000 neurons per column (Cherniak 2012; Herculano-Houzel 2012).

On the basis of this data regarding cortical volume and NPD, we can calculate the *number of cortical neurons* in mammals. The results are given in Table 14.1. Due to their large cortex volumes, their small neurons and high NPD, primates have considerably more cortical neurons than expected on the basis of absolute brain size. The relatively small New World squirrel monkey has 450, and the much larger Old World rhesus monkey about 840, the New World white-fronted capuchin 720, gorillas 4,300, chimpanzees about 6,200, and humans about 15,000 million (or 15 billion) cortical neurons. The largest number of cortical neurons in *non-primate* mammals is found in the false killer "whale" (dolphin), with 10,500 and the African elephant with 11,000 millions, which is less than the number found in humans, despite their much larger brains. The reason is that their cortices are much thinner, their cortical neurons are much larger, and accordingly their NPD is much lower.

Chimpanzees have brains that include one-third of the cortical volume of that of humans, its cortex is as thick as the human one and the size of their pyramidal cells is comparable to that of humans. Because their NPD is somewhat higher than in humans, they have little less than half of the cortical neuron number found in humans. Cats have much smaller brains than dogs, but a much higher NPD, and therefore they have almost twice as many cortical neurons as dogs. Particularly impressive are the results from a comparison between horses and chimpanzees: the latter have a smaller brain, but five times more cortical neurons than the former.

Estimates on cell numbers are strongly influenced by the methods applied. Herculano-Houzel et al. (2007) report 1,100 million cortical neurons for the rhesus monkey, which appears far too high given the robust data by Haug on cortex volume and NPD in that species. Estimates in humans also vary widely in the literature, between 10,000 and 22,000 million, the latter being reported by Pakkenberg and Gundersen (1997), which again appear to be too high even when calculated on the basis of the highest measured human NPD. Herculano-Houzel and colleagues (Azevedo et al. 2009), with their isotropic fractionator method, arrive at 16,000 million cortical neurons in humans, which is close to the 15,000 million neurons calculated by Roth and Dicke (2012).

With respect to the information processing capacity (IPC) of the cortex, the number of synapses could be of importance. However, this topic is controversial. Some authors, like Schüz (2001), state that the number of cortical contacts per neurons is constant throughout mammals, while others like Changizi (2001) assume that it increases with cortical volume and neuron size with an exponent of 0.33. Thus, larger cortical neurons should have more synapses, but this increase in the number of synapses is believed to be compensated by a decrease in NPD, so that in mammals, cortical synapse density would remain constant. Unfortunately, exact data on a number of synapses is largely lacking. The number of synapses per neuron in the human cortex likewise is controversial; Cherniak (1990) reports 1,000–10,000, and Rockland (2002) nearly 30,000 synapses per neuron on average. If we, somewhat arbitrarily, assume 20,000 synapses per neuron for the human cortex, this would yield a total number of 3×10^{14} synapses, which at first glance seems incredibly high, but probably is quite realistic.

Besides the number of cortical neurons and synapses, another factor that is important for cortical IPC is *processing speed*, which, in turn, critically depends on (1) interneuronal distance, (2) conduction velocity, and (3) synaptic transmission speed. Interneuronal distance is determined by NPD: the higher the NPD, the shorter, trivially, is the interneuronal distance. We easily see that animals with large brains but low NPD might have severe problems in this respect. Conduction velocity rather strictly depends on the diameter of mostly myelinated axons, i.e., axons with a thin myelin sheath (or none at all) have low, and those with a thick myelin sheet have high conduction velocities. In mammals, axon diameter varies little from 0.5 μm in the mouse to 1 μm in monkeys (Schüz 2001). Apes are reported to have thicker axons than other mammals, and for fibers connecting cortical and subcortical areas in the brain, velocities of 10 m/s are reported, while peripheral nerves (e.g., the schiatic nerve) may reach 150 m/s. On the other hand, the axons of cetaceans (whales and dolphins) and elephants have thin myelin sheaths and consequently relatively low conduction velocities (Changizi 2001; Zhang and Sejnowski 2000;

Rockland 2002). Finally, the speed of synapse transmission is assumed to be constant among mammals and primates, but exact data are lacking.

Thus, in large-brained animals like cetaceans and elephants we find an unfavorable combination of high interneuronal distance plus low conduction velocity, which strongly slows down neuronal IPC. In the human brain, in contrast, we find a reasonable interneuronal distance plus very high conduction velocity, and this alone may result in processing speed that may be about five times higher compared to that found in cetaceans and elephants.

Another factor important for cortical IPC is the mode of connectivity among cortical neurons. In order to approach that problem, let us imagine full reciprocal connectivity among cortical neurons, which at first glance seems optimal: every neuron is reciprocally connected with every other neuron via at least one synapse. In such a case, the connections within the cortex would grow according to the formula $c = n \times (n - 1)$ or $n^2 - n$, where c is the total number of connections and n the number of neurons. In the case of larger n (roughly from 1,000 on), we can reduce the formula to $c = n^2$, which means that with a linear increase in the number of neurons we get a square increase in the number of connections. With 1,000 neurons we would get 1 million (10^6) connections, with 1 million neurons we would already have one trillion (10^{12}) connections, and if we assume the number of cortical neurons in humans to amount to 15 billion, we get the astronomical number of more than 10^{20} cortical connections. Our cortexes would be gigantic and consist mostly of myelinated axons. Already for metabolic reasons, this is impossible.

Luckily enough, a full connectivity pattern is not realized in the mammalian, including human, cortex, which—besides space problems—would be highly noneconomical. Most importantly, there is no full connectivity in the sense that each neuron is connected with each other, but there is the principle of "dense local and sparse global connectivity" or "small-world connectivity" known from the social network theories (cf. Cherniak 2012; Hofman 2001, 2012; Sporns 2010). This means that within a restricted area, every neuron is connected with nearly every other neuron (via an average of 4–6 synapses, as said before), forming a *functional assembly*, while within such assemblies only few neurons have connections with more distant neurons, which again are part of other functional assemblies. The number of connections among neurons would roughly increase with the natural logarithm of the synapses, which is dramatically less than in the case of a square increase. This principle appears to be optimal for all large and complex networks for information processing and relies on "compartmentalization." In other words, within such information processing networks, local working groups communicate intensely with each other, while across working groups it is mainly the group leaders who communicate. In this way, the communicative connections are reduced from full connection to optimal connection by several orders of magnitude, in the human cortex by 5–6, i.e., from 10^{20} to 10^{15}–10^{14} connections. It is clear that this makes a dramatic reduction in brain size possible.

14.3.2 Modularity of the Cortex

Anatomically and functionally, this "small-world" principle is realized in the mammalian isocortex by parcellation into functionally different (sensory, motor, and integrative-associative) areas. During the past few years, scientists such as the American neurobiologist Jon Kaas (cf. Kaas 2007) have studied the evolution of modularization of the cortex in detail.

In the small brains of "insectivore-like" mammals, the olfactory system is the dominating sensory system, and, accordingly, these brains have relatively large olfactory bulbs, a large hippocampus, which is the seat of the olfactory memory in these animals, and an olfactory (piriform) cortex. In the hedgehog cortex, for example, this olfactory cortex is at least three times as large as the non-olfactory cortex. In the tenrec (the Madagascar "hedgehog" *Tenrec ecaudatus*), the non-olfactory cortex consists of one visual and one auditory cortex, two somatosensory areas, and one primary motor area. In the cortex of the European hedgehog (*Erinaceus europaeus*), we find—besides a rather large olfactory bulb and cortex—one primary and one secondary visual and somatosensory area plus a parietal area, one auditory area and one motor area. The rat cortex has a relatively large olfactory bulb, while the olfactory cortex is relatively smaller, and there are two visual areas, one auditory area and a relatively large somatosensory area, in which the fore- and hindpaws, face, and vibrissae are represented. In addition, there is a relatively small secondary somatosensory and secondary auditory area (which is surprising, given the strongly developed auditory system of rats) as well as a large primary and a smaller secondary motor area.

A phylogenetic reconstruction suggests that besides a large olfactory area, the cortex of ancestral placental mammals, on average, had at least two somatosensory and two visual areas, but only one auditory area and one motor area, adding up to a total of seven to ten primary sensory and motor areas, without signs of integrative-associative areas (Kaas 2007). The main evolutionary pathway now led to the extant placental mammals with at least 4 visual and somatosensory and 2–3 auditory areas plus one specialized area for taste separated from the somatosensory cortex, as well as several limbic cortical areas, with 1–2 areas in the pre- or orbitofrontal cortex, 2–4 in the anterior cingulate cortex, and additional areas in the insular and entorhinal cortex. The human cortex is assumed to possess 150 areas and 60 connections per area, resulting in 9,000 area–area connections (Changizi and Shimojo 2005). The number of cortical areas is believed to increase with cortex volume at an exponent of 0.33 in most mammals and all primates, but not in elephants or cetaceans (see below). At the same time, the relative sizes of cortical areas are supposed to decrease. This has been interpreted as a tendency to maintain an optimal connectivity at increasing cortical volume and consequent number of neurons and areas, which is realized via the "small-world" principle of dense local connectivity within cortical areas and columns and sparse global connections across cortical areas, as described above.

This evolutionary pathway, however, is not the only possible one. As in evolution in general, many animals remained essentially unchanged until the present, and this means for the current context that the small "insectivores" existed for almost 200 million years with a relatively and absolutely small brain dominated by the olfactory system. It would be wrong to view them as "undeveloped," because they have successfully survived until the present by relying on olfaction and a predominantly nocturnal lifestyle. Another alternative was realized by small mammals that stayed with a small brain and specialized in a single sensory system at the expense of other sensory systems. One example is the enormous enlargement of the auditory system in bats in the context of the evolution of an echolocation system, or the enormous increase of the tactile-somatosensory system in many rodents (mice, rats, moles, etc.) or of the visual system in other small mammals, like squirrels.

Still another evolutionary alternative was realized in whales and elephants, where a strong increase in brain and cortex size was *not* accompanied by a strong parcellation of the cortex into many unimodal and multimodal-associative areas. Here, only one or two sensory systems strongly increased in size, but did not undergo substantial parcellation. In whales, this is the auditory system in the context of echolocation, and in elephants the somatosensory-vibratory system. Accordingly, these large-brained mammals appear to have many fewer cortical areas than primates. In these animals, besides a very large limbic cortex, enormous temporal and parietal cortical areas are found, but no substantial nonlimbic prefrontal cortex (Hart and Hart 2007; Hart et al. 2008).

14.3.3 Specialties of the Cytoarchitecture of the Mammalian Cortex

In comparative and evolutionary neurobiology, there has been an ongoing debate about whether or not across mammalian taxa the cortex has to be considered rather homogeneous or heterogeneous, with specialties found in different taxa. While authors previously tended to emphasize the homogeneity, today there is more search for heterogeneity and specialties. Clearly visible are the just-mentioned differences in size and number of sensory, predominantly visual, somatosensory, and auditory cortical areas. In "insectivores," we find the dominance of the olfactory system, while an olfactory cortex is absent in cetaceans, which instead have a large auditory cortex. These animals have a relatively small hippocampus (Hof et al. 2006), which, as already mentioned, was the site of olfactory memory in the primitive state of mammals. In addition, the cetacean cortex lacks a prominent layer IV, which is called "granular layer," because of the presence of many small-sized neurons. In most mammals, this layer IV is thick to very thick, particularly in the primary visual cortex, where we find a double layer IV, and it is the input layer of visual afferents from the thalamus. In cetaceans, instead, layer II is relatively thick and contains large pyramidal cells oriented upside down. The reasons for these specialties are unknown, because the cortex of even-toed mammals (artiodactylids), which presumably gave rise to the cetaceans, has a "normal" cortical cytoarchitecture with a well-developed layer IV.

Neuroanatomists like Preuss (1995) and Wise (2008) argue that only primates have a prefrontal cortex in the strict sense, together with its specific functions including control of attention, working memory, action planning, and decision making. Accordingly, lesions of the granular (pre)frontal area in primates have dramatic consequences for the mentioned functions, which is not the case in rats, when their dorsal frontal cortex is lesioned. A specialty of the frontal cortex of primates is the presence of a granural prefrontal area, which is characterized by a layer IV containing many small neurons. The frontal cortex of other mammals (e.g., rodents) lacks such a granular area, and is therefore called *agranular*. According to Elston et al. (2006), neurons in the prefrontal cortex of humans exhibit a higher degree of branching, an increased number of neurons, and a number of dendritic spines per neuron leading to a higher number of spine synapses, and wider cortical columns compared to nonhuman primates. The authors interpret these findings as proof of a dramatic increase in IPC of the human prefrontal cortex.

One alleged peculiarity of the cortex of hominid primates (including humans) that is often discussed these days is the presence of spindle-shaped neurons in layer Vb of the medial frontal and anterior cingulate cortex, which are four times as large as the other pyramid cells and are said to have extraordinarily widespread connections with other parts of the brain (Nimchinsky et al. 1999; Elston 2002). However, such "von Economo cells" have recently been found in some cetaceans and in elephants as well, but not consistently in all large-brained mammals (Hof and van der Gucht 2006; Hakeem et al. 2009). Whether this mosaic existence of "von Economo cells" is due to independent evolution or, when absent, to secondary loss, is unclear, as is their specific significance for cognition (Sherwood et al. 2008). Furthermore, it is unlikely that superior mental abilities are based on the presence of a single type of neuron.

14.4 Bird Brains and Mesonidopallium

In the preceding chapter we have seen that corvids and parrots turned out to be the most intelligent birds, and their intelligence has been considered equal to that of primates. Their brains are relatively large, their telencephala occupy 70–80 % of total brain mass, and the mesonidopallium (MNP) as well as the hyperpallium are extraordinarily large in a relative sense (Iwaniuk and Hurd 2005, for psittacids; Mehlhorn et al. 2010, for New Caledonian crows). However, these brains are small in an absolute sense, with a range of 8–12 g in corvids and up to 24 g in psittacids, which are equal to the lowest sizes found in monkeys (see above). A capuchin monkey with a comparable degree of intelligence has a brain of 26–80 g.

Particularly interesting in this context is the fact that the anatomy and cytoarchitecture of the MNP of birds as the "site" of intelligence has no resemblance to the mammalian cortex, i.e., no lamination or presence of pyramid-shaped cells, but a rather diffuse structure, where substructures are difficult to recognize (cf. Chap. 9). Birds generally have very small neurons, and these appear to be tightly packed inside the MNP, but unfortunately there are no quantitative data, and the same is true for the diameters of myelinated fibers in that region. Therefore, no direct comparisons between these important parameters in birds and mammals are possible. If we, very speculatively, start from the situation found in the cortex of small monkeys characterized by small and densely packed cells and assume an even higher packing density in birds, because their neurons are even smaller, then large-brained corvids and parrots might have around 200 million MNP neurons. In addition, it could well be that due to an extremely high packing density, information processing of these animals is considerably higher than that of monkeys, particularly because the metabolism of birds is higher. However, these speculations need to be tested by detailed empirical-experimental studies.

14.5 What Does All This Tell Us?

We began this chapter with a comparison of the brains of vertebrates and particularly of mammals (detailed data is available only from this group) and a discussion of the significance of absolute and relative brain size. We recognized that small vertebrates on average have small brains and large animals large brains in absolute terms, and the reason for this is that brain size is determined by roughly 90 % by body size. Whales/dolphins and elephants have the largest brains with weights up to 10 kg; the human brain, with an average weight of 1.350 kg, is of moderately large size. At the same time, brain size relative to body size tends to decrease with an increase in body size, resulting in the fact that small animals have relatively large and large animals relatively small brains. This is called *negative brain allometry*, which, for example, in mammals leads to dramatic differences. In shrews, brains include 10 % or more of body volume, while in the largest mammal and animal, the blue whale, the brain occupies less than 0.01 % of the body. In this context, the value of 2 % for the human brain is very high, given the fact that *Homo sapiens* belongs to the larger mammals. This becomes evident when we calculate the "encephalization quotient" (EQ) or residuals of brain-body regression, which for a given taxon indicates how much the actual brain size of a species deviates from the average brain-body relationship in this taxon. It turns out that humans have a brain that is roughly eight times larger than expected from the average mammalian brain-body relationship, closely followed by some dolphins, which have a fivefold larger brain than expected.

If we compare the values of absolute or relative brain size of a vertebrate or mammalian taxon with their intelligence as described in the preceding chapter, it becomes evident that there is *no* clear correlation between absolute or relative brain size and intelligence. Assuming that absolute brain size is decisive for intelligence, then whales or elephants should be more intelligent than humans, and cows more intelligent than chimpanzees, which definitely is not the case. If instead it were relative brain size that counted for intelligence, then shrews should be the most intelligent mammals, which nobody believes. Taking the EQ into account removes some inconsistencies, because then finally humans are on top, but many other inconsistencies remain, for example, that gorillas have a rather low EQ, but are considered highly intelligent, while capuchin monkeys and dolphins have unusually high EQs but are not considered to be as intelligent as gorillas. Thus, other factors have to be considered.

The cerebral cortex is considered the "seat" of intelligence and mind. During mammalian evolution, there was a dramatic increase in cortical surface with increasing brain size, while the thickness of the cortex increased only slightly. Among large-brained mammals, primates have the thickest cortices of 3–5 mm, while those of cetaceans and the elephant are surprisingly thin (1–1.8 mm). With increasing cortical volume, NPD usually decreases, but primates have unusually high and cetaceans and elephants unusually low packing densities. All this adds up to the fact that the human brain has the largest number of cortical neurons, (about 15 billion) despite the fact that its brain and cortex are much smaller in size than those of cetaceans and elephants (with 10–12 billion cortical neurons).

However, this alone cannot explain the unquestionable superiority of human intelligence. Here, differences in the speed of intracortical information processing come into play. We have reason to assume that in humans, cortical information processing is much faster than in the large-brained elephants and cetaceans. Of course, the speed of information processing probably is faster in much smaller brains with still much higher neuronal packing densities, but these brains have many fewer neurons. Thus, it is the combination of very many cortical neurons and a relatively high IPC that appears to substantially contribute to high nonverbal (and maybe even verbal) human intelligence.

Despite intense research, so far we have found no anatomical or physiological properties that would qualitatively distinguish the human brain from other mammalian or, in general, animal brains. All existing differences are quantitative in nature. There remains the question of whether human language represents such a qualitative step. This will be discussed in the next chapter.

There is still the question of why corvids and parrots, with very small brains compared to those of most mammals, including primates, are so intelligent. Presumably, because of the extremely high packing density of neurons in their MNP, they have an unusually high number of pallial neurons, probably several hundred million, despite the small size of their brains. This could result in a very high IPC. Most astonishing is the fact that the "seat" of avian intelligence, the nidopallium, exhibits an anatomy that differs radically from that of the mammalian isocortex. This could indicate that high intelligence can be realized by very different neuronal architecture. I will come back to this important point in the last chapter.

Chapter 15
Are Humans Unique?

Keywords Evolution of *Homo sapiens* · Australopithecines · *Homo habilis* · *Homo erectus* · *Homo neanderthalensis* · Enlargement of human brain · Human language · Language/speech centers · Animal language · Social behavior of humans

In the introductory chapter, I addressed the central question of the "uniqueness" of humans. Since Darwin, it has become increasingly clear that with respect to our biological nature, there is no uniqueness: we are descendants of chimpanzee-like ancestors, and genetically we are more closely related to chimpanzees than chimpanzees to other non-human apes. As a consequence, the defenders of the "uniqueness view" concentrated—and still concentrate—on the search for certain cognitive or communicative abilities that would underline the uniqueness of humans—abilities that are not found in non-human animals even in rudimentary forms. However, during the past 50 years of extensive comparative behavioral, psychological, and neurobiological research, the once long list of alleged "unique" properties that included tool use and tool making, mental maps, action planning, imitation, mirror self-recognition, theory of mind, teaching, cultural transmission of knowledge, consciousness, self-reflection, a syntactical-grammatical language, a "theory of mind," religion, morality, science, and art has become very short, and the defenders of human "uniqueness" are struggling for any feature that stands for a qualitative rather than quantitative difference between humans and non-human vertebrates.

In his book "*Human: The science behind what makes us unique*" (2009), the neuropsychologist Michael Gazzaniga stated that humans and non-human animals, including apes, are "hugely different" or that the two are "light years apart," but the evidence he cites in favor of such an emphatic statement is very weak at best, or could not be confirmed experimentally, e.g., regarding the human variants of "Microcephalin (MCPH1)" and "abnormal spindle-like microcephaly associated (ASPM)" genes (cf. Evans et al. 2005; Mekel-Bobrov et al. 2005; criticism by

Timpson et al. 2005; Yu et al. 2007). Other genes that are often cited to have played a role in human evolution like FoxP2 are discussed further below.

Certainly, humans differ from their closest biological relatives, i.e., chimpanzees and gorillas, in a number of features. Among the most conspicuous ones are the gracile body, the upright bipedal walk and the ability for sprints as well as for long runs, an arched food without a grasping toe, loss of climbing abilities, relatively minor size differences between the sexes (at least compared to gorillas), orthognath instead of prognath dentition, lack of protruding canines, reduced body hair, strongly increased number of sweat glands, and, consequently, an increased ability to perspire. There is, in addition, in female humans, an estrus that is not externally visible, a strongly prolonged childhood of altricial young and intensive parenting, extensive meat eating and extractive foraging for tubers, nuts, and other high quality food, accurate and powerful throwing of projectiles and—last but not least—by far the largest brain among primates in absolute as well as relative terms and, as a consequence, a strongly increased "general intelligence."

These and many more features that distinguish humans from non-human primates together form no "round" picture, but rather give the impression of mosaic evolution, i.e., the merging of many independent evolutionary events. No single "key" evolutionary event can be identified in the evolutionary line leading to *Homo sapiens*. Let us take a closer look at the evolution of our species.

15.1 How Did *Homo sapiens* Evolve?

Despite details on research, the evolution of *Homo sapiens* are only partially known. It is assumed to have begun about 85–65 mya with the divergence of proto-primates from other mammals. Earliest fossils are from 55 mya. Simiiforms (or *Anthropoidea*), i.e., monkeys and apes, appeared about 40 mya, and the split between New World monkeys and Old World monkeys (including apes) occurred about 30 mya. Apes or *Hominoidea* originated about 25 mya, the divergence of gibbons from the *Hominidae* (orangutan, gorilla, chimpanzees) occurred about 15–18 mya, and that of orangutans from the *Homininae* (gorilla, chimpanzees) about 13 mya. About 10 mya, there was the divergence of the line leading to gorillas from that leading to the *Hominini*, i.e., chimpanzees and direct human ancestors, and about 7–5 mya there was the split between the ancestors of chimpanzees and the ancestors of australopithecines, the hominins in a narrow sense. However, in the literature there are great discrepancies concerning the nomenclature.

During the past two decades, the evolutionary history of the australopithecines, including our direct ancestors, has been studied extensively, and a present view is given in Fig. 15.1. A putative ancestor of the australopithecines, *Ardipithecus ramidus*, lived 5.8–4.4 mya in the region of Aramis in today's Ethiopia. He exhibited many characteristics that clearly distinguish him from extant chimpanzees as well as gorillas. *Ardipithecus* was 120–130 cm tall and already capable of upright, bipedal walk, but apparently could still brachiate in the trees and still

15.1 How Did *Homo sapiens* Evolve?

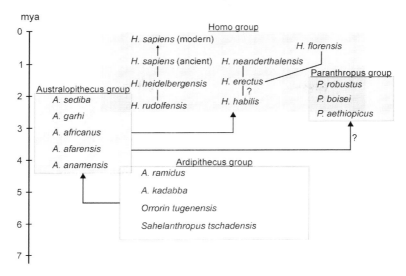

Fig. 15.1 The putative evolution of hominins. About 4 mya, the *Ardipithecus* group split into the Australopithecus group and the *Paranthropus* group. Members of the former, perhaps *A. africanus*, gave rise about 2.5 mya to the *Homo* group. Here, one line led to the ancient and modern *Homo sapiens*, another line to *Homo erectus*, *H. Neanderthalensis*, and the dwarfed *H. florensis*

possessed a grasping toe, like chimpanzees and gorillas. The brain of *Ardipithecus* was only slightly larger than that of extant chimpanzees, while the canines were not as protrusive. This may indicate that the common ancestor of chimpanzees and australopithecines was not as chimpanzee-like as previously thought. Perhaps chimpanzees and gorillas represent cases of independent evolution toward a higher specialization for arboreal life.

The next level of hominization is represented by the genus *Australopithecus*, which lived 4.4–1.9 mya in the vicinity of Lake Turkana (formerly Lake Rudolph) and Lake Victoria in East Africa as well as in South Africa. This genus comprised, among others, *A. anamensis*, which lived 4.2–3.9 mya in the region around Lake Turkana, was about 120 cm tall, and already exhibited a good upright walk; *A. afarensis* (with "Lucy" as the most famous representative), which lived in Tanzania, Kenya, and Ethiopia 3.8–2.9 mya and was about 105 cm tall (young female); *A. africanus*, which lived 3.5–2.5 mya in South Africa and was 110–140 cm tall; and *A. garhi*, which lived in Ethiopia around 2.5 mya. They all had brain sizes ranging from 350 to 550 ccm, which is equal to or slightly larger than brain sizes found in extant chimpanzees and gorillas. There are no signs of stone tool use. The evolution has probably gone from *A. anamensis* via *A. afarensis* to *A. africanus* and *A. garhi*. The genus *Paranthropus* probably split from *A. afarensis*. Its species, *P. aethiopicus* (2.8–2.3 mya) and *P. boisei* (2.3–1.4 mya) in Tanzania, were 140 cm tall with a brain of 485 ccm, and *P. robustus* (2.0–1.5 mya) in South Africa, which was 140 cm tall and had a brain size of 493 ccm.

The recently described *Australopithecus sediba* (cf. Carlson et al. 2011) lived about 1.9 mya at the boundary between australopithecines and the genus *Homo*, with unclear phylogenetic position. He was about 130 cm tall and had a brain volume of about 420 ccm. Of interest is his enlarged ventral frontal brain. His hand morphology indicates at least partial arboreal locomotion, but reveals a long thumb, which is human-like, and a predisposition to stone tool production. *A. sediba* clearly was bipedal and had a more inferred foot, but like all australopithecines, apparently possessed no modern human locomotion (Zipfel et al. 2011).

From one of the australopithecines, representatives of the new genus *Homo* developed including *H. habilis* in regions south of the Sahara. Traces of *H. habilis* are found in Ethiopia, Tanzania, Kenya, and South Africa from around 2.4–1.5 mya. He was about 140 cm tall and had a brain volume of 550–780 ccm, which is well above the values for extant great apes as well as the australopithecines. He used spears for hunting and stone tools for cutting meat and hammering. *H. rudolfensis* lived 2.5–1.8 mya in the region of Lake Turkana (Lake Rudolph) in South Ethiopia and of Lake Malawi. He was 155 cm tall and had a brain volume of 600–700 ccm. Primitive tools were found with him, e.g., sharp-edged stones for cutting or abrading meat from carrion. Many experts believe that *H. rudolfensis* and perhaps *H. ergaster* left Africa 1.8 mya as the first member of the genus *Homo*. From an early representative of that genus, perhaps *H. ergaster*, developed *H. erectus*, probably 1.8 mya, as well as (perhaps via an intermediate species *H. antecessor*) *H. heidelbergensis*. However, the precise relationships between *H. habilis, H. rudolfensis, H. ergaster*, and *H. erectus* are still unclear (Pickering et al. 2011). There may have been a parallel occurrence of *H. habilis, H. ergaster*, and *H. erectus* for about 500,000 years. The latter survived for a long times as *H. soloensis* (until about 100,000 years) and the miniaturized *H. floresiensis* (possibly until 12,000 years ago).

Homo ergaster/heidelbergensis lived from 1.8 until about 200,000 ya in Europe (Germany, France, Northern Spain, and Balkan), in the Caucasus, in Morocco, and all of East Africa, while *Homo erectus* was found in Southeast Asia, in China, and in East and South Africa. The brain volumes of these species had a range of 700–1,250 ccm and thus, at least partly, reached the volumes of extant humans. They knew how to use fire and stone axes. The first settlement of South Europe by *H. ergaster/heidelbergensis* occurred around 800,000 years ago (some experts assume 1 my or even more), but became stable only after 500,000 years. *H. heidelbergensis* probably gave rise to both *H. neanderthalensis* and *H. sapiens*. The former lived from 220,000 until 27,000 ya in Israel, at the Black Sea, in East Turkey, Iran and Afghanistan, Spain, France, Germany, and England. *H. neanderthalensis* was up to 160 cm tall and was characterized by a massive bone structure and a musculous body; his head possessed strong supraorbital ridges and a fleeing chin. Neanderthals buried their dead together with grave goods and fabricated finer tools. They had brain volumes of 1,400–1,900 ccm (another size range reported is 1,125–1,740 ccm), which is more than the average brain volume of modern *Homo sapiens* (1,300–1,400) and is the largest brain of all hominins and primates.

Our direct ancestors, *Homo sapiens*, originated in its archaic form around 500,000 ya and in its modern form, *Homo sapiens* 200,000–150,000 ya in East

Africa. From there he spread all over the entire world in a form that did not substantially differ from that of extant humans. South Africa was invaded about 150,000, Northern Africa and Asia Minor about 100,000 years ago. In Asia Minor, there are no signs of any conflict between modern humans and Neanderthals. They produced and used similar kinds of tools, and appear to have mixed genetically in moderate terms (Green et al. 2010).

From Asia Minor, *H. sapiens* spread via Afghanistan and Northern India to China and Southeast Asia. Modern humans arrived in Australia in several waves about 60,000 years ago, invaded Northeast Asia and from there entered North and South America 30-15,000 ya. They settled in Southern Europe about 45,000 years ago, which is relatively late. Here, they met *H. neanderthalensis*, who had lived there for 200,000 years. What happened between *H. sapiens* and the Neanderthals living there is unknown, but 13,000 years later, i.e., 27,000 ya, the latter became extinct. Some authors speculate that *H. sapiens* exterminated them actively, but there is no evidence for that, and the same holds for the assumption that infectious diseases were the cause of the disappearance of *H. neanderthalensis* in Europe. Also, as opposed to the situation in Asia Minor, genetic mixture has not been discovered in Europe. Recently, Mellars and French (2011) found that there was a tenfold population increase in Western Europe at the Neanderthal-to-modern human transition, and this population explosion could be due to improved hunting and food-processing technology, food storage, enhanced mobility and transportation technology, increased social integration and cohesion.

With the end of the last glacial period, about 10,000 years ago, earlier agriculture and earlier settlements with more than a thousand inhabitants occurred in Asia Minor. The first areas of high population density formed in regions where desertification as a consequence of climate warming brought people more closely together, or in landscapes with highly favorable climatic, geological, and botanic-zoological the conditions like in China, at the Indus River, in Mesopotamia, and at the Nile. There, the first advanced civilizations appeared with written language, efficient administration, and a basis for astronomy and mathematics as well as art and culture.

Exactly which *biological* factors may have favored this development is unclear. The human brain and its functions probably did not substantially change over the past 30,000 years. The cave paintings from 40,000 ya on, e.g., in Altamira and Lascaux, are of such mastery, as are the tools and pieces of art of that time that a fundamental increase in cognitive and manipulatory functions in the meantime is rather unlikely.

15.2 Leaving the Jungle and Its Consequences

The first big step in the evolution toward *Homo sapiens* was the complete exodus from the tropical rain forest and a continuous life in the much dryer savanna or open grassland, where trees do not form canopies. Of the two species of chimpanzees, only the bonobos (*Pan paniscus*) are exclusive forest dwellers, while the common chimpanzee (*Pan troglodytes*) also lives within the transient zone between rain forest and savanna, but cannot survive permanently in the dry and hot savanna. This has—among others—to do with their lower tolerance for heat and with their diet. Chimpanzees are omnivorous, feeding mostly on fruits, nuts, leaves, and flowers of trees. Like monkeys, they regularly eat insects and any kind of small mammals and go out for monkey hunting, but this kind of "meat" represents only a small portion of their diet. These kinds of food are found predominantly at the fringe of the tropical forest and wetter parts of the savanna.

Decades ago, primatologists like the Dutch Adriaan Kortlandt, proposed that the evolution of australopithecines and humans was intimately connected with large-scale geological changes in East Africa during the late Pliocene, around 3.6–2.6 mya, when Africa collided with Europe and the Mediterranean Sea was formed (Kortlandt 1968). Due to this collision, high mountain ranges and deep valley systems were formed, constituting the East African Rift System, which extends from the "Afar Triple Junction" southward across eastern Africa and splits the African Plate into the western Nubian and the eastern Somalian plates. The Rift system includes several large and very deep lakes like, Lake Tanganyika and Lake Victoria. Together with the formation of the Rift, the climate became cooler and dryer and the rain forests shrunk.

This geological system has long been considered the cradle of humanity, because the Rift Valley in East Africa has been a rich source of fossils related to human evolution, especially since the rapidly eroding highlands filled the valley with sediments, which created a favorable environment for the preservation of bones and other remnants. Here, Lucy (*Australopithecus afarensis*) and other putative ancestors of modern humans have been found. For chimpanzees of that time, the Rift system formed an insuperable barrier because they would not find enough fruits and could not swim, and therefore they died out in East Africa and survived only in Western Central Africa. Our ancestors, in contrast, managed to survive under these new conditions. In addition, the open savanna gave rise to incredible numbers of ungulates that wandered through the grassland in large herds and represented a virtually unlimited source of meat—i.e., together with nuts, the most nutritious food. The only problem then was to get that meat.

There were two possibilities. One was to feed on dead or dying animals that had been abandoned by the herd. This was not without danger, because in the savanna there were and still are many larger carnivores, like lions, leopards, or hyenas that had the same interest. So, early humans had to find ways to fight them. The other was hunting, which, however, in addition to the appropriate hunting techniques, requires the ability for enduring runs—impossible for the great apes. The evolution of upright walking and bipedalism was one of the key events. This was made possible by substantial changes in the skeleton and the related muscular apparatus, more precisely, the formation of a multi-curved vertebrate column (cervical, thoracic, lumbar, and pelvic curves forming a double-S shape) essential for balancing the large-brained head in the upright position, of an arched, rather than flat foot, changes in the hips for better stabilization of ball-and-socket joints and in the knee joints for better stabilizing the body by bringing the legs under the body, and,

15.2 Leaving the Jungle and Its Consequences

finally, lengthening of the legs. The upright walk freed the forearms from locomotion functions and favored further specialization of the hands. By walking upright, our ancestors could better survey the neighborhood, and bipedal locomotion enabled them for quick runs, e.g., to escape from large carnivores or to snatch a piece of food from a competitor (animal or conspecific), as well as for long walks.

Some of the greatest differences between humans and chimpanzees concern the anatomy and function of legs and feet. While the foot of chimpanzees as well as of gorillas is a typical instrument for climbing and grasping, it is poorly adapted for longer bipedal walking. Both taxa of great apes exhibit "knuckle-walking," when moving quadrupedally, because during walking, the fingers of the forearms are partially flexed and the animals actually walk on their knuckles. It has been discussed that at least in chimpanzees, knuckle-walking, which may have evolved independently from that of gorillas, originated after the split between human-like hominids and ancestors of extant chimpanzees, and *Ardipithecus ramidus* showed no signs of knuckle-walking. The human foot, with short toes and a big toe that has only limited grasping abilities, is specialized for walking and running.

In contrast to the foot, the human hand differs only insignificantly from that of chimpanzees. It is generally smaller, and the fingers are shorter, except for the thumb. Extensive hand use and fine motor skills of the hand are already found in chimpanzees, which can exert the same "precision grip" as humans, as can be seen in Fig. 15.2, which shows the chimpanzee Julia using a screwdriver. What has changed in humans, however, is the much more sophisticated neuronal control of the hand during tool use and tool fabrication. Recent studies demonstrate that extended frontal, premotor, and parietal cortical networks, mostly within the left hemisphere, are involved in this task. Interestingly, here two different networks can be distinguished, which can be impaired and consequently "dissociate" relatively independently, i.e., a "semantic" one concerning the concept of a specific tool use including the underlying principle, recognition and naming of the tool, and a "ideomotor" one involved in the sensorimotor control including practical experience (Johnson-Frey 2003).

There were and still are negative consequences of upright walking, predominantly osteological malfunctions in the lower back and the joints due to the increased body weight they had to support, and already our early hunter-gatherer ancestors suffered from arthritis. However, experts tell us that bipedal locomotion is economically optimal and enabled our ancestors to follow herds for days and weeks in the hope of ill, old, or dead ungulates. To attack and kill was difficult without claws and a prognath dentition with large canines and incisors (weapons for hunting were invented much later). Accordingly, it is assumed that besides meat from dying or dead animals or fishes caught in rivers or in shallow waters of the lakes, roots, tubers, and fruits still made a substantial contribution to the diet.

Another big problem was the heat in the savanna. It is assumed that exposition to the sunlight was substantially reduced by the upright walk, and heat management was improved by the strong reduction of body hair and increase in the

Fig. 15.2 Chimpanzee Julia guides the insertion of a screwdriver into a small screw with the index finger of her left hand. From Rensch (1968)

density of sweat glands. Still another severe consequence of upright walking was the reorganization of the pelvis, which now assumes the function of carrying the body organs, while still making fast walking and giving birth to a child possible. A strong prenatal brain growth, together with upright walking, created the major complications for childbirth in human—problems that do not exist in the rest of the animal kingdom. The female pelvis is a compromise between two functions, i.e., upright walking, which favors a narrow pelvis, and a wide birth canal for birth giving to large-brained fetuses. Humans are born when, despite all complications, the majority of mothers and children will survive the birth. This moment is somewhat earlier than in other primates, but the much-cited statement of the Swiss anthropologist Adolf Portmann, that human beings, compared to other primates and mammals, are born extremely prematurely and accordingly are extremely helpless, is incorrect since at least chimpanzee babies are born equally helpless. What is correct, however, is that in humans, postnatal brain growth is much stronger and lasts much longer than in other primates, including chimpanzees. This long period of absolute and relative helplessness of babies and young children, however, is a key prerequisite for the peculiar sociality of humans (see below).

15.2 Leaving the Jungle and Its Consequences

In the context of hunting, a key step was the invention of the spear about 400,000 years ago. Most efficient was this weapon in the form of a spear-thrower, by which a wooden spear or dart can achieve a velocity of about 150 km/h (93 mph). Such spear-throwers are believed to have been in use by humans since the Upper Paleolithic, i.e., around 30,000 years ago. Bows and arrows may have been invented earlier, i.e., 70–60,000 years ago (earliest possible arrowheads are about 64,000 years old). Stone axes have been in use since Mesolithic (10,000–6,000 ya) or even Paleolithic times (2.6 mya–10,000 ya). In addition, cooperative hunting techniques were invented like those used by lions.

15.3 Enlargement of the Brain and Its Consequences

The evolutionary line from *Australopithecus afarensis* and *A. africanus* to the Neanderthals and modern human is characterized by an increase in brain volume within roughly 3 million years. As can be seen in Fig. 15.3, this increase is assumed to have an exponent of 1.73, which means that it was very positively

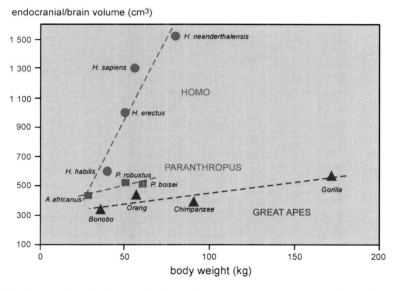

Fig. 15.3 The relationship between body weight and brain or endocranial volume (in extinct species) in the great apes (bonobo, orangutan, chimpanzee, gorilla), australopithecines (*Australopithecus africanus, A. robustus, A. boisei*) and hominins (*Homo habilis, H. erectus, H. sapiens, H. Neanderthalensis*) (data from Jerison 1973). While in the great apes as well as in *A. robustus* and *A. boisei*, probably not among our ancestors, brain/endocranial volume has increased only slightly with body size; in the genus *Homo*, a steep increase in brain/endocranial volume has occurred over about 2.5 million years, culminating in the brain of the extinct *Homo neanderthalensis*, which was considerably larger than that of the modern *Homo sapiens*. After Pilbeam and Gould (1974), modified

allometric (Pilbeam and Gould 1974). The first human-like dwellers of the open savanna, like *Australopithecus africanus*, had a brain volume of 350–550 ccm, in *Homo habilis* it was already 550–780 ccm, and *Homo erectus* arrived at 1,000 ccm and beyond. The subsequent separate evolution toward *Homo neanderthalensis* led to a brain volume of 1,400–1,900 ccm and toward *Homo sapiens* with an average of 1,350 ccm. The fact that *Homo sapiens* had (and has) a considerably smaller brain than *H. neanderthalensis* remains unexplained until today and demonstrates that a big brain alone does not prevent a species from becoming extinct—for whatever reason.

Figure 15.3 also shows the situation regarding brain size in the great apes, i.e., bonobos, orangutan, chimpanzee, and gorilla, as well as in those australopithecines, here *Paranthropus robustus* and *P. boisei*, which do not belong to our ancestors. In both groups, a substantial evolutionary increase in body size is paralleled by only a minor increase in brain size, with an exponent of 0.34 and 0.33, respectively, as is typical of brain-body allometries found at "low" taxonomic levels like families or genera.

The fast and steep increase in brain size from about 400 to about 1.350 ccm (or even much more in the case of the Neanderthals) leads to two questions: First, what were the exact causes for it, and second, which were the benefits and which were the costs?

The latter question can be answered more easily. A large brain creates two major problems, a metabolic and a developmental one. As to the first, we have to recall that a large brain is enormously costly in metabolic terms, and this is particularly true for the human brain. In adults, it occupies about 2 % of body volume, but already in its resting states it consumes about 20 % of glucose and oxygen metabolism, and when it comes to intense mental activities, like in a state of high concentration, this rate greatly goes up. We become aware of that fact when we stop any physical activity while being mentally engaged in something important and complicated, and after a few minutes of strong mental concentration, we feel exhausted.

This means that in order to afford a large brain, an animal or human has to be in a position to nourish it, and the best way to do so is to feed on highly nutritious food, i.e., meat, roots, nuts, etc. Additionally, there have been substantial savings in the metabolic costs of other body organs, because they, too, may be metabolically costly. The digestive tract, heart, liver, and kidney, together with the brain, consume about 70 % of adult metabolism. According to the *Expensive Tissue Hypothesis* developed by the anthropologist Leslie Aiello and colleagues (Aiello et al. 2001), a substantial reduction of gut length occurred during the early evolution of humans, which reduced the metabolic costs of the gut and could be compensated by more nutritious food. According to the authors, this favored the further increase in brain size.

However, even if metabolic and other problems (like those mentioned above) are solved, a brain cannot simply grow in size, but appropriate genetic-epigenetic mechanisms are needed to accomplish this. In this context, it is important to determine whether during human evolution the entire brain or only parts—and if so, which parts—increased in size. As already mentioned, all parts except the olfactory system and the medulla oblongata increased in size, but the cerebellum as well as the cortex, and within the latter the frontal cortex, grew somewhat faster in a positively allometric fashion. According to experts, this makes it likely that this nearly uniform increase in brain size was due to changes in the genetic control of brain growth, e.g., by changes in regulatory genes (Finlay and Darlington 1995; Rakic and Kornack 2001).

Changes in such genetic-regulatory mechanisms may affect cell division rate or a prolongation of the time of production of neuronal precursor (or progenitor) cells. At the beginning of cortical development there is first a symmetric division of precursor cells close to the ventricles of the telencephalon, which means that each precursor cell divides into two precursor cells, leading to an exponential growth. This is followed by a phase of asymmetric cell division, where a precursor cell gives rise to a further progenitor cell and a nerve cell. While the former keeps dividing asymmetrically, the newly formed nerve cell migrates outward and forms the so-called cortical plate, from which the cortex originates. Thus, the most efficient way for fast increase in the number of neurons is an increase in symmetric cell divisions.

According to the *Maternal Energy Hypothesis* developed by brain scientist Robert Martin (Martin 1996), another "bottleneck" affecting the evolutionary increase in brain size is the prenatal and early postnatal brain growth. Both are very costly, because they consume about 60 % of resting metabolism of the fetus or baby. The dramatic prenatal brain growth continues for a while after birth and increases until the end of the seventh year. Only this way—according to Martin—does the human brain reach its enormous size in absolute and relative terms. This has two important consequences: On the one hand, it puts a heavy metabolic load on the expectant mother, which means that she continuously needs high-caloric food. On the other, this requires intense additional care after birth, e.g., by grandmothers, sisters, or neighbors, and not least by the partner of the mother and his family. This help presumably was facilitated by increased verbal communication—a topic that I will address now.

15.4 Language and the Brain

15.4.1 Animal Language

Many non-human animals possess complex intraspecific communication. The signals used may be either auditory-vocal, somatosensory (e.g., vibratory, tactile), electrical, visual (body signs, hand gestures, facial expressions), or olfactory (pheromones etc.) (cf. Chap. 11). A fascinating phenomenon is *bird song*. Detailed studies (for overview see Mooney 2009; Beckers 2011; Woolley and More 2011) demonstrate striking parallels between bird song and human vocal language, e.g., song learning (cf. Scharff and Petry 2011) and demonstrate that many characteristics of human speech perception are not uniquely human. Songbirds originated some 50 million years ago and evolved vocal abilities, which may even be considered superior to those of humans. Furthermore, there are signs of syntax and grammar, but they are not linked to the semantic components (i.e., the level of meaning) of the song (cf. Berwick et al. 2011).

Among mammals, complex vocal communication systems are found in prairie dogs of the genus *Cynomys* (Slobodchikoff 2002) and in vervet (or green) monkeys (Seyfarth and Cheney 2008; Seyfarth et al. 1980). Vervet monkeys have about ten different intraspecific vocal sounds which do not simply encode the actual emotional-affective state of an animal (arousal, joy, anger, pain, etc.), but also serve as alarm calls signaling the presence or approach of a predator (leopard, eagle, snake, baboon, human, etc.) and deliver information about their most important features like size, sex, and the degree of threat, about relations and even objects that are not present, for example, in order to deceive. Here, different sounds reliably elicit different behavioral responses, e.g., climbing trees or searching the sky. Many of these calls must be learned by the young monkey, and there are—like in songbirds—dialects (Ghazanfar and Hauser 1999). Nevertheless, in a study by Fitch and Hauser on the linguistic abilities of cotton-top tamarins (*Sanguinus oedipus*), the animals did not exhibit any signs of complex syntactic abilities termed "phase structure grammars" (Fitch and Hauser 2004).

In the context of "language in animals," one generally has to distinguish between the ability to communicate via vocal language and the ability to understand language—more precisely, to follow verbal commands. As to the latter, some mammals are capable of following up to several hundred commands of humans. Some years ago, Juliane Kaminski, Julia Fischer, and Joseph Call from the Max-Planck Institute for Evolutionary Anthropology in Leipzig obtained astonishing results while working with the border collie, Rico (Kaminski et al. 2004). When Rico was given the names of toy animals or these were shown to him, he was able to select the correct animal from a collection of more than 200 toy animals in a neighboring room. He could also pick out an animal unknown to him out of the otherwise familiar toy animals. It is, of course, difficult to decide whether Rico showed any understanding of the meaning of words or was just trained to respond to certain sounds with a certain behavior.

In the past, there were several attempts to teach human language to apes (cf. Rütsche and Meyer 2010). In the 1950s, Keith and Catherine Hayes tried to teach the chimpanzee Viky to imitate human vocal language, but the success was minimal, because Viky was only able to vocalize something like "mama," "papa," "cup," and "up" (Hayes and Hayes 1954; Premack and Premack 1983). The reason is that chimpanzees are physically unable to produce the full range of sounds of human language, particularly vowels (see below).

More successful were the attempts to teach apes a non-vocal language system, e.g., plastic tokens, as in the case of *Sarah*, American sign language (ASL) in the case of *Washoe*, and computer keyboards in the case of *Kanzi*. In David Premack's study, the chimpanzee *Sarah* learned to select objects and express intentions and desires by using plastic tokens as signs that were completely different in appearance from the things they stood for. Also, Sarah learned concepts like negation, name-of, more-or-less, and if-then, and she was able to follow rather complex commands like "Sarah put banana pail apple bowl" by putting the banana into the pail and the apple into the bowl, although both fruits would fit in both containers (Premack 1983).

Washoe, a common chimpanzee caught in the wild at an age of about 10 months, was reared by Beatrix and Allen Gardner in their family under the same conditions as a human infant with deaf parents (Gardner et al. 1989). The Gardners and their assistants communicated with Washoe by ASL, minimizing the use of spoken language. There was impressive progress in the acquisition of words, and Washoe became able to form two- to three-word sentences like "open food drink" meaning "open the refrigerator" or "please open hurry" meaning "please open (it) quickly," to use the sign "more" in many different situations or the sign for "flower" to express the idea of smell and to generalize the meaning of words, e.g., for "hat." Remarkably, Washoe also taught other chimpanzees some ASL without any help from humans.

Soon after publication of these findings, linguistic critics argued that the Gardners had simply conditioned Washoe to use ALS in certain contexts in order to reach certain goals, but that the ape had not learned the same linguistic rules that humans innately know. In order to respond to that criticism, the chimpanzee *Nim Chimpsky* (alluding to the name of the linguist *Noam Chomsky*) was taught by Herbert S. Terrace to communicate via ASL. In about 3 1/2 years, Nim Chimpsky learned 125 signs, but according to Terrace, his use was purely symbolic and lacked the grammar and syntax that characterize humans' language, as well as a deeper understanding of linguistic communication in the same way, as there is tool use without understanding the underlying principles among primates (see Chap. 12). Furthermore, Nim Chimpsky's vocabulary learning rate was roughly 0.1 words per day, while a human learns roughly 14 words per day between ages 2 and 22, which, according to Terrace, indicates fundamental differences in language acquisition (Terrace et al. 1979).

These findings were, in turn, criticized by pointing out that Nim Chimpsky's learning rate was considerably lower than that of Washoe, and that the results of the experiment led by Terrace were not fully representative. Another criticism was that Terrace had adopted classical behavioral methods to condition the ape to use certain hand signs to name certain objects—exactly the way he interpreted what Premack and the Gardners had done. However, for the critics of Terrace it was clear that both Sarah and Washoe used symbols "referentially," i.e., standing for another thing, as well as "communicatively," i.e., to express intentions and desires. Nevertheless, there remained the critical argument that all apes investigated so far had been explicitly *taught* the use of "words," while human children learn to speak rather *spontaneously*.

Such spontaneous acquisition of language was found in *Kanzi*, a male bonobo that was brought to the Georgia State University Language Research Centre with his mother when he was 6 months old. When Sue Savage-Rumbaugh and Duane

Rumbaugh began to teach his mother, Kanzi was present. After being separated from his mother at age 2 1/2, he started his own training, but already on the very first day he produced 120 utterances and used all 12 symbols on the keyboard, and on the second day he spontaneously produced the message "melon go" indicating that he wanted to go outdoors and eat melon. Thus, in contrast to Sarah and Washoe, he did this by pure observation or imitation of the training of his mother. Rumbaugh immediately refrained from teaching language to Kanzi and let him find out the meaning of new vocabulary that appeared on the keyboard. Kanzi also used the keyboard to talk to himself. A few years later, he had learned 256 words as well as vocalizations, gestures, and a combination of these three. While the messages that he produced did not reveal clear signs of syntax, he could at least *understand* the relevance of word order. When Rumbaugh used the computer to ask him "Can you make the dog bite the snake?"—a sentence he had never heard before—he found a toy dog and a toy snake and put the snake into the mouth of the dog and with his fingers closed the dog's mouth over the snake. When he was 7 1/2 years old, he was able to answer correctly 74 % of 400 complex questions.

In summary, Kanzi learned language by himself without explicit teaching or conditioning and started conversations with humans as well as with conspecifics in the absence of rewards, which speaks against operant conditioning. In addition, at least in language comprehension, he revealed an understanding of simple syntax (Savage-Rumbaugh 1984).

The vocabulary of Kanzi amounts, as in Rico, to about 200 words or concepts. *Koko*, a female gorilla, was said by her trainer, Francine Patterson, to understand and follow about 1,000 ASL signs. However, most experts, including Savage-Rumbaugh, agree that despite greatest efforts and training that a chimpanzee or gorilla can learn over the years, as well as the length and structure of sentences, they are highly limited in great contrast to the virtually unlimited nature of human language. Apes are capable of forming new sentences from known words, but whether these sentences reveal at least a rudimentary syntax is debated—perhaps in understanding. Generally, the great apes appear to be unable to go beyond the linguistic abilities of a human child aged 2–3 i.e., a stage characterized by two- to three-word sentences without clear signs of grammar and syntax (Savage-Rumbaugh 1984, 1986).

15.4.2 The Evolution of Human Language

There is much controversy about the question of whether the evolution of human language was a slow and continuous process extending over many preliminary stages in non-human apes and the more direct ancestors of *Homo sapiens*, or a fast, "saltatory" event without precursors, as Thomas Huxley, Darwin's famous "combatant," believed and the linguist Noam Chomsky still believes. A major problem consists in the number, complexity, and diversity of cognitive, motor, and linguistic prerequisites of human language that must have come together in an apparently very short evolutionary time (cf. Pinker 1995; Pinker and Jackendorf 2005). Also, it is hotly debated whether human language originated from vocal communication or from gestural communication in non-human primates, or both (see below).

Of special interest is the evolution of the vocal apparatus necessary for the production of a large number of consonants, and particularly vowels, which required substantial modifications of the mouth (lips, tongue, and velum), nose, and throat region, including the larynx. One key event appears to be the evolution of upright-bipedal walking and the reduction of dentition. The former allowed for the more L-shaped vocal tract and relatively lower larynx necessary for the production of many sounds, especially vowels, while the latter probably was made possible due to changes in nutrition habits, which required less biting and chewing. As already mentioned, non-human primates have a relatively high larynx, which makes the production of vowels very difficult. When during evolution the descent of the larynx occurred is unclear; most experts assume that the Neanderthals did not possess a fully lowered larynx and accordingly not a fully articulated speech as found in *Homo sapiens* (Corballis 2010). There was likewise a modification of the inner ear concerning both its vestibular and auditory functions: The vestibular system had to keep up with new challenges regarding balance during upright walking, and the auditory system had to specialize on the recognition of the sounds of human voice, which is characterized by much higher frequencies than that of non-human primates.

The second important factor was a novel control of this new vocalization system. In non-human primates, vocalization is exerted mostly by limbic cortical areas, such as the cingulate gyrus and subcortical centers, like the periaqueductal gray and the nucleus ambiguus, and there is no direct cortical, but only subcortical limbic control of the nucleus ambiguus (Heffner and Heffner 1995). Accordingly, most vocal sounds represent expressions of emotions like pain, arousal, alarm, or threatening calls. Direct cortical pathways exist for the control of facial muscles, lips, and jaws, but they are not used for volitional control of the larynx (see below). In humans, isocortical control areas like Broca's speech area in the lateral frontal cortex are added, making volitional control of speech production possible.

The human brain possesses a number of frontal, temporal, and parietal areas related to language and speech located in the left hemisphere (cf. Friederici 2011; Vigneau et al. 2011; Price 2012), plus a number or areas related to contextual information, and emotional aspects of speech (prosody) without phonological components in the right hemisphere (Vigneau et al. 2011). With many subdivisions, these include areas A 44/45 (the classic Broca's area) and the inferior frontal sulcus in the frontal lobe, the frontal operculum above the insular cortex, areas A 42/22 of the superior temporal gyrus (i.e., the anterior and posterior portion of the classical "Wernicke area"), parts of the middle temporal gyrus, and the inferior parietal and angular gyrus. According to recent evidence, the mentioned parietal and temporal areas are connected, probably bi-directionally, with the frontal areas by two "dorsal" and two "ventral" pathways. The "dorsal pathway I" connects the superior temporal gyrus with the premotor cortex responsible for speech production, and the "dorsal pathway II" connects the dorsal temporal gyrus with Broca area A 45. Both are part of the arcuate and superior longitudinal fascicles. The "ventral pathway I" connects the anterior temporal cortex with A 45 and A 47, and the "ventral pathway II" the anterior temporal cortex with the frontal operculum plus medial and orbital frontal regions, both included in the uncinate fascicle (Friederici 2011). According to Weiller et al. (2011), the dorsal pathway system is involved in speech production (articulation, dorsal I) and precise and rapid analysis of serial sequences as the basis for syntax, i.e., complex phonological segmentation (dorsal II), in close interaction with the working memory located in the dorsolateral prefrontal cortex. The ventral pathway system, instead, is generally involved in speech comprehension based on meaning and its contextual cues. Thus, inside the human brain, speech and language are based on a complex system with centers distributed over frontal, temporal, and parietal areas in both hemispheres processing the syntactical, lexical, semantic, emotional, and pragmatic components, with the left hemisphere dominating in syntactical and grammatical aspects.

There is extensive discussion about the question of whether, in addition to "Wernicke-like" temporal areas, at least parts of Broca area are already present in non-human primates. According to Corballis (2010), the posterior part, A 44, bordering the regions of the primary motor cortex controlling hand, face, lips, and mouth muscles is not only involved in the control of these muscles during speech production, but also in the recognition of manipulation and grasping hand and arm movements. A number of authors believe that A 44 is homologous to the F5 region in monkeys, where the mirror neurons mentioned in Chap. 12 are located, and a close relationship between these mirror neurons and the evolution of human language is assumed (Rizzolatti and Arbib 1998). However, connecting mirror neurons to human language creates a number of problems (cf. Aboitiz et al. 2006; Corballis 2010). First, neural control of vocalization in non-human primates is not exerted by prefrontal cortical regions. Second, at least monkeys, where the F5 neurons or cortical areas are found, unlike humans, do not naturally imitate gestures or facial expressions like mouth or hand opening except lip smacking in neonates, and mirror neurons appear *not* to be involved in imitation, as opposed to similar neurons found in humans. However, monkeys can be trained to follow attentional cues provided by humans, which means that there may be an "exaptation" in the sense of a latent ability (cf. Chap. 3). Furthermore, mirror neurons respond only to *transitive* movements, i.e., reaching for goals, but not to intransitive ones, i.e., arm and hand movements *without* involving objects. Finally, monkeys do not appear to possess a theory of mind as an important prerequisite for human linguistic communication.

Thus, while it may be that the evolution of the Broca speech area is somehow linked to the F5 mirror neuron region, one must assume a dramatic change in the function of that region and associated temporal and parietal regions, like the superior temporal sulcus (STS) and the inferior parietal lobule (area PF), plus substantial changes in projections of these areas to frontal cortical areas (cf. Aboitiz and Garcia 1997). In humans, and in contrast to monkeys, these areas, in combination with the Broca area, are involved in imitation, and this new function is viewed by some authors as one of the starting points of human language evolution (cf. Rütsche and Meyer 2010; Fitch 2011a, b). These authors reject the idea of an evolution of human speech from vocal expression systems in non-human primates because of the mentioned differences in neural control and the rigidity of their sounds. They rather assume the evolution from manual gestures, which in non-human primates are much more flexible, more sophisticated and highly intentional and are learned to a large degree. While in non-human primates, manual gestures are not coupled to communicative sounds, such a coupling to word comprehension and production is very tight in humans, as we all know, and appears very early in ontogeny, i.e., from 11 months on after birth. Even more, manual gestures can fully replace spoken language, as is the case in any sign language, which are considered fully developed linguistic systems.

However, in order to make such a scenario plausible, we have to assume two fundamental evolutionary steps. The first step concerns a coupling between manual and facial gestures, including mouth movements, and the second includes the coupling between mouth gestures and volitional speech production. It is speculated that the first step was linked to hand-mouth coordination, and then mouth movements gradually assumed dominance over hand movement and were eventually accompanied by movements of the tongue and the vocal tract and resulting sounds (Rütsche and Meyer 2010; Corballis 2010). Recently, the "lip smacking" found in non-human primates during communication may have been such a "prelude" for vocal language in humans (Fitch 2011a, 2011b). In addition, the Broca region must have acquired an involvement in speech-associated gestures, which may improve the understanding of verbal communication by reducing semantic ambiguity and syntactical complexity. Thus, there may have been an evolutionary transformation of Broca region A 44 from classical "mirror neuron" functions regarding the recognition of transitive, goal-directed movements, to intransitive, meaningful gestures to the additional comprehension of facial expressions and, eventually, sounds. In this sense, words and sentences are nothing but "speech gestures" (Corballis 2010). However, in the present context of the origin of human language, we must leave the question of "gestures first" or "vocalization first" undecided; perhaps both processes took place in parallel.

Another important step for the evolution of human language appears to be the formation of new connections between the anterior and the posterior language zone as described above. According to Friederici (2009), the dorsal pathway system appears to be evolutionarily younger than the ventral pathway system, and the dorsal pathways mature relatively late in the sense of myelination and only after the 7th year. This is consistent with the fact that up to that age, children still make typical mistakes in the comprehension of syntactically complex sentences (Friederici 2009). Aboitiz and Garcia (1997) likewise assume that the ventral system is phylogenetically older and that the already existing general function of the dorsal system, i.e., temporal ordering of events and acts, was linked to language during hominid evolution. In this context, the evolution of Brodmann areas A 39 and 40 at the junction of the parietal, temporal, and occipital lobe, apparently unique to humans, was a decisive step by increasing the connectivity of superior temporal auditory regions with parietal regions involved in action planning and execution via connections to frontal premotor areas (Aboititz and Garcia 1997).

A final important factor was the substantial changes in cognitive-executive functions. These changes were either preceded or paralleled by substantial modifications of the frontal cortex, particularly of the dorsal and lateral prefrontal (A 9, 46) and the frontopolar cortex (A 10), which is considerably enlarged in absolute and relative terms in humans (Semendeferi et al. 2002; cf. Chap. 14). This part of the brain, together with temporal and parietal association areas, is assumed to be the "seat" of all faculties that form the cognitive basis of language, i.e., thoughts, imaginations, memories, wishes, and goals including an efficient working memory with a likewise efficient access to the verbal as well as non-verbal long-term memory. One of the central functions of the prefrontal and frontopolar cortex is the *temporal segmentation* of events as one major task of working memory. We can assume that this function increased with the increasing demand for an exact short-term recall and prediction of the sequence of events and actions, e.g., in the context of remembering the number and kind of conspecifics, prey, or predators coming and going, of steps in the fabrication and application of tools or at building a shelter or house. Most primates, including monkeys and even apes, are not good at these mental abilities, as the "delayed-match-to-sample" experiments demonstrate (cf. Chap. 12), but the capacity of working memory strongly expanded during human evolution (Aboititz and Garcia 1997; Aboititz et al. 2006).

The increase in the ability for temporal segmentation was put into the service of mental manipulations, i.e., thinking and action planning, and finally of language. In that way, syntax and grammar could evolve or at least substantially improve. This, however, required that the lateral prefrontal cortex was connected with the vocalization apparatus (mouth, lips, larynx, etc.) via the "dorsal pathway system," which is missing or present only in rudimentary form in the non-human primate brain. This evolutionary process apparently led to the formation of the "phonological loop" of the working memory as an essential prerequisite of human language (Aboititz et al. 2006).

15.4.3 The Tempo of the Evolution of Human Language

While it is now widely accepted that human language evolved gradually, albeit with substantial changes and additions of functions, the entire process, nevertheless, took place in a relatively short time. Most experts assume that australopithecines did not differ much from chimpanzees in sound production, and *Homo habilis* may not have possessed much better linguistic capabilities, although he already had a much larger brain and better abilities for cooperative hunting and food gathering. Presumably, with his 700 ccm large brain, he still remained mostly below the linguistic "cerebral rubicon", i.e., a cranial capacity which—according to the Scottish anthropologist Arthur Keith—lies around 750 ccm. *Homo erectus* reaches this cranial volume at age 6, whereas a *Homo sapiens* child reaches it already after 1 year. We know nothing about the linguistic abilities of *Homo erectus*, and the same is true for *Homo neanderthalensis*, despite the fact that he had a brain that was larger than *Homo sapiens*. Presumably, his larynx differed from that of modern humans and appears to have had a restricted sound production capacity, especially with respect to vowels. If we assume that the split between the older *Homo sapiens* and *Homo erectus* occurred 600–500,000 years ago in East Africa, and that the modern *Homo sapiens* evolved in the same region 200–150,000 years ago, then human language probably evolved between 150,000 and 80,000 years ago. A recent study by Atkinson based on an analysis of phonemic diversity suggests that modern human language probably predated the African exodus and paralleled the earliest archeological evidence of symbolic culture in Africa 80–160,000 ya (Atkinson 2011).

Some years ago, the discovery of the so-called FOXP2 transcription factors excited evolutionary linguists (Enard et al. 2002). These factors were first discovered in a family whose members had severe language deficits in combination with non-linguistic cognitive impairments and a significantly lowered intelligence quotient. Genetic studies showed that these multiple deficits are causally linked to a defect in the gene family FOXP2. These genes encode a transcription factor controlling the expression of perhaps a larger number of other genes. Structural deficits concern a decrease in the volume of the Broca area and the left caudate nucleus, which is highly active during speech production. The evolutionary geneticist Svante Pääbo and his colleagues from Leipzig were able to demonstrate that the human FOXP2 gene differs from the non-human primate gene only in two amino acids (Enard et al. 2002), and the authors speculate that the last FOXP2 mutation in humans occurred between 100,000 and 10,000 years ago. There was much debate about the function of that gene/transcription factor, from "speech gene" to even a "grammar gene." Since then, it became evident that FOXP2 is widely distributed among vertebrates and linked to sound production like echolocation in bats or bird song (cf. Scharff and Petri 2011). In *Homo neanderthalensis*, the same allele as the human ones was present. This latter finding could be either due to a "contamination" from the human genome, or indicate that Neanderthals already had something like a "modern" language, or that even in the genus *Homo*, FOXP2 has a more general function beyond being a "grammar gene."

Besides all this discussion around the evolution of human language, it is clear that this step had huge consequences for the further development of mankind. As a mighty "intelligence amplifier," it certainly boosted the development of more refined tool fabrication, cooperative hunting, culture in the modern sense, civilization and art, as represented in the impressive cave paintings dating back about 40,000 years ago.

15.5 Do Humans Exhibit a Special Social Behavior?

In recent years, many psychological-anthropological investigations aimed at the question of in what respect human social behavior differs from that of non-human animals, especially regarding our closest relatives, the chimpanzees. I have already mentioned some aspects, for example, imitation, theory of mind, and knowledge attribution, and it appears that they are already found in chimpanzees, but that humans have substantially further developed these abilities. Humans imitate other humans from birth on and do this extensively, as opposed to apes (cf. Chap. 12). Furthermore, they readily distribute new experience to conspecifics. Human empathy and theory of mind far exceed what is found in chimpanzees. However, all this does not disrupt the picture of a quantitative rather than qualitative evolution.

But aren't there truly qualitative differences in complex social behavior, e.g., in the context of cooperativity? Michael Tomasello of the Leipzig Max-Planck Institute and Felix Warneken of the Harvard University (Tomasello and Warneken 2009) recently distinguished three ways of cooperative or altruistic behavior that is shown (1) to help others reach a certain goal, (2) to share goods, e.g., food, and (3) inform others about things that are possibly important for them. There are a number of studies comparing the behavior of chimpanzees and young children. These studies demonstrate that children help other individuals from age of 14–18 months on, even when not rewarded. They spontaneously pick up things dropped by adults, or open a door for them, even if their hands are full. The same behavior is found in chimpanzees, although to a lesser extent. In children as well as in chimpanzees, there seems to be some inborn cooperativity or at least some intrinsic reward, because they do that without extrinsic reward, and subsequent rewarding does not increase this altruistic tendency. Instead, at least in children, being rewarded decreases the inclination for helping, giving the impression that the previous intrinsic reward was replaced or at least reduced by an extrinsic one.

The differences between children and chimpanzees were larger when *sharing* was considered. While children generally like to share things from early childhood on, and even valuable things, chimpanzees do that only with things that have little value. They do not participate in joint activities, where others get the same share as they do, especially regarding food, while children do that spontaneously. Even if chimpanzees are "begged for it," they share only in the case of things of low value, e.g., tasteless food. The most striking differences are observed in the case of mutual information, which in the eyes of Tomasello and Warneken are a special kind of altruism and cooperativity.

Quite normally, humans inform other people without having any profit from doing so. In chimpanzees as well as in all other mammals except humans, all kinds of information about food places or threats by predators or enemies—according to the argument of the authors—are ego-centered. Even when chimpanzees address humans, this normally aims at getting something. In contrast, people help other people who are not their relatives or whom they do not even know and without having any profit ("charitable altruism"; cf. Harbaugh et al. 2007). The conclusion is that humans are characterized by special pro-social behavior exceeding the widespread reciprocal altruism ("I will help you if you help me") and suggests a truly "unselfish" behavior. Recent investigations, however, showed that even in these cases of "charitable" altruism, the cerebral self-reward system located in the mesolimbic system rewards the actor by releasing endogenous opioids leading to a state of "feeling good" in the do-gooder. There is also the phenomenon of "altruistic punishment," i.e., that individuals punish individuals behaving egoistically, although the punishment is costly for them and yields no material gain (cf. Fehr and Gächter 2002).

Of special interest is the fact that very young children exhibit such an altruistic behavior to a higher degree than older ones, and this phenomenon is explained by the influence of negative experience with ungrateful peers. Tomasello and Warneken assume an inborn prosociality, which is modulated by later social experience (cf. also Henrich et al. 2006; Almas et al. 2010). They see the roots of this disposition in the specific ecological and social conditions under which our ancestors had to live. Survival in the savanna was promoted by increased cooperativity in the form of mutual help, sharing and eventually exchange of information. This concerned not only cooperative hunting, but particularly the joint raising of children.

15.6 What Does All This Tell Us?

The evolution of the australopithecines and eventually *Homo sapiens* after the "exodus" from the tropical forest and invasion of the dry savanna is characterized by fast changes in locomotion (upright walk), feeding habits and diet, hunting style and, finally, with the appearance of *Homo habilis* about 2 mya (or somewhat earlier), a strong increase in brain size from about 600 to 1,350 ccm in *H. sapiens* and around 1,700 ccm in *H. neanderthalensis* paralleled by the invention of fire use, weapons, and other tools. The reasons for this rapid evolution are unclear, but it is likely that it was made possible by changes in genes controlling general brain growth, because increases in human brain size occurred along general trends including positively allometrical growth of the cortex, which more or less automatically led to a relatively larger prefrontal cortex including an increase in working memory.

In practically all cognitive abilities, humans exceed all other animals. This holds for all types of learning and memory formation as well as for all so-called higher cognitive functions like thinking, abstraction, categorization, mirror self-recognition, deception and counter deception, empathy, theory of mind, knowledge attribution, and metacognition. Similarly, humans exhibit an increase in prosocial behavior, i.e., cooperativity and "unselfish" altruism. Differences between humans and non-human animals are particularly large regarding the capacity for intermediate and long-term action planning and syntactical-grammatical language. Apes, and particularly chimpanzees, are capable of planning actions a few hours in advance, but all other animals investigated so far exhibit no longer action-planning. Apparently, this has to do with the difficulty of "keeping in mind" goals and plans for more than a few minutes or even seconds. In most humans, such an ability is essentially linked to the "phonological loop" of our working memory, letting us mentally list things like numbers, names, or places, and this enormously enlarges the span of our working memory. The planning of future actions beyond one day usually requires a calendar.

Syntactical-grammatical language appears to play a crucial role in cognitive achievements of humans because it makes a way of thinking and reasoning possible, which is impossible or at least difficult to perform non-linguistically. Human language is based on the general ability to process mental events in a *temporal sequence*, and this general ability is essentially a-modal, i.e., it may concern sounds, words, thoughts, or images. Deaf people usually have intact "Wernicke" and Broca speech areas, and deficits in these areas lead to comparable impairments as in individuals with vocal language. This demonstrates that there is an amodal cognitive ability of temporal segmentation and handling, which was put into service of language probably 100–50,000 years ago. Although there are clear forms of proto-language in apes, these animals never pass the barrier of an essentially agrammatical and asyntactical language consisting of 2- to 3-word sentences, even if they are taught (or teach themselves) the use of non-vocal language.

In summary, even after intense search, we find no "truly unique" characters in humans compared to other animals—at least in the cognitive domain. There is nothing in the evolution of humans that does not have pre-stages or could not have served as "exaptation" for further evolution. Rather, humans appear to be characterized by a unique combination of traits, which were—at least in rudimentary form—already present in their ancestor, like hand use, upright walking, a large brain enabling high general intelligence, and a highly efficient way of verbal communication.

Chapter 16
Determinants of the Evolution of Brains and Minds

Keywords Determining factors of brain evolution · Evolution of cognitive-mental functions · Relationship between intelligence and brains · Ecological intelligence · Social intelligence · General intelligence

I will begin this chapter with a summary of the data regarding the putative evolution of nervous systems and brains and ask whether a general pattern of that process emerges. Then I will summarize the insight into differences in intelligence among extant animals, and I will ask to what extent these observed differences can be related to patterns of brain evolution. Finally, I will ask what are the driving forces behind this co-evolution of brains and minds.

16.1 Patterns of the Evolution of Nervous Systems and Brains

The basic organization of organisms for the control of a behavior that promotes survival and reproduction is as old as life itself. Already at the levels of bacteria and eukaryotic unicellular organisms (protozoans), we find the fundamental organization of behavioral control into a sensory, integrative, and a motor part. This includes a short-term memory, and with this, a minimum of information processing. In multicellular animals above the level of sponges, true nerve cells and diffuse nerve nets originated. From there two basic lines of development diverged. The first and minor one is the evolution of ring-shaped nervous systems in cnidarians and ctenophorans (the former "coelenterates"); the other, and dominating one, leads to nervous systems of bilaterally organized animals, with a supra- or circum-esophageal ganglion located in the head and nerve cords being highly variable in number and extending throughout the body of the animals. This evolutionary bifurcation between cnidarians-ctenophorans and bilaterians took place about 600 million years ago or even earlier. At this time, we already find near ion channels, neuroactive substances (transmitters, neuropeptides, neurohormones), electrical and chemical synaptic transmission mechanisms and simple ways of learning and memory formation, together forming the "language of neurons."

From the first bilaterally organized animals, again two major evolutionary lines originated around 540 mya with the beginning of the Paleozoic era, one leading to the protostomes (or "invertebrates"), comprising the largest number of species and the greatest diversity in form and lifestyle, and the other leading to the deuterostomes including craniates-vertebrates. Protostomes, in turn, split into two major lines, the lophotrochozoans and the ecdysozoans. The basic organization of the nervous system of the last common ancestors of both protostomes and deuterostomes is debated. Traditionally, it was assumed that the CNS of this common ancestor had a very simple organization resembling the diffuse nerve net found in *Hydra*, and from there a *multiple* and mostly *parallel-independent* evolution of complex sense organs and nervous systems/brains took place. According to that view, this happened within the lophotrochozoans in predatory flatworms and polychaetes and the likewise predatory cephalopods. Here, powerful eyes and visual systems evolved as well as multilobed supraesophageal ganglia, and the brain of the mollusk *Octopus* is regarded as the largest and most complex one among the invertebrates. Within the ecdysozoans, the arthropods, as the largest animal group, likewise developed complex sense organs of great diversity and a multilobed or tripartite brain, which in spiders, crustaceans, and insects has taken specific modifications. The brains of flies and hymenopterans, including bees and wasps, are another prominent example of high neural complexity among protostomes-invertebrates.

This contrasts with a more recent view, that the evolution from a diffuse nerve net into a tripartite brain already took place *before* the split between protostomes and deuterostomes about 600 mya. Accordingly, the division of the arthropod brain into a protocerebrum, deutocerebrum and tritocerebrum as well as that of the chordate-vertebrate brain into a prosencephalon, mesencephalon, and metencephalon are due to "deep homology" and not a product of convergent evolution. This view is based on the presence of homologous genes involved in the anteroposterior organization of brain like the Hox, Pax, and otd/Otx genes (Chap. 10) in distantly related taxa like insects (*Drosophila*), ascidians (i.e., "primitive" chordates), frogs (*Xenopus*), and mice (Hirth and Reichert 2007). If this view is correct, then many protostome taxa with simple brains represent many cases of secondary simplification, and the same must have happened in some deuterostome taxa with likewise simple brains (e.g., in the echinoderms and hemichordates). It could also be, however, that the above-mentioned developmental control genes *pre-dated* the realization of complex brains. At the same time, there can be no doubt that despite the existence of a relatively ancestral complex brain in the last common ancestor of bilaterians, there was a further independent increase of brain complexity in many lines of protostomes and deuterostomes.

The evolution of the deuterostomes likewise exhibits two major pathways, one leading to the echinoderms, which, like cnidarians and ctenophorans, are characterized by a ring-shaped, radially symmetric and de-centralized nerve net of unknown phylogenetic origin—probably as the result of secondary simplification.

The other pathway leads to the chordates comprising the simply organized (or simplified) uro- and cephalochordates and the more complex craniates which comprise hagfish and vertebrates. All craniates, like most invertebrates, possess a tripartite brain, which then divides into five parts (myelencephalon, metencephalon, mesencephalon, diencephalon, and telencephalon). This basic and highly conserved organization of the vertebrate brain appears to have originated about 500 million years ago. However, there are differences in brain size over 8 orders of magnitude, as well as in the relative enlargement of parts of the brain, i.e., some parts, like the telencephalic pallium or the cerebellum and the valvula cerebelli, became enormously large and complex.

In summary, although there is a basic organization of a tripartite vertebrates brain, possibly homologous to that of protostomes, in many vertebrate lines, increases in anatomical and functional complexity took place many times independently, mostly in the context of a specialization, *de novo* formation or re-invention of sensory systems like gustation, electrosensation, infrared and echolocation system, visual and auditory system. However, even among vertebrates, there are a number of cases of secondary simplification, definitely in all amphibians and probably in hagfish. Finally, there were independent changes within the diencephalon and telencephalon, particularly of the dorsal thalamus and the closely connected dorsal telencephalon, the pallium. This pallium underwent dramatic evolutionary modifications in cartilaginous and bony fishes, in sauropsids and in mammals. In the latter, a six-layered cortex developed from the dorsal pallium with primary, secondary and associative visual, auditory, somatosensory, and vestibular areas receiving afferents mostly from the dorsal thalamus. In sauropsids, presumably the ventral pallium transformed into the dorsal ventricular ridge or mesonidopallium, which likewise became the targets of primary sensory afferents from the dorsal thalamus. Thus, among terrestrial vertebrates and starting from the situation found in amphibians, there have been two parallel and independent evolutionary lines leading to the "exploitation" of the pallium/cortex for the formation of sensory maps, integrative areas, extended memory, and motor systems, i.e., one in mammals and the other in the sauropsids culminating in birds.

What about specialties of the human brain? Humans neither have the absolutely nor relatively largest brains. However, if we apply Jerison's encephalization quotient (EQ) indicating to what degree brain size of a given species is below or above average brain size of the respective higher taxon (here mammals), then humans are on top with a brain that is about eight times larger than expected. Nevertheless, at the same time the human brain follows ordinary laws of brain allometry, because its isocortex, including the frontal cortex, increases slightly positively allometrically, as in all mammals. Thus, humans do not have an unusually large isocortex or prefrontal cortex; what is unusual is their large brain given a relatively large body, as a consequence of a strongly positively allometric brain growth. The causes for this process are unclear, but again this was not a unique event, but happened in dolphins as well, although to a somewhat lesser degree.

Two major conclusion can be drawn. The first is that the evolution of nervous systems and brains was not a linear process "from worm to man," but one that from an initial state of a diffuse nerve net or primitive bilateral nervous system (a simple and perhaps already tripartite supraesophageal ganglion and ventral nerve cords) diverged into several large and many smaller developmental lines. As with biological evolution in general, the evolution of the nervous systems is a *tree-like*, i.e., multiply branching process. This process is neither linear nor goal-directed: *Homo sapiens* and his brain is not the ultimate goal of evolution, but a momentary terminal point of one of countless evolutionary processes.

The second fact is that phylogenetic history is *not* identical with evolution in the sense of increase in complexity ("anagenesis") of form and function. No matter how impressive the improvements of sense organs, nervous systems, and brains in the different lines of invertebrates and vertebrates appear, they represent only a very small part of the entire phylogeny of animals. Far more animal taxa remained with their sense organs, nervous systems, and brains at a level of relatively low complexity or modified them only in tiny steps, and hundreds of thousands of species even became simpler, mostly in the context of transition to sedentary or parasitic lifestyles.

Thus, during the phylogeny of animals we recognize three major "strategies." The first and dominant one is "Remain as you are—no further experiments!" The second is: "Simplify your life and brain, whenever this is possible!" Only the third and least frequently realized one is "Become more complex, whenever this is necessary and advantageous!" *An evolutionary increase in complexity is an exception, not normality.*

Most impressive is the large number of cases of parallel evolution of complex forms and functions in sense organs and brains, e.g., the formation of lens or compound eyes or auditory organs, "rope-ladder" nervous systems, tripartite brains, mushroom-body-like structures, pallium- or cortex-like structures as seats of intelligence. As already discussed, there is no consensus among experts about how to interpret this situation. Traditionally, and in the context of Neo-Darwinism, this is seen as cases of truly independent and *convergent* evolution of similar forms and functions as "adaptations" under similar selective pressures of certain living conditions. However, together with the discovery of ancient developmental genetic mechanisms, there is growing insight that these evolutionary phenomena are not as independent as they appear. Rather, the presence of developmental genetic mechanisms makes the evolution of certain forms and functions like a lens eye or a tripartite brain more likely (see above). Of course, both interpretations are not mutually exclusive: given certain developmental-genetic mechanisms, organisms will follow certain environmental challenges more easily than others and will develop similar forms and functions.

16.2 The Evolution of Cognitive-Mental Functions

All animals exhibit some behavioral flexibility, and all eukaryotic animals can learn, although to very different extents. Thus, we can test the degree of *behavioral flexibility* which animals under study reveal when confronted with new problems either under natural or laboratory conditions with respect to ecological, social, physical-instrumental, or other abstract challenges.

Bacteria exhibit a system of behavioral control that definitely exceeds the level of a mere "reflex machine." All true multicellular organisms reveal behavioral plasticity in the form of habituation and sensitization and simple classical conditioning. Among invertebrates, we find complex cognitive abilities in many lines culminating in cephalopods and insects, especially in hymenopterans like bees and wasps and even in tiny fruit flies (*Drosophila*). *Octopus* and the honeybee keep up with many "smart" vertebrates, like corvids or primates, with respect to navigation, learning, and memory formation.

Among vertebrates, we find highly developed mechanisms for spatial orientation and recognition of prey, food, enemies, and conspecifics, which are paralleled by sophisticated sensory systems. Some bony fish, like cichlids and weakly electric fish, exhibit complex communication systems. Among terrestrial vertebrates, birds and mammals generally exceed both amphibians and "reptiles" with respect to cognitive abilities. Among birds, corvids and parrots stand out in this regard with behavioral flexibility, innovation rate, tool use and tool fabrication, and also with respect to truly mental abilities such as logical reasoning and mirror self-recognition—at least in one corvid species.

Several groups of mammals, like dolphins and whales, dogs, elephants, and bears (just to mention a few) show signs of high intelligence at least in some cognitive domains. Primates on average exhibit an intelligence superior to all other mammals. Among primates, there is a rather clear-cut ranking order in intelligence, from the prosimians to monkeys and to the great apes. Except a few species (e.g., capuchins), the great apes exhibit at least some aspects of cognitive and mental abilities not found in monkeys regarding tool fabrication, insight into causal mechanisms, mirror recognition, theory of mind, knowledge attribution, metacognition, and consciousness.

However, humans, even under the most critical aspects, are superior to other animals in all cognitive functions, no matter how astonishing the achievements of the latter may be. The most clear-cut differences between humans and non-human primates lies in two abilities that are interconnected: planning abilities and a syntactic-grammatical language. When we compare the cognitive-mental abilities of the most intelligent non-human animals with those of humans, then we find that they roughly correspond to the abilities of children aged 2 1/2-5. As for linguistic abilities, chimpanzees and gorillas equal a 3-year-old child, while with respect to psychosocial abilities (empathy, theory of mind, etc.) they may be equivalent to those of a 5-year-old child. In light of these empirical findings, the standard question of whether human intelligence differs qualitatively or only quantitatively from the non-human one, may ironically be transformed into the question about whether, with respect to cognitive functions, an adolescent or adult is qualitatively or only quantitatively superior to a 3–5-year-old child. Besides maturation of social competences, the most decisive feature that distinguishes humans from non-human animals is the appearance of a syntactical-grammatical language at age 2 1/2, which is paralleled by an enormous increase in the capacity of working memory and, consequently, intelligence, i.e., novel problem solving.

Therefore, the "rubicon" between animal and human intelligence seems to be the evolution of the syntactical-grammatical language, which is essentially bound to an increase in the ability to mentally manipulate processes (first actions, then thoughts, then words) in the temporal domain. Once evolved, human language served as a mighty "intelligence amplifier," as was later development of writing and invention of the computer.

16.3 How Do Differences in Intelligence Relate to Differences in Brain Structures and Functions?

If we make comparisons *within* the major evolutionary lines, e.g., lophotrochozoans, ecdysozoans, and craniates-vertebrates, we get a rather clear-cut correlation between the complexity of nervous systems and brains on the one hand and degree of intelligence in the sense of learning capabilities, behavioral flexibility, innovation rate, etc., on the other. The highest levels of complexity of brains and intelligence are invariably found in those taxa that are predatory and/or live in complex, relatively unpredictable and therefore challenging natural or social environments. Difficulties arise when we make comparisons *across* major evolutionary lines, i.e., compare the honeybee with *Octopus*, or a corvid bird with a monkey or even with a chimpanzee. If we consider the intelligence of a honeybee to be comparable to that of *Octopus*, then the differences in brain size are remarkable, and the same situation holds for the comparison of a New Caledonian crow and a macaque monkey or a chimpanzee. When we take the number of neurons into consideration, rather than mere absolute brain size, the situation does not become better for the honeybee and *Octopus* (1 million versus 40 million neurons, which is a ratio of 1:40). A similar situation is found when we compare the New Caledonian crow, with up to 200 million neurons (gross estimate), with the chimpanzee with more than 6 billion neurons (which is a ratio of 1:30). Here we have to take into consideration additional factors relevant to the information processing capacity (IPC), like cell-packing density/interneuronal distance and conduction velocity, which usually are more favorable in small brains with tiny and tightly packed neurons compared to larger brains with larger and more loosely packed neurons.

The best correlation between brain properties and levels of intelligence is obtained when we focus our attention on primates. Here, as mentioned in Chap. 14, prosimians and tarsiers have relatively small brains with an average of 7 g, followed by New World monkeys with an average of 45 g and Old World monkeys with an average 115 g. Among apes, gibbons have brain sizes around 120 g, which lies within the range of Old World monkeys, and the great apes, i.e., orangutans, gorillas, and chimpanzees, have brain weights between 300 and 600 g. Far on top are humans, with brain weights around 1.350 g. As discussed in Chap. 14, this corresponds relatively well with the ranking order of intelligence of the primate taxa, including humans. Exceptions are the New World capuchin monkeys, which have relatively small brains, but are considered at least as intelligent as an Old World macaque or baboon, and the gorilla, which has a considerably larger brain than a chimpanzee, while it appears to be somewhat less "talented" than chimpanzees. At least this latter exception can be explained by determining the neuron number: It turns out that chimpanzees have considerably more neurons than gorillas, because within their brains, the neurons are smaller and more densely packed.

Taking into consideration neuron number, neuronal distance and, finally, cortical conduction velocity, we likewise can remove inconsistencies in comparing primates to non-primate mammals. Many of the latter have considerably larger brains than the former: ungulates, cetaceans, and elephant, just to mention a few, but it turns out that they have fewer neurons than primates with the same or even smaller brain size. Only dolphins, whales, and elephants have more neurons than the great apes, but here the large interneuronal distance and a relatively slow cortical conduction velocity are unfavorable factors for neuronal information processing capacity. The great advantage of primates over all other mammals regarding IPC lies in the fact that they generally have larger brains relative to body size, smaller and more densely packed neurons, and a higher cortical conduction velocity.

However, before we can draw a general conclusion about the co-evolution of brains and minds, we have to further address two unsolved problems. The first is which factors ultimately drove the increase in complexity of brains and accordingly intelligence? Maybe the term "intelligence" means very different things that cannot be compared directly. The second conundrum concerns the puzzling fact that animals with rather tiny brains (honeybees, corvid birds) may be rather smart, while those with relatively large brains may be of only moderate intelligence (ungulates, cetaceans, elephants).

16.4 Which Are the Ultimate Factors for the Evolution of Brains and Minds?

Until recently, three major factors determining the evolution of brains and intelligence have been discussed: (1) *ecological intelligence*, i.e., mastering challenges of an environment, (2) *social intelligence*, i.e., mastering the challenges of social life and survival, and (3) *general intelligence*, i.e., efficient information processing. Recently, Bates and Byrne have argued in favor of *physical intelligence*, which includes, among others, tool use, innovation rate, and causal understanding and reasoning, and they distinguish "lower" social intelligence regarding factors such as home range, group size, degree of social interaction, deception, ranking order, simple forms of cooperativity (cooperative hunting), and social communicative systems from "higher" ones including individual recognition, sophisticated social tactics, coalitions, theory of mind, knowledge attribution, and self-recognition (Bates and Byrne 2010).

16.4.1 Ecological Intelligence

Regarding the hypothesis of *ecological intelligence* as the major factor driving an increase in brain size or in relevant parts of the brain like the cortex or pallium, frontal cortex, etc., the relationship between spatial orientation and spatial memory on the one hand and size of the hippocampus on the other has been scrutinized. In a much-cited article, Krebs and coworkers (1989) found a significant correlation between the ability of birds to cache and/or recover hidden food and the size of their hippocampus. Plowright and colleagues (1998) found similar correlations in mynah birds (*Gracula religiosa*). Sherry (2011) found that the hippocampus of the storing chickadees (*Poecile atricapillus*, a tit species) was much larger than that of the non-storing canaries. The fact that elephants, with their astonishing spatial memory (Chap. 12), have an unusually large hippocampus (Hart and Hart 2007) points in the same direction. However, whales likewise have excellent navigation abilities, but a surprisingly small hippocampus (Hof and van der Gucht 2007). This latter finding could rather be a consequence of the fact that in "primitive" mammals the hippocampus is the site of olfactory memory (even in primates, the hippocampus is in close proximity to the olfactory and entorhinal cortex), and that cetaceans have almost completely lost their olfactory system (Hof and van der Gucht 2007).

In a meta-analysis published in 2001, MacPhail and Bolhuis came to the conclusion that empirical evidence for a correlation between spatial orientation and hippocampal size is weak at best and reaches significance only in birds, but not mammals. This has been confirmed more recently by Lefebvre and Sol (2008). In addition, Cnotka and colleagues (2008b) found that in homing pigeons, the size of the hippocampus is influenced by experience. This could also be the case in the famous correlation between spatial orientation abilities and hippocampal size in London taxi drivers (Maguire et al. 2000).

Within the past decade, the relationship between climatic changes, innovation rate, and behavioral flexibility on the one hand, and brain features on the other has been studied in birds (cf. Burish et al. 2004; Iwaniuk and Hurd 2005). In a number of bird taxa, Lefebvre and colleagues (2004) determined the degree of "behavioral innovations" in correlation with the ability to cope with seasonal changes in the environment. They found that in the bird species investigated, the degree of this ability was significantly correlated with the corrected relative size of the "hyperstriatum ventrale" (today called "hyperpallium") and of the "neostriatum"(today called "mesonidopallium"). They came to the conclusion that within birds, this correlation must have evolved six times independently. In another study based on a large number of different bird taxa, Sol and colleagues (2005) found that birds with corrected relatively larger brains are better capable of coping with new environments than those with relatively smaller brains. However, there was no such correlation with respect to absolute brain size. In a study on a large number of species of neotropical parrots, Schuck-Paim and colleagues (2008) studied the relationship between climatic variability and absolute as well as corrected relative brain size. These authors, too, found a stronger correlation with corrected relative, but not with absolute brain size.

The arguments of the authors in these studies are based on the general assumption originally made by Jerison (1973) that higher cognitive abilities result from more "extra-neurons" found in the relatively larger brains of some taxa compared to either related taxa or the average of the entire higher taxon (e.g., family). However, this argument is valid only when animals (here birds) of the same body size, but different brain sizes (or sizes of relevant brain areas like the mesonidopallium) are compared and other important variables like cell density and neuron size are taken into consideration. This, however, was not done by the authors.

As to the relationship between "ecological intelligence" and brain size in mammals, sufficient data is available only for primates. In a recent meta-analysis, Lefebvre (2012), on the basis of 26 primates, found that "more encephalized" primates (i.e., those with a larger corrected brain size) eat a higher quality diet, have larger home ranges, and are more arboreal and more frequently live in closed forests than "less encephalized" ones. This confirms the earlier findings of Clutton-Brock and Harvey (1977) that frugivorous primates have larger brains than do folivorous ones. Barton (1996) found that besides social group size, the percent of fruit in the diet predicts relative cortex size. The explanation for this finding is that inside the forest, the spatial and temporal distribution of fruit is more difficult to track and over a wider range than that of leaves, and this requires higher cognitive abilities. However, a later study by Walker et al. (2006) showed that there is a significant relationship between residual brain size and home range, but not with the percent of fruit in the diet, whereas Dunbar and Shultz (2007) as well as Reader et al. (2011), using still other statistical methods, confirmed the correlation between diet and residual brain and cortex size. Thus, in primates the findings regarding the correlation between "ecological" factors like diet and home range are equivocal, and the strength of correlation depends on the methods used.

16.4.2 Social Intelligence

The hypothesis of *social intelligence* or the "social brain" was originally proposed by Robin Dunbar (1995) and Richard Byrne (1995), who assumed that at least in primates, the size of the cortex is determined more by the complexity of social relationships than by environmental complexity. Dunbar (1998) found a significant correlation between cortex size and size of social groups as well as complexity of social interactions in primates. In the opinion of the author, the best example for this relationship is found in baboons, who have the largest isocortex among monkeys and a high degree of sociality. Byrne and colleagues found a significant correlation between cortex size and the degree of "tactical deception" (or "Machiavellian intelligence"; cf. Byrne und Whiten 1988, 1992; Byrne 1995). The underlying assumption is that a larger neocortex can process a larger amount of information important for social life, including alliance networks, dominance relationships, anticipating the behavioral responses of conspecifics, and manipulating them. A number of studies discussed by Lefebvre (2012) showed that neocortex size is associated with group size, number of females in the group, grooming clique size, frequency of coalitions and network connectivity, but Lindenfors et al. (2007) found that this holds only for females, and not for males, which instead exhibit a correlation between size of the limbic system and "social life."

Some years ago, Holekamp (2006) questioned the "social intelligence hypothesis," at least in its generalized form. His main arguments are based on his studies with spotted hyenas, which exhibit high sociality comparable to that of primates, but much lower cognitive abilities, and they do not have an increased absolute or relative brain size. In contrast, bears conduct a solitary life, but are highly intelligent and have relatively large brains compared to other carnivores, e.g., dogs. Furthermore, Old World monkeys like baboons or macaques with high sociality are poor at tool-use. At the same time, social life of the large-brained great apes is no more complex than in monkeys, while their cognitive abilities of the former are much more evolved. For Holekamp, group size, which is central for Dunbar and colleagues, does not correlate well with social complexity. Also, more gregarious birds are not more, or are even less intelligent than non-social birds. He argues that tool use must be interpreted independent of social intelligence.

In summary, in birds there seems to be some correlation between brain size and "ecological intelligence," which is stronger for corrected relative than for absolute brain size, while there is little or no evidence for the "social brain hypothesis." Conversely, in primates data supports at least some aspects of the social intelligence hypothesis, while there is only scanty data on the "ecological intelligence." Lefebvre and Sol (2008) argue that "ecologically intelligent" animals mostly are "socially intelligent," too, for example, as regards innovation rate, tool use, and socially complex behavior. Such a coupling has been demonstrated by Reader and Laland (2002) in primates and Lefebvre and colleagues (2004) as well as Bouchard and colleagues (2007) in various groups of birds. Thus, we have to consider two alternatives: either "ecological intelligence" and "social intelligence" are two independent variables, which may or may not significantly correlate with absolute or corrected relative brain size, or there is "general intelligence" that underlies, in various forms, both types of more special intelligence.

16.4.3 General Intelligence

Several years ago, Gibson and colleagues investigated the relationship between general cognitive abilities and brain features in primates (Gibson et al. 2001). In order to estimate the level of cognitive functions at reward learning, the authors used a "transfer index (TI)" expressing the ability to switch from one strategy to another. TI can assume positive as well as negative values and indicates how much a given animal lies, with its cognitive abilities, above or below the average of a taxon—here, primates. The result was that prosimians generally had a negative TI (i.e., below average) and simians generally a positive TI (i.e., above average), with a value of 9 in macaques and beyond 10 in the great apes (chimpanzee and orangutan 12, gorilla 14), while gibbons had a remarkably low TI of 0.9.

This TI ranking was significantly correlated with both body and absolute brain weight, while there was no significant correlation either with uncorrected or corrected relative brain weight and Jerison's EQ. In this context, it is interesting that the authors found a very modest TI of 0.5 for the capuchin monkey (*Cebus*) for which Jerison had found an astonishingly high EQ of 3.5–4.8. On the other hand, the gorilla turned out to have the highest TI among all non-human primates, while Jerison had found a surprisingly low EQ of 1.76 (Chap. 13). A ranking order similar to that of the TI is achieved if other cognitive functions like tool use and tool fabrication (which Bates and Byrne now call "physical intelligence"), and forms of "higher" social cognition, like Theory of Mind and mirror self-recognition, are taken into account (Chap. 12). Prosimians occupy the lowest ranks, although they exibit some of these abilities. Monkeys are better on average, and the great apes are far better. Here, too, Gibson and colleagues found the best correlation of these performance levels and absolute, but not relative size of cortex, cerebellum, striatum, diencephalon, and hippocampus.

By using the term "physical intelligence" and adopting the distinction between "lower" and "higher" social intelligence or cognition by Bates and Byrne (2010), we can remove a number of inconsistencies within the mentioned findings. First, "lower" social intelligence, including home range and group size, is not or only weakly correlated with absolute or corrected relative brain size and cannot explain the large differences in brain size and intelligence between non-primate mammals and primate mammals as well as between monkeys and apes, the latter of which have roughly equal group sizes and home ranges. For example, cooperative hunting is widespread, but, according to the authors, does not indicate an understanding of the strategy of the others; the same is true for tactics of social manipulation, which does not need an insight into how it works (Bates and Byrne 2010). "Higher" intelligence, according to the authors, is almost exclusively found in the great apes, above all understanding the intentions and other mental states of the others and eventually leading to an understanding of oneself. For Bates and Byrne, "lower" social cognition, found in many mammals, is a matter of degree, while "higher" is essentially restricted to the great apes and perhaps some cetaceans and elephants. In the great apes, such "higher" social cognition is strongly correlated with higher physical intelligence, i.e., understanding the principles of tool use and tool fabrication. Here, corvid birds excel, while hints for "higher" social intelligence are sparse—but this may be a consequence of the lack of intense research.

The arguments proposed by Bates and Byrne eventually lead to the insight that behind the different kinds of intelligence, i.e., environmental, physical, lower and higher social intelligence, there is just one decisive parameter: *general intelligence*. The significance of general intelligence for cognitive functions in primates was analyzed a few years ago by Deaner and colleagues (2007). The authors compared "general intelligence" with absolute brain size, Jerison's EQ, and corrected relative brain size. They found—in contrast to the above-mentioned data on birds—that absolute brain size correlates better with general intelligence than Jerison's EQ, as well as corrected relative brain size. This again speaks in favor of the assumption that general cognitive abilities, such as quick problem solving, can be universally used in any context, whether ecological or social (Hofman 2003; Lefebvre and Sol 2008). The great apes, including humans, are the best examples of that view, because they occupy top positions in technical or "physical" intelligence, e.g., tool fabrication and use, as well as in "higher" social intelligence, i.e., theory of mind, knowledge attribution, and mirror self-recognition.

General intelligence is intimately bound to information processing capacity, which on the one hand depends on the basic efficiency of the cortex or pallium at processing detailed and complex information, but more specifically is related to the efficiency of working memory and, accordingly, "mental manipulation" abilities (cf. Marois and Ivanoff 2005). In primates, this predominantly takes place in the prefrontal and frontopolar cortex and is based on the ability to handle the sequence of events, whether actions, imaginations, memories, thoughts or words, which eventually leads to the evolution of human language.

The question of why animals with absolutely small brains can be relatively smart must remain unanswered, because we do not really know the properties that determine their IPC. Certainly, the relationship between number of neurons, packing density/interneuronal distance and conduction velocity on the one hand, and IPC/intelligence on the other is nonlinear: small brains like that of the honeybee, with high packing density and very short interneuronal distances, may have a high IPC despite an extremely low number of neurons, because information processing may be based more on dendritic than axonal (spiking) processing. It could also be that in such tiny animals and brains, intelligence is much more restricted to a few domains, e.g., spatial orientation and odor-object associative learning, although honeybees exhibit some astonishing abilities regarding categorical learning (Chap. 8). Finally, it could be that sophisticated social communication systems like the "bee language" served as a strong "intelligence amplifier," as is the case with the human language.

16.5 Basic Mechanisms of the Evolution of Brains and Cognitive Functions

So far, we have discussed only correlational data between absolute or relative size of the brain or cortex and ecological, social or general intelligence. However, correlations tell us nothing about *causal relationships*. If A is ecological, social, physical or general intelligence and B brain or cortex size, then higher demands for A could have caused an increase in B. This is the usual concept of Neodarwinian adaptionism. However, an increase in B for reasons *unrelated* to ecological, social or cognitive demands could, as a side-effect or as an indirect consequence, for example of increase in body size, have ultimately led to an increase of A.

As already stated, in the framework of Neodarwinism it is assumed that in the context of the "struggle for survival" and greater reproductive success, selective forces from the environment drive the concerted evolution of brains and cognitive abilities. One could argue that the astonishing complexity of the brain and cognitive abilities of the honeybee or *Octopus* were highly adaptive, as are those of birds and primates. However, such an adaptionist scenario has to struggle with several basic problems—besides empirical evidence.

First, there is the question of why certain taxa developed larger and/or more complex brains, while many others did not, if larger or more complex brains are as highly adaptive as is commonly assumed. This question is not new in biology. Why did eusociality evolve in some insects (hymenopterans, termites, and a few more groups), but not in all, if—as is often stated—eusociality is highly adaptive? Why did eu-teleosts lose an electroreceptive system, if it is highly advantageous, and why did only very few groups re-evolve them? We could make a long list about spectacular adaptations, particularly in the domain of sense organs and sensory information processing (Chap. 11), and the question always is, why—if they are highly adaptive—did the other ones not evolve these mechanisms? Of course, there are genetic and phenotypic limitations of adaptation: not all species can equally adapt to certain environments. But many of those that did not adapt did not become extinct at all—apparently because they, too, are *sufficiently* adapted to their environment, while remaining "primitive."

David Wake and I, together with colleagues, carefully studied prey-capture mechanisms in amphibians and found that under very similar ecological conditions many frog and salamander species co-exist, either with primitive or highly sophisticated feeding mechanisms (or intermediate ones, too). Some of them developed very fast and precise projectile tongues, and it was argued by Neodarwinian colleagues that these mechanisms must have developed "under strong, albeit unknown selective forces." In a recent study, Wake and colleagues were able to show that among the lungless salamanders, family Plethodontidae, projectile tongues have developed at least four times independently, always exhibiting slightly different mechanisms. Lunglessness has been shown by us to be a prerequisite for the evolution of a projectile tongue (Roth and Wake 1989), but not all salamander taxa that became lungless likewise evolved a projectile tongue. Careful investigations reveal that there are a number of reasons why some groups did and the others did not. An important one is an increase in genome and cell size (cf. Roth et al. 1997, and Chap. 3), causing a strong decrease in metabolic rates, and due to this the inability to move quickly. Salamanders that are unable to move fast during foraging are easily caught by predators. Thus it was favorable for them to switch to "ambush" feeding rather than hunting. But in order to become a good "ambush" feeder, a fast and precise feeding mechanism combined with excellent depth perception mechanisms is needed (cf. Roth 1987).

Thus, whenever we carefully study evolutionary processes leading to more complex or more sophisticated neural and non-neural mechanism, we find that the taxa under consideration had opportunities which were absent in the others. However, what in most cases did *not* happen was that those taxa that *did not* evolve these mechanisms, died out. Rather, they survived well in their traditional habitat, while the others could move into new "ecological niches" or developed new lifestyles, new feeding habits, etc. Thus, rather than *winning* competition, they were capable of *escaping from* it. For example, many tongue-projecting salamanders specialize in collembolans or catching insects on the wing, which for other salamanders are too fast to be caught. Many toads have specialized in ants after becoming immune to formic acid, which many other frogs and salamanders reject as food. Some teleosts have re-evolved electrosensation, and this enabled them to become nocturnal predators or live in muddy waters, where other fishes cannot survive.

In this context, Bates and Byrne (2010) discuss the question of why the great apes did not become extinct. In many aspects of "lower" social intelligence, movement abilities, and feeding habits, e.g., with respect to the ability to digest coarser material and less ripe fruit, monkeys appear to be better adapted than the great apes. The answer of the authors is that the great apes were able to *avoid competition* with the monkeys by access to food that monkeys cannot reach, e.g., extracting insects, honey, and seeds by means of tools, or dealing with plant defenses (spiny rattans and palms, etc.). Sophisticated tool use and tool fabrication was possible only after strongly increased mental abilities as well as complex social interactions including cultural transmission, e.g., of tool use and tool fabrication.

In this context, let us consider the evolution of *Homo sapiens*, as described in the preceding chapter. Apparently, one key event was that our ancestors left the tropical rain forest completely 7–5 mya. As already said, of the two chimpanzee species, only the bonobos (*Pan paniscus*) are exclusive forest dwellers, while the common chimpanzee (*Pan troglodytes*) lives mostly within the transient zone between rain forest and savanna, but cannot survive permanently in the dry and hot savanna. It is reasonable to assume that the last common ancestor of chimpanzees and australopithecines did the same and made increasingly larger excursions into the dryer savanna. Among the often cited traits that might have favored such behavior is the tendency toward bipedal locomotion, which enabled our ancestors to run quickly, e.g., to escape from large carnivores, such as leopards, as well as for long walks, better thermoregulation mechanisms and fishing in lakes. This might have enabled ancestral australopithecines to get better food, i.e., fish from lakes, more fruit, roots, and also meat from dead or dying animals. This new lifestyle certainly enabled our ancestors to avoid competition with chimpanzees and to settle in the savanna.

At the same time, they still had a brain with a size in the range of those of great apes, i.e., 350–550 ccm, or slightly more. Maybe that little bit of additional brain mass was advantageous. However, such brain size remained essentially unchanged until the appearance of *H. habilis*, who had a brain size up to 780 ccm. Thus, for at least 2 million years, australopithecines lived in the savanna with a chimp-like brain; there was no stone tool use or fabrication, no use of fire, and definitely no language in the sense of modern humans. Thus, a large brain was *not* necessary for successful survival in the new habitat. What has caused the considerable increase in brain size between Lucy-like australopithecines and *H. habilis* is unknown, and the same is true for the next "jump" (if there was any) in brain size from *H. habilis* to *H. erectus/ergaster* up to 1000 ccm, and finally from the latter to the brain size around 1,350 ccm in modern *H. sapiens* and to 1.400–1.900 ccm in *H. neanderthalensis*. Since during this process of "encephalization" brains became generally larger, it is likely that this event was due to changes in the genetic control of brain growth, e.g., by changes in regulatory genes (Finlay and Darlington 1995; Rakic und Kornack 2001), instead of more specific increases in the sizes of certain parts of the brain. However, while increasing in general, the cortex, including the frontal cortex, grew positively allometrically. As already mentioned, inside the frontal cortex, the dorsolateral prefrontal and frontopolar part, involved in cognitive and executive functions, became particularly large at the expense of the ventral and more limbic parts.

Thus, much of the evolution of humans occurred *without* a large brain, and when the brain became large from *H. habilis* on, it simply followed the lines of general brain allometry. This leads us to speculate that a general increase in brain size in the ancestor of *H. habilis* provided him with a larger cortex including prefrontal cortex, and this may have happened as a result of *neutral, non-adaptive variability* of brain size as an exaptation sensu Gould and Vrba (Chap. 3). However, once *H. habilis* had such a brain, he could do things which neither his ancestors nor his competitors could do, e.g., using spears for hunting and stone tools for cutting meat and hammering. Finally, the enlarged brain and higher intelligence enabled *H. rudolfensis* and perhaps *H. ergaster* to leave Africa as the first members of the genus *Homo* 1.8 mya.

If such a scenario is correct, then in the evolutionary line leading to *H. sapiens* and *H. neanderthalensis*, substantial brain growth was not the result of strong ecological selection pressure occurring in the savanna in the first place. But even if there was strong ecological selection pressure, for a long time our ancestors did not or could not respond to it with an increase in brain and cortex size, and yet they survived quite successfully.

Thus, we have to substantially modify our view of the evolutionary process leading to larger and/or more complex brains. With no doubt, the two most important factors are (1) genetic variability of certain traits relevant for survival and (2) scarcity of vital resources. If there is sufficient genetic variability, then what is mostly observed is the *avoidance* of competition: animals develop forms and functions that enable them to feed on new kinds of food or invade habitats inaccessible to the competitors, new lifestyles, etc. Only if escape from competition is impossible, then is there a struggle for existence in the same habitat, and the "better adapted" will eventually win. This explains that under *artificial selection pressure* using animals with fast generation succession, one can demonstrate adaptive changes in sense organs, feeding behavior, predator defense, escape reactions, better insulation mechanisms, etc., but this should not be mistaken as normal evolution. In the wild, struggle for existence mostly leads to *stabilizing selection*, i.e., the continuous excision of less favorable characters, but not necessarily to an improvement of these characters, an increase in complexity or to the disappearance of simpler mechanisms.

In addition, we have to bear in mind that the way forms and functions evolve is *strongly canalized*, besides genetic variability, by two major factors. One of them is mass extinctions, through which relatively abruptly large biotopes were "freed" from competitors, as was the case of the disappearance of dinosaurs from the oceans and lakes, from land and air about 65 mya, which gave way to the evolution of modern mammals, fish, and birds. The other is *increasing coupling* of the development of structures and functions with increasing complexity: the longer evolutionary lines persist, the more restricted the "degrees of freedom" of further modifications at a given structural or functional level appear to be. Consequently, the origin of fundamentally different "fundamental plans" occurred very early, i.e., with the "Cambrian explosion," and further and increasingly minor changes occurred at increasingly lower taxonomic and complexity levels. Whenever terrestrial vertebrates evolved wings (dinosaurs, birds, mammals), they modified their existing limbs (mostly forelimbs), but did not evolve an extra pair of limbs. This means that they did not become angels, which probably would have been a great advantage. The same true holds for the vertebrate brain: whatever the life conditions were for fish, amphibians, sauropsids or mammals, all adaptive processes occurred within the genetic-developmental framework of a five-fold brain and its major subdivisions. The vertebrate brain is a structurally and functionally coupled system par excellence, and the general rule is that structures and functions appearing later during ontogeny are more likely to undergo adaptive modifications than earlier ones, because they are less coupled than the other ones.

16.6 What Does All This Tell Us?

So far, attempts to explain increases in absolute or relative brain size or in brain complexity as well as in cognitive functions by referring to either ecological, or physical-instrumental, or "lower" and "higher" social selection pressures have yielded mixed results: in some taxa of birds, mammals or primates, there are correlations with either absolute or corrected relative brain or cortex size, or size of

16.6 What Does All This Tell Us?

other parts of the brain, like the hippocampus, and the results heavily depend on the statistical methods used by the various authors. The relatively strongest correlations are obtained in birds regarding environmental factors, and in mammals and primates regarding social factors and in corvid birds and primates, especially apes, regarding factors working on physical-instrumental intelligence and on "higher" social cognitive functions.

The most convincing explanation for these findings is that behind the more special forms of intelligence, there is "general intelligence," i.e., the ability to quickly process complex and detailed information. This points directly to just one dominating factor, i.e., *neuronal information processing*, depending partly on rather general factors like number of neurons, interneuronal distance, conduction velocity aiming predominantly at the function of short-term and working memory, and partly on more specific factors like patterns of connectivity (e.g., "small-world" connectivity), a high degree of functional modularity and parallel processing, the formation of hierarchies, etc., enabling the brain to form second- and third-order hierarchies. I will come back to that question at the end of the final chapter.

Chapter 17
Brains and Minds

Keywords Dualism · Strong emergentism · Reductionism · Anatomy and physiology of mind · Structural basis of intelligence—birds · *Octopus* · Honeybee · Multiple realization of mind · Artificial mind/intelligence · True nature of mind

At the end of this book, I will ask to what extent all the data and concepts presented here will help us further clarify the "big question" of the mind-brain relationship in a scientific as well as philosophical context. The central question will be whether from an evolutionary perspective, a plausible naturalistic and physicalistic concept of mind and consciousness is possible.

17.1 The Problems of Dualism

Explicitly or implicitly, dualistic positions are far more widespread that one might think as a scientist, and even among scientists themselves. One reason already mentioned in Chap. 2 is that mind-body dualism comes naturally from everyday psychology, and one of the most frequently used arguments against identism or naturalism is that a natural origin of mind from the brain is "inconceivable," and therefore scientifically inexplicable. Philosophically, this is of course a naïve attitude, even when put forward by well-known philosophers or philosophizing scientists. Many things in our universe can be explained scientifically, although they are not conceivable, e.g., quantum-physical or relativistic phenomena. But even more specifically, nobody can realistically imagine how one million or even billions of neurons interact in order to guide our behavior, yet they do it without any mysticism.

First, I will ask how plausible a dualistic position can be in the light of evidence from evolutionary and comparative neurobiology, as presented in this book. A hard-core dualist, when accepting this empirical evidence, will run into two major problems. First, he cannot make plausible why the mind, as an independent immaterial entity, should have evolved in parallel to nervous systems and brains *at all*. Why should mind "need" brains? One solution to this problem is to declare the apparent parallelism between the mind and the material world an *illusion* created by God, as Leibniz did. Such a radical solution, however, is scientifically uninteresting. Another solution would be to assume that animals—except humans—have no mind, but only "natural intelligence," and that mind "emerged" at some point in time during human evolution or does so during ontogeny. But even then for a dualist there remains the question of why the human mind cannot guide our behavior without the brain. If, however, the dualist agrees that mind "needs" the brain in order to become effective in the material world (as did for example, Eccles), then he inevitably runs into the second problem, that of mental causation, i.e., the question of how the "immaterial" mind can act upon "material" brain processes without violating the laws of nature.

Descartes left this problem unsolved, but his "Cartesianist" followers, like Arnold Geulincx, adopted a position called "occasionalism" saying that true causal relationships between mind and matter are impossible, and that only God himself can truly cause events. Several hundred years later, John Eccles seriously dealt with the problem of mental causation and took an explicit evolutionist view in his book "*Evolution of the Brain: Creation of the Self*," first published in 1989, as well as in his article "The evolution of consciousness," published in 1992. Interestingly, Eccles accepted that there may be simple forms of consciousness, for example in birds and mammals, while *self-consciousness* is bound to the evolution of the human brain and especially the cortex.

Eccles, an "interactive dualist" and at the same time a leading neurobiologist, was looking for a possibility, how an interaction between "immaterial" mind and "material" brain could actually take place happen *without* violating the laws of physics and particularly the law of conservation of energy. He argued that at the level of quantum physics, the transfer of information is possible *without* the transfer of energy. Accordingly, he looked for a mechanism inside the brain, where such energy-free transfer of information could occur. In his view, this is the case at synapses of cortical pyramidal cells forming so-called dendrons, i.e., bundles of shafts of pyramidal cells. According to Eccles, 40 million dendrons with up to 100,000 spine synapses exist in the human cortex and each dendron is the basis of one mental event. For him, the precise mechanism of mind-brain interaction is the process of release ("exocytosis") of one "quantum" of transmitter substance contained in a synaptic vesicle in a cortical synapse. By a mere play on words, Eccles calls this definitely macromolecular release of one transmitter vesicle a "quantum process" in the sense of quantum physics. He assumed that the release of one synaptic vesicle is a probabilistic process like those known from quantum physics. His basic idea, partially developed together with the German physicist Friedrich Beck from Darmstadt Technical University, was that the immaterial mind influences the *probability of vesicle release* at pyramidal synapses, more precisely through "quantum tunneling" of electrons between the lipid bilayers of the synaptic vesicles and the presynaptic membrane of the synapse, which then triggers exocytosis. Although this effect would be minimal, just because of the immense number of cortical synapses, Eccles assumed it to become strong enough to influence cortical activity at a macroscopic level.

17.1 The Problems of Dualism

Most importantly, Eccles assumed that such a "mechanism" for mind-brain interaction would not violate the law of conservation of energy, which, however, is wrong. Not only the transmitter molecules released, but also elementary particles appear to "obey" the laws of conservation of energy. In addition, neurophysiologists are uncertain about whether or not the release of one synaptic vesicle is truly random. It could be that the observed non-predictability of the release is a consequence of the enormous complexity of processes involved in exocytosis (cf. Chap. 5). Finally, even truly random processes at a single synapse need not accumulate in a certain direction, but could also average themselves out at higher levels. Thus, the "old" problem of an energy-free mind-brain interaction of dualism has become obsolete.

17.2 Problems of Strong Emergentism

Strong emergentism like that recently proposed by Terrence W. Deacon (1997, 2011), while not being explicitly dualistic, has to struggle with the question of what exactly is meant by "irreducible" differences, for example, between humans and even their closest non-human relatives, with regard to mental capabilities, culture, language, etc. Almost all properties in nature are in a certain sense "emergent," from the properties of atoms and molecules to snowflakes, superconduction, the organization of living beings, and, finally, brains (cf. McLaughlin 1997). System properties often appear to be "irreducible" simply because they are not found at the level of single components, as happens in the famous examples of the water molecule or the sodium chloride molecule. However, in many cases, system properties can be causally linked to those of the components and can even be predicted on the basis of their knowledge (e.g., the properties of sodium chloride). In the framework of "supervenience theory" (cf. Chap. 2), a system has certain properties only because its components have certain properties leading to certain forms of interaction between them. If the components have other properties, then they will interact differently, and as a result the system likewise will have other properties.

This is also true for the nervous system and brain: a single neuron reveals neither cognition nor intelligence, but—as I have tried to demonstrate—cognitive functions and intelligent behavior in animals and humans arise only because there are membranes with ion channels, neurons and synapses, graduated and action potentials, the formation of brains with nuclei and layers, etc. The fact that in most cases we cannot precisely predict the properties of the brain on the basis of the properties of the components is due first to the incredible complexity of the brain, second to the strong limitations of mathematics, and third to the fact that many, if not most components inside the brain at least partially change their properties while interacting (cf. Chap. 4).

Hard-core emergentists, however, will insist on the fact that mind and consciousness are so *radically different* from any other phenomenon in nature that phenomenally they cannot be linked to any known physical phenomenon. Also, they will emphasize that consciousness is a completely private experience, i.e., accessible only for those who have it. However, these "fundamental gap" arguments are unconvincing. Of course, the privateness of consciousness implies that the presence of consciousness in any other human being cannot be directly experienced, but we can rather reliably infer consciousness without any hesitation from his or her behavior, including verbal reports. Regarding consciousness in animals, the situation is essentially the same: we are willing to attribute consciousness to animals in the same measure as they exhibit kinds of behavior which in humans require consciousness, as described in Chap. 15. Furthermore, the (relative) "inaccessibility" of conscious experience follows from the vast degree of *internal connectivity* of the cortex compared to the number of input and output lines leading to an almost infinite variety and number of "internal" states. But as every neural network expert knows, this is nothing mystical but occurs in any network with so-called hidden layers, and what such a network is doing between the input and output layer often cannot be precisely reconstructed mathematically.

The weakest point in "strong emergentism" regarding human mind and consciousness is the fact that its defenders must leave open at *exactly which moment* in human evolution mind and consciousness "fulgurated." Was it at the origin of australopithecines, after having settled in the savanna? Or was it at the origin of the genus *Homo*, or of the ancient or modern type of *Homo sapiens*, or together with the appearance of syntactical-grammatical language? All available data suggests that human evolution was a slow process with many intermediate steps, including the evolution of human language. As mentioned in Chap. 15, even here many events had come together during a period of 100,000 years (or even more) in order to make human language possible, for instance, walking upright, the descent and new innervation pattern of the larynx, reorganization of the mouth, nose and throat region, of the inner ear, further development of the prefrontal cortex and of the language centers, transition from gestural to vocal language, etc. The outcome, syntactical-grammatical language, certainly has had enormous consequences functioning as a strong "intelligence amplifier" and as a basis for new kinds of social interaction. Despite its complexity, nothing of that is enigmatic.

Most importantly, an "emergence" of the human mind and consciousness occurs in every *ontogenetic development* of the brain. Important steps of that process are the formation of the neural tube, the development of the three, and later five, major parts of the brain and sub-divisions of these parts, differentiation and migration of neurons, formation of subcortical and then cortical parts of the telencephalon (Bystron et al. 2008), and, finally, the maturation of the limbic cortex and isocortex, including its myelination and cell differentiation (Huttenlocher and Dabholkar 1997; Sowell et al. 1999). There is a strict correspondence between the process of synapse elimination and myelination on the one hand and the appearance of cortical including cognitive functions on the other. Most importantly, the beginning of syntactical-grammatical language in a 2 1/2-year-old child nicely coincides with the maturation of the Broca area (cf. Chap. 15). The same holds true

17.2 Problems of Strong Emergentism

for the substantial increase in working memory capacity and related complex cognitive-executive functions on the one hand and the maturation of the dorsolateral prefrontal cortex on the other. No other fact can better demonstrate the unity of brain development and the maturation of cognitive-mental function.

17.3 Problems of Reductionism

Does that mean that a *reductionist* view is more appropriate? It cannot be denied that in some simple cases, a nearly complete reduction of the properties of a system to those of its components, including the mechanisms of interaction between the components are possible, but nobody has ever been able to demonstrate such a thing with respect to brains and their cognitive functions. For a complete reduction, it is characteristic to be able to construct system properties "bottom up:" by knowing all the properties of nerve cells and their interactions in space and time, it should be possible to construct the properties of whole brains. This is far from being possible even in the simplest nerve cell assemblies like the famous stomatogastric ganglion found in decapod crustaceans (Selverston et al. 2007).

What one can try, however, is an explanation *in retrospect*, i.e., studying the properties of a system including its behavior and then search for correlations and hopefully causal relationships with properties of components. In the same way, we will never be able to really predict the course of the evolution of nervous systems and brains and their functions "bottom up," because we do not know the initial conditions; but while looking back we can possibly identify regularities and perhaps even laws. Although this procedure has gained very interesting results, it is by no means a reduction. As to the brain, neurons are not intelligent or "mindful." We can list many apparently necessary conditions (neurons with ion channels, synapses, transmitters, the formation of nuclei, layers and areas, and even the entire brain and nervous system), but all that does not explain their functions, unless we did not refer to a specific behavior we have already studied as well as the conditions under which organisms and their brains exist.

What remains as the most plausible mind-brain concept is a *non-reductionist physicalism* that avoids the inherent difficulties of both dualisms and strong emergentism on the one hand and of reductionism on the other. With respect to the origin of intelligence, mental states, and consciousness, we recognize that despite all the huge differences occurring between bacteria and *Homo sapiens*, there is no true "leap," nothing that apparently violates laws of nature, including those of thermodynamics. At the same time, such a non-reductionist physicalism concedes that at new system levels of the brain, certain properties may arise that are not found as such at the level of components. This, however, is a ubiquitous phenomenon in nature and in no way specific to the relationship between brain and mind.

17.4 The Anatomy and Physiology of the Mind

In recent times, experimental neurobiologists, psychologists, theoretical neuroscientists, and even philosophers have tried hard to elucidate the neural conditions under which mental states, including the various kinds of consciousness, originate in the brains of humans and at least in some animals. As to mammals, and particularly primates including humans, it turns out that "higher" mental states and different kinds of consciousness are invariably bound to the activity of the thalamo-cortical system, in cooperation with the activity of many other brain systems like the reticular formation, the basal forebrain, and cortical and subcortical limbic centers (cf. Chap. 10). By combining electroencephalography (EEG), magnetoencephalography (MEG) and functional magnetic resonance imaging (fMRI), scientists are able to demonstrate that in humans as well as in non-human primates, states of consciousness are always preceded by unconscious processes usually lasting between 200 and 400 ms (Noesselt et al. 2002; Seth et al. 2008; Soon et al. 2008). Also, by means of direct cortical stimulation, one finds out that there is a minimum time of 100 ms and minimum intensity of cortical activity necessary for the occurrence of conscious states (Libet 1978, 1990; Cleeremans 2005). On the basis of such experiments, one can predict with a probability of 60-100 % conscious mental states on the basis of certain brain processes, and vice versa (Haynes and Rees 2005, 2006; Bles and Haynes 2008; Bode and Haynes 2009).

It is believed that during conscious states, processes of a "re-wiring" of existing neuronal networks, e.g., short-term modifications of synaptic coupling take place, predominantly within the dorsolateral prefrontal cortex as the seat of working memory in interaction with associative parietal and temporal regions, where meaningful information processing occurs and respective memories are located. In this process, neuromodulators play an important role, especially dopamine and acetylcholine in the context of attention, evaluation, and goal-setting. Such fast rewiring of synaptic coupling is metabolically highly demanding and leads to a significant rise in local glucose and oxygen consumption, which triggers an increase in local cortical blood flow (Logothetis et al. 2001).

As mentioned before (cf. Chap. 13), a number of neurobiologists assume that oscillatory activity and synchronization of cortical networks, more or less directly, is linked to consciousness by "binding together" neuronal activity to meaningful entities (Engel et al. 1991; Crick and Koch 2003). However, evidence for a *direct* link between cortical oscillation-synchronization and consciousness is sparse. Rather, these phenomena appear to be involved, among others, in the guidance of attention (Kreiter and Singer 1996; Crick and Koch 2003; Taylor et al. 2005), or are one *unspecific* precondition for consciousness (cf. Seth et al. 2008).

An important role for the origin of conscious sensory experience in the primate brain appears to be the *sequential activation* of primary and secondary sensory plus associative cortical areas in a specific manner by a *combination of ascending and recurrent-descending pathways* between these areas (Edelman and Tononi 2000; Lamme 2000; Lamme and Roelfsema 2000). The idea is that certain sensory experiences remain unconscious, as long as they activate only ascending connections and not descending-recurrent ones back to the primary sensory areas. This assumption was verified in a study in which MEG and fMRI technique was applied under otherwise identical experimental conditions (Noesselt et al. 2002). Here, the presentation of visual stimuli first activated the primary visual and secondary cortex (V1, V2) after about 100 ms, followed by an activation of higher order visual areas (V4) after 200-250 ms. Then, after a short delay, the primary and secondary visual cortex was activated *again* around 300 ms, and this was the moment at which the stimuli were reported to be consciously perceived. This coincides well with the appearance of the P3 or P300 wave in event-related EEG, which is often viewed as the moment at which unconscious processes become conscious.

The interpretation of these findings is that in V1 and V2, visual stimuli are processed *unconsciously* according to basic and initially meaningless visual properties (e.g., contrasts, wavelengths, orientation of edges, direction of movement, disparity), and the results of this processing are sent to associative visual cortical areas. Here, with the help of other cortical and subcortical areas and—most importantly—by an adequate readout of memories, they are further processed according to their *meaning*. This global and meaningful "interpretation" is then *sent back* to the primary and secondary visual cortex, and the content becomes conscious. Such a sequence of information processing solves the fundamental problem of cortical recognition processes that activity of primary sensory areas leads to *details without meaning* and the activity of "higher" associative areas to *meaning without details*. Only by recurrent pathways and by fusion of activity of primary and associative visual cortical areas, detailed *and* meaningful conscious perception arises.

For such a parallel-divergent-convergent transformation of meaningless sensory signals into meaningful information, certain network properties of the cortex are crucial (cf. Schüz 2002). These properties include (1) laminar arrangement of a very large number of neurons (from many millions to billions), in which input from different sources (sensory, limbic, intrinsic) and output to different targets are processed partly in parallel and partly in a convergent and divergent manner; (2) a high degree of intrinsic connectivity according to the principle of dense local and sparse global connections ("small-world" organization); (3) vertical-columnar and modular organization with a dominance of excitatory projection neurons (pyramidal cells) and a minority of excitatory or inhibitory interneurons; (4) excitatory and inhibitory synaptic contacts capable of fast and temporary changes in transmission efficiency (short-term and working memory) as well as slower and longer lasting changes (long-term memory); (5) anatomical and functional segregation of (a) primary and secondary sensory and motor areas, (b) unimodal and multimodal associative areas, (c) integrative–executive areas, (d) premotor and motor areas, and (e) limbic-evaluative areas; (6) parallel as well as hierarchical, i.e., ascending and descending-recurrent connection and interaction between these areas; (7) strong dominance of short-range and long-range intracortical connections over extracortical afferents and efferents—in humans at a ratio of about 100,000:1 (Roth 2003).

Thus, in humans and other large-brained mammals, the cortex is a gigantic *associative network* capable of detailed processing of unimodal sensory information (topographic and non-topographic), comparison and integration into multimodal representations, short-term manipulation of information and consequent learning, intermediate and long-term storage of learned information and experience, categorization and abstraction, and higher order, e.g., "purely mental" representations of sensory events as a kind of self-description of the cortex.

Of great importance is the interaction of the cortex with limbic evaluation centers, because evaluation is the process that ultimately generates meaning. All the contents of the non-limbic cortex processes become meaningful only to the degree that they are *evaluated* according to individual as well as social experience identifying what is positive or negative for biological, psychic, mental, and social survival. This interaction is strongest between subcortical limbic centers like the amygdala and the mesolimbic system (VTA, nucleus accumbens, substantia nigra) and the associative, predominantly prefrontal, orbitofrontal, and ventromedial cortex, which in turn interacts with the cognitive-executive cortex.

17.5 Brains and Minds in Birds, *Octopus* and the Honeybee

As described in the preceding chapters, in mammals and particularly primates, the level of intelligence and of "higher" cognitive functions can be correlated with properties of the cortex as an associative network, particularly with respect to the size, number of neurons and synapses, synaptic plasticity and storage capacity, its information processing speed, to the capacity of short-term and working memory functions, the degree of parcellation, and the formation of functional hierarchies.

The question that arises is whether we will find similar properties when looking into the brains of birds that are comparable with respect to "physical-cognitive" intelligence to monkeys or even to great apes. As stated in Chaps. 10 and 14, at first glance the anatomy and cytoarchitecture of the mesonidopallium (NMP), including the entopallium (the former "ectostriatum") of birds, considered the "site" of their intelligence, have no resemblance to the mammalian cortex: there is no lamination, no presence of pyramid-shaped cells, but a rather diffuse structure where substructures are difficult to recognize. The bird brain generally contains very small neurons, and these appear to be tightly packed inside the MNP, while nothing is known about the actual information processing speed. The main type of MNP neurons are medium-sized projection neurons with dendrites that are moderately to heavily covered with spines like the mammalian pyramidal cells. Interneurons have rather smooth dendrites (cf. Fig. 17.1). The total number of MNP neurons may, grossly astimated, range between 50 and 200 million.

Unimodal and multimodal thalamic afferents with very thick diameters enter the entopallium ventromedially and quickly divide into secondary dendrites (Fig. 17.1).

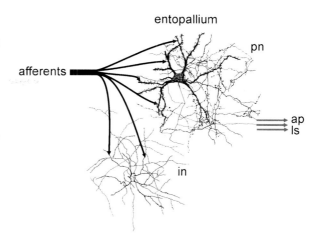

Fig. 17.1 Connectivity in the entopallium of the chicken as revealed by Golgi staining. Afferents, mostly from the nucleus rotundus, make contact both with projection neurons (*pn*) and interneurons (*in*), which in turn contact with fine local axons the projection neurons. These project to the arcopallium (*ap*) and lateral striatum (*ls*). After Tömböl et al. 1988, redrawn

These secondary processes extend straight forward, divide again, and form a very regular fiber network resembling the network of thalamic afferents to the mammalian cortex and making contact with projection neurons as well as with interneurons. There is another type of afferents with smaller diameter, which also run straight forward. This rather regularly arranged system of incoming fibers does not meet a regular, laminated arrangement of cells as in the mammalian cortex, but a seemingly irregular, perhaps globular or nuclear organization of projection neurons and interneurons.

Apparently, there is a hierarchy of processing areas, with the nidopallium caudolaterale as the most important convergence center equivalent to the prefrontal cortex and including working memory. Here, too, the neurotransmitter-neuromodulator dopamine appears to play an important role (Güntürkün 2005). The presence of spine synapses indicates that the MNP is the primary site of both fast short-term and long-term learning and memory in birds. This suggests that despite the gross-anatomical differences, the cortex of mammals and the MNP of birds follow similar principles.

What about intelligent invertebrates like the *Octopus* and the honeybee? Will we find a similar principle? As described in Chap. 7, inside the *Octopus* brain the vertical lobe is considered the "seat" of intelligence and memory of the *Octopus*. It is composed of five lobuli, similar to the gyri of the cortex of mammals, and contains about 26 million neurons, which is more than half of the neurons inside the brain. Like the cortex of mammals and the MNP of birds, it consists only of two major types of neurons, i.e., nearly 26 million tiny interneurons representing the smallest ones inside the *Octopus* brain, and 65,000 large projection neurons, and the former converge on the latter (Fig. 17.2). The difference to the mammalian cerebral cortex consists of the fact that the ratio between projection neurons and interneurons is inverse in the vertical lobe of *Octopus*, i.e., the interneurons are far more numerous than the projection neurons. Unfortunately, information regarding limbic-evaluative afferents is lacking.

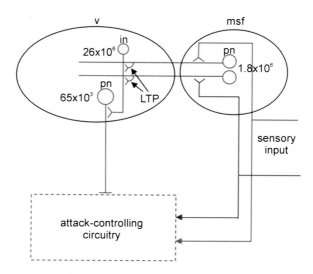

Fig. 17.2 Highly schematic wiring diagram between "higher" lobes of the *Octopus* brain. Sensory (visual, tactile) afferents reach the median superior frontal lobe (*msf*). Neurons of the msf project, via a special tract composed of 1.8 million fibers, to the *vertical lobe* (*v*) making "en-passant" contacts with 26 million interneurons in a *rectangular fashion*. These converge onto 65,000 *vertical lobe* projection neurons which in turn project, via the subvertical lobe (not shown), to attack-controlling centers of the brain. From Shomrat et al. 2008, modified

The vertical lobe receives sensory (mostly visual and tactile) afferents predominantly from the median superior frontal lobe. These afferents form a distinct tract composed of 1.8 million fibers, which terminates in the rind of the vertical lobe. Processes of the nearly 26 million interneurons located there penetrate the tract in a rectangular fashion and form "en passant" contacts. This is considered the site of long-term potentiation and formation of long-term memory. The vertical lobe is closely connected, via the projection neurons, to the subvertical lobe, which contains about 800,000 neurons, and the interaction of both lobes is based on the work of an impressively regular network of millions of crossing fibers.

Finally, let us have a look at the brain of the honeybee. Here, the paired mushroom bodies (MB) are considered the "seat" of intelligence (Menzel 2012) (Fig. 17.3). The calyces exhibit three ring regions: the *lip ring* region processing olfactory input, the *collar ring* region processing visual input, and the *basal ring* region processing mixed olfactory and mechanosensory input. As described in Chap. 7, the somata of about 150,000 neurons of each MB, the "Kenyon cells," are the smallest ones found among insects, and their packing density appears to be much higher than the highest ones found in the vertebrate brain. They receive input from the 800 projection neurons of the antennal lobe via about 1 million presynaptic contacts, plus about 10 postsynaptic contacts (Menzel 2012). These synaptic contacts, together with synapses from inhibitory neurons, recurrent axons, and afferents from VUMmx1 neurons (see below) form microglomeruli

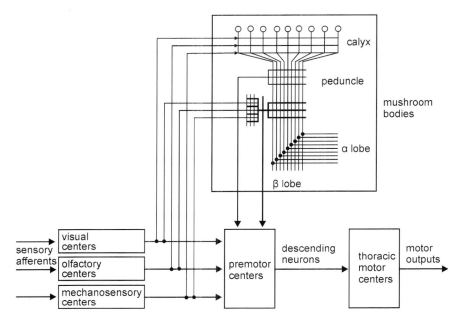

Fig. 17.3 Wiring diagram of the insect mushroom body (MB) and its connections with other supraesophageal brain areas. Sensory afferents (*left*) supply visual, olfactory, and mechanosensory neuropils (antennal lobes, optic lobes, etc.) that connect to premotor neuropils from which pathways descend to thoracic motor neurons and interneurons. Sensory neuropils likewise project to the calyx of the MB where sensory afferent maps are transformed into distributed and partially overlapping domains that provide convergence onto many thousands of intrinsic neurons. These cells, the *Kenyon cells*, supply with their axons the peduncle and the lobes of the MB. Dendrites of output neurons intersect and are contacted by Kenyon cell axons. Output neurons likewise receive afferents from the sensory neuropils with their dendrites. These neurons project to premotor centers. Recurrent axons from the MB lobes back to the calyces occur, but are not shown. After Breidbach and Kutsch 1995

representing "microcircuits". Recurrent pathways of efferent MB neurons run back to the lip region of the calyx. In the honeybee and in hymenopterans in general, the MB represents a highly complex multimodal center that forms the neural basis of processing and integrating olfactory and visual information and enables learning (mostly olfactory and visual) complex cognitive functions and complex behavior.

In the brain of a honeybee, an equivalent of the limbic system exists in the form of neurons situated in the subesophageal ganglion. One of them is the VUMmx1 neuron which projects, among others, to the lip region of the MB calyces and is involved in the formation of the olfactory engram on the basis of reward learning (Hammer 1993; Menzel and Giurfa 2001). Interestingly, these kinds of neurons are characterized by the transmitters dopamine and octopamine (the equivalent of noradrenaline in insects). In the honeybee brain, like in the mammalian brain, the interaction between the perceptive-cognitive and the limbic system leads to the *anticipation of reward* on the basis of reward-mediated learning (Menzel 2012). Another striking similarity between the mammalian-primate cortex and the MB is the occurrence of the principle of sparse coding, probably in the context of the high metabolic needs of such associative matrices (Creutzfeldt 1993; Menzel 2012).

Thus, despite gross differences in neuroarchitecture, size, and number of neurons among the primate cortex, the bird MNP, the *Octopus* vertical lobe and the honeybee mushroom bodies, important commonalities can be recognized. First, there is a highly ordered input system of unimodal and multimodal afferents establishing multiple connections with a large to very large network of excitatory projection neurons and excitatory or inhibitory interneurons. Afferent and intrinsic fibers together form a regular matrix that primarily serves as an *associative network*, i.e., an instrument for bringing the most diverse kinds of input into the *same data format* and integrate the respective kinds of information. The synaptic contacts of this matrix are highly plastic, i.e., they can modify their coupling strength, via LTP or other mechanisms, within seconds or even less, which is necessary for the function of a working memory. The result of the activity of the working memory is then sent for consolidation to an intermediate memory and from there to a long-term memory (this may also happen in parallel). The site of long-term memory is identical with the associative network proper.

Due to the very large number of intrinsic connections, these associative networks, while being the site of short-term as well as of long-term memory, are capable of forming *internal representations* of outside events in a detailed, i.e., topological, as well as abstract, symbolic fashion. Certainly, among animal groups, there are differences in the number and kind of such representations. In honeybees and *Octopus*, unimodal and multimodal representations predominate, but a level of more abstract cognitive functions exist. Perhaps in birds, and definitely in apes, including humans, the cortex has *many levels of representation*: primary and secondary unimodal, associative unimodal and associative multimodal up to global models of their own bodies, the environment and, eventually, a conscious self.

17.6 Is Mind Multiply Realized and Artificially Realizable?

My assumption is that the network properties described above essentially contribute to the formation of states of high intelligence, including forms of consciousness, at least in birds and mammals/primates and self-consciousness in humans and perhaps in the great apes. If this is the case, then these functions and properties must have evolved *independently*, since phylogenetically honeybees, *Octopus*, corvid birds, and primates including humans are only very distantly related. This means that during evolution, "mind," in the defined sense, has been realized independently at least several, and perhaps many times.

From this it follows that mind, including consciousness, can be realized, or (as philosophers prefer to say) "instantiated" in *different* architectural ways, but always along the same or very similar functional principles. Some features, like a

17.6 Is Mind Multiply Realized and Artificially Realizable?

six-layered cortex or pyramidal neurons which often have been assumed to be necessary prerequisites for consciousness, appear to be among the sufficient, but not necessary ones. This speaks in favor of a functional understanding of consciousness as a specific mode of information processing in the context of detailed perception and cognition that are both new and important in the light of past experience. Thus, the human mind would be just one, albeit highly efficient, kind of mind.

This leads us to the question of whether or not it is possible that mind in the sense of high intelligence, perhaps combined with consciousness, can be realized *artificially*. If we adopt the position of naturalism-physicalism, then the answer would be "in principle, *yes*." So far we have found nothing that could not be realized artificially in principle, because nothing of the neuronal preconditions of mind appears to lie outside nature. Rather, the principles we have identified are well known in the context of "artificial intelligence," at least in theory, although so far technicians have been largely unable to realize them.

However, it could turn out that there are at least two basic obstacles against a technical realization of mind and consciousness. The first could be that an artificial system that possessed the mentioned properties necessary for the origin of mind and consciousness can be built only from material that is quasi-biological, for example, something that resembles membranes containing ion channels, forming nerve cells, synaptic structures, etc. Progress in that direction is very slow. Second, even if there were such quasi-biological material with self-organizing and self-maintaining properties, its assembling into a hypercomplex system like the brain of a honeybee, let alone of humans, is impossible by conventional methods, i.e., by an external agent putting them together like a watchmaker. During ontogenetic development of the human brain, there are periods in which millions of synaptic contacts are formed every second. Also, in order to do so, we would need a full understanding of the connectivity of brains and the underlying principles in order to construct them, but such a full understanding does not exist. There would be only the possibility to equip these artificial systems with *self-connecting* or *self-wiring* properties like the natural ones described above. However, here again the necessary algorithms are not yet fully known.

Furthermore, even after having developed such artificial self-producing and self-organizing systems, for their further evolution they would probably need sensory experience and realistic interaction with certain environments over a very long time. Maybe this could happen in a highly abbreviated manner lasting only some hundred rather than several billion years. Thus, we conclude that artificial systems possessing intelligence and conscious minds comparable in function to those of birds, primates, and eventually humans, while being possible in principle, may be extremely difficult or too expensive to produce.

It could, however, turn out that all this is possible, but that we *do not want* highly intelligent, conscious and eventually self-conscious artificial systems—perhaps not so much for ethical reasons, but because such systems most probably would develop their own experience, motives, and will. We would then have the same difficulties with them as we do with our conspecifics, and we would have to struggle with the fact that their intentions are not necessarily the same as ours.

17.7 What Is the True Nature of the Mind?

As already mentioned, one of the most popular arguments in the philosophy of mind for an alleged uniqueness of mind is its phenomenal difference to anything else in nature—nothing "out there" seems to resemble mind and consciousness. However, this "fundamental gap" argument is based on a basic epistemological error.

Neurobiologists, psychologists, and many philosophers accept the fact that all conscious perceptual, cognitive, emotional, etc., experience results from the activity of neurons and neuronal ensembles. The brain has no direct contact with the external world, because all events that stimulate sensory receptors are transformed into neuronal signals as the "language of the brain," as described in Chaps. 5 and 11. Therefore, consciously experienced phenomena have to be considered *constructs* of the brain on the basis of these neuronal signals, and consequently these constructs should not be mistaken as being "real" entities of the external world. This is commonly accepted in the case of colors, where every expert hastens to admit that colors "do not exist in the outside world," but rather are constructs of the brain based on neuronal activity. But this is equally true for the entire phenomenal difference between "mental" phenomena, bodily sensations and "material" events of the outside world, i.e., these *phenomenal differences* are likewise constructs of the brain formed in early childhood. Every child has to *learn* that "material" things are not "mental" things.

The first fundamental distinction a developing brain has to make is that between "body" and "environment," based on the criterion whether or not a certain event can be *controlled directly* by the brain and gives *sensory feedback*. As is known from patients with disturbances in this sensory feedback, e.g., from limbs, everything that does not give sensory feedback is subjectively experienced as "non-body" and belonging to the external world. Thus, sensory feedback is essential for the development of the body scheme. The second fundamental distinction is that between the external world and body as something "material" on the one hand, and sensations like dreams, thoughts, imaginations, memories, and emotions as something "mental-immaterial" on the other. This distinction is vague and instable in early infancy, and it takes the human brain several years to make it robust, i.e., being able to reliably distinguish between something that has "really" happened and something else that was only imagined or remembered.

Even in adulthood it may happen that we confound these two in domains of conscious experience, for instance at low perceptual intensities, strong emotional states, or at strong wishful thinking. Children as well as adults have great difficulties describing these mental states, and psychologists tell us that the distinction between "real sensations" and "mental events" is based on a variety of criteria including *clarity* and *detailedness* of experienced states (e.g., colors or objects), *consistency* with previous experience and *logical coherence* of what happens. The more these criteria are fulfilled, the more we are willing to accept that something "really" happened, and the less they are given, the more they appear as "purely mental" or "imagined."

17.7 What Is the True Nature of the Mind?

The crucial point is that the difference between a "real" or "material" world and a "mental" world as we experience it consciously is a *difference formed by our brains*. From a philosophical and epistemological perspective, we might be absolutely convinced that a "material" world or "reality," as I have called it (Roth 1996), exists independent of our conscious experience. At the same time, we have to accept that the consciously experienced world is a mere construct of our brain (called "actuality") and the question of the relationship between "reality" and "actuality" has been discussed since antiquity and is at the center of modern philosophy since Kant. Together with Kant, we have to accept that the "true" nature of the world is inaccessable, but the same goes for the "true" nature of mind. Therefore, we can by no means reveal the "true" or "real" nature of mind.

Thus, what we are studying while investigating the mind-brain problem is *not* the relationship between "real" neural and "real" mental events and their relationship, but between *brain constructs* experienced as external or physical-neural events with other brain constructs experienced as immaterial-mental events. From this follows trivially that the "material" brain constructs *do not* give rise to the "mental" brain constructs, but that this happens only in a mind-independent world. This world, however, is inaccessible for us. From this likewise follows that the question of the "true" nature of mind; how it "really" arises from brain states or processes is epistemologically mindless, because it can never be answered.

What we can do and what I have tried to do in this book, however, is to develop *most plausible models* for the mind-brain relationship in our phenomenal world (actuality)—the only world we have conscious access to. Thus, from an epistemological point of view, the alleged "fundamental gap" between the material and the mental likewise is a construct created by our brain, albeit a very useful one. Thus, the philosophizing brain falls into its own trap.

Literature

Aboitiz F, García RR (1997) The evolutionary origin of the language areas in the human brain. A neuroanatomical perspective. Brain Res Rev 25:381–396
Aboitiz F, García RR, Bosman C, Brunetti E (2006) Cortical memory mechanisms and language origins. Brain and Language 98:40–56
Aiello LC, Bates N, Joffe T (2001) In defense of the expensive tissue hypothesis. In: Falk D, Gibson KR (eds) Evolutionary anatomy of the primate cerebral cortex. Cambridge University Press, Cambridge, pp 57–78
Albert JT, Göpfert MC (2013) Mechanosensation. In: Galizia G, Lledo P-M (eds) Neurosciences Springer, Berlin
Alberts B, Johnson A, Lewis J, Raff M, Roberts K, Walter P (2002) Molecular biology of the cell. Garland Science Textbooks, London
Almas I, Cappelen AW, Sorensen EO, Tungodden B (2010) Fairness and the development of inequality acceptance. Science 328:1176–1178
Amici F, Aureli F, Visalberghi E, Call J (2009) Spider monkeys (*Ateles geoffroyi*) and capuchin monkeys (*Cebus apella*) follow gaze around barriers: evidence for perspective taking? J Comp Psychol 123:368–374
An der Heiden U, Roth G, Schwegler H (1984) System-theoretic characterization of living systems. In: Möller DPF (ed) Systemanalyse biologischer prozesse. Springer, Heidelberg
An der Heiden U, Roth G, Schwegler H (1985a) Die Organisation der Organismen: Selbstherstellung und Selbsterhaltung. Funkt Biol Med 5:330–346
An der Heiden U, Roth G, Schwegler H (1985b) Principles of self-generation and self-maintenance. Acta Biotheor 34:125–138
Anderson PAV, Greenberg RM (2001) Phylogeny of ion channels: clues to structure and function. Comp Biochem Physiol Part B 129:17–28
Arbib MA (2005) From monkey-like action recognition to human language: an evolutionary framework for neurolinguistics. Behav Brain Sci 28:105–167
Azevedo FAC, Carvalho LRB, Grinberg LT, Farfel JM, Ferretti REI, Leite REP (2009) Equal numbers of neuronal and nonneuronal cells make the human brain an isometrically scaled-up primate brain. J Comp Neurol 512:532–541
Armus HL, Montgomery AR, Jellison JL (2006) Discrimination learning in paramecia (*Paramecium caudatum*). Psychol Rec 56:489–498
Atkinson QD (2011) Phonemic diversity supports a serial founder effect model of language expansion from Africa. Science 332:346–348
Baddeley AD (1986) Working memory. Clarendon Press, Oxford
Baddeley AD (1992) Working memory. Science 255:556–559

Baron G (2007) Encephalization: comparative studies of brain size and structure. In: Kaas J, Krubitzer L (eds) Evolution of nervous systems. A comprehensive review, vol 3, Mammals. Academic (Elsevier), Amsterdam, Oxford, pp 125–136

Barbas H (2007) Specialized Elements of orbitofrontal cortex in primates. Ann NY Acad Sci 1121:10–32

Barth FG (ed) (1985) Neurobiology of arachnids. Springer, Berlin

Barth FG (2012) Sensory perception: adaptation to lifestyle and habitat. In: Barth FG, Giampieri-Deutsch P, Klein H-D (eds) Sensory peception. Mind and matter. Springer, New York, pp 89–107

Barth J, Call J (2006) Tracking the displacement of objects: a series of tasks with great apes (*Pan troglodytes, Pan paniscus, Gorilla gorilla,* and *Pongo pygmaeus*) and young children (*Homo sapiens*). J Exp Psychol Anim Behav Process 32:239–252

Barton RA (1996) Neocortex size and behavioural ecology in primates. Proc R Soc Ser B Biol Sci 263:173–177

Bates LA, Byrne RW (2010) Imitation: what animal imitation tells us about animal cognition. Wiley Interdisc Rev Cogn Sci 1:685–695

Bates LA, Sayialel KN, Njiraini NW, Moss CJ, Poole JH, Byrne RW (2007) Elephants classify human ethnic groups by odor and garment color. Curr Biol 17:1938–1942

Beckermann A, Flohr H, Kim J (eds) (1992) Emergence or reduction?. de Gruyter, Berlin

Beckers GJL (2011) Bird speech perception and vocal production: a comparison with humans. Hum Biol 83:191–212

Beran MJ (2007) Rhesus Monkeys (*Macaca mulatta*) enumerate large and small sequentially presented sets of items using analog numerical representations. J Exp Psychol 33:42–54

Berg HC (2000) Motile behavior of bacteria. Phys Today 53:24–29

Berwick RC, Okanoya K, Beckers GJL, Bolhuis JJ (2011) Songs to syntax: the linguistics of birdsong. Trends Cogn Sci 15:113–121

Bird CD, Emery NJ (2009) Rooks use stones to raise the water level to reach a floating worm. Curr Biol 19:1410–1414

Blaisdell AP, Sawa K, Leising KJ, Waldmann MR (2006) Causal reasoning in rats. Science 311:1020–1022

Bles M, Haynes J-D (2008) Detecting concealed information using brain-imaging technology. Neurocase 14:82–92

Bloch JI, Boyer DM (2002) Grasping primate origin. Science 298:1606–1610

Bode S, Haynes J-D (2009) Decoding sequential stages of task preparation in the human brain. Neuroimage 45:606–613

Boesch C, Boesch H (1990) Tool use and tool making in wild chimpanzees. Folia Primatol 54:86–99

Botvinick MM, Cohen JD, Carter CS (2004) Conflict monitoring and anterior cingulate cortex: an update. Trends Cogn Sci 8:539–546

Bouchard J, Goodyer W, Lefebvre L (2007) Innovation and social learning are positively correlated in pigeons. Anim Cogn 10:259–266

Boycott BB, Young JZ (1955) A memory system in *Octopus vulgaris* Lamarck. Proc R Soc Lond B Biol Sci 143:449–480

Bräuer J, Call J, Tomasello M (2005) All great ape species follow gaze to distant locations and around barriers. J Comp Psychol 119:145–154

Breidbach O, Kutsch W (eds) (1995) The nervous system of invertebrates: an evolutionary and comparative approach. Birkhäuser

Brembs B, Heisenberg M (2000) The operant and the classical in conditioned orientation of *Drosophila melanogaster* at the flight simulator. Learn Mem 7:104–115

Brenner S (1974) The genetics of *Caenorhabditis elegans*. Genetics 77:71–94

Brodmann K (1909) Vergleichende Lokalisationslehre der Großhirnrinde. Barth, Leipzig

Bshary R, Wickler W, Fricke H (2002) Fish cognition: a primate's eye view. Anim Cogn 5:1–13

Bugnyar T (2010) Knower-guesser differentiation in ravens: others' viewpoints matter. Proc R Soc Lond B 278:634–640

Bullock TH, Horridge GA (1965) Structure and function in the nervous system of invertebrates. Freeman, San Francisco

Burish MJ, Kueh HY, Wang SSH (2004) Brain architecture and social complexity in modern and ancient birds. Brain Behav Evol 63:107–124

Burkart JM, Heschl A (2007) Understanding visual access in common marmosets, *Callithrix jacchus*: perspective taking or behaviour reading? Anim Behav 73:457–469

Byrne R (1995) The thinking ape. Evolutionary origins of intelligence. Oxford University Press, Oxford

Byrne R, Bates L, Moss CJ (2009) Elephant cognition in primate perspective. Comp Cogn Behav Rev 4:1–15

Byrne RW, Whiten A (1988) Machiavellian intelligence: social expertise and the evolution of intellect in monkeys, apes and humans. Clarendon Press, Oxford

Byrne RW, Whiten A (1992) Cognitive evolution in primates: evidence from tactical deception. Man 27:609–627

Bystron I, Blakemore C, Rakic P (2008) Development of the human cerebral cortex: boulder committee revisited. Nat Rev Neurosci 9:110–122

Call J, Tomasello M (2008) Does the chimpanzee have a theory of mind? 30 years later. Trends Cogn Sci 12:187–192

Call J, Hare B, Carpenter M, Tomasello M (2004) 'Unwilling' versus 'unable': chimpanzees' understanding of human intentional action. Dev Sci 7:488–498

Camerer CF (2003) Behavioral game theory: experiments in strategic interaction. Princeton University Press, Princeton

Carlson KJ, Stout D, Jashashvili T, de Ruiter DJ, Tafforeau P, Carlson K, Berger LR (2011) The endocast of MH1, *Australopithecus sediba*. Science 333:1402–1407

Cattell RB (1963) Theory of fluid and crystallized intelligence: a critical experiment. J Educ Psychol 54:1–22

Chalmers DJ (1996) The conscious mind. In search of a fundamental theory. Oxford University Press, Oxford

Changizi MA (2001) Principles underlying mammalian neocortical scaling. Biol Cybern 84:207–215

Changizi MA, Shimojo S (2005) Parcellation and area-area connectivity as a function of neocortex size. Brain Behav Evol 66:88–98

Chappell J, Kacelnik A (2004) Selection of tool diameter by new Caledonian crows *Corvus moneduloides*. Anim Cogn 7:121–127

Cherniak C (1990) The bounded brain: toward quantitative neuroanatomy. J Cogn Neurosci 2:58–66

Cherniak C (2012) Neuronal wiring optimization. Prog Brain Res 195:361–371

Churchland P (1995) The engine of reason, the seat of the soul: a philosophical journey into the brain. MIT Press, Cambridge Mass

Churchland PM (1997) Die Seelenmaschine. Spektrum Akademischer Verlag, Berlin

Cleeremans A (2005) Computational correlates of consciousness. Prog Brain Res 150:81–98

Clutton-Brock TH, Harvey P (1977) Primate ecology and social organization. J Zool 183:1–39

Cnotka J, Möhle M, Rehkämper G (2008a) Navigational experience affects hippocampus sizes in homing pigeons. Brain Behav Evol 72:233–238

Cnotka J, Güntürkün O, Rehkämper G, Gray RD, Hunt GR (2008b) Extraordinary large brains in tool-using New Caledonian crows (*Corvus moneduloides*). Neurosci Lett 433:241–245

Collier-Baker E, Davis JM, Nielsen M, Suddendorf T (2006) Do chimpanzees (*Pan troglodytes*) understand single invisible displacement? Anim Cogn 9:55–61

Corballis MC (2009) The evolution of language. Ann N Y Acad Sci 1156:19–43

Corballis MC (2010) Mirror neurons and the evolution of language. Brain Lang 112:25–35

Cowey A, Stoerig P (1991) The neurobiology of blindsight. Trends Neurosci 14:140–145

Creutzfeldt OD (1983) Cortex Cerebri. Leistung, strukturelle und funktionelle Organisation der Hirnrinde. Springer, Heidelberg
Crick FHC, Koch C (2003) A framework for consciousness. Nat Neurosci 6:119–126
Darwin C (1859) On the origin of species by means of natural selection, or the preservation of favoured races in the struggle for life, 1st edn. John Murray, London
Darwin C (1871) The descent of man and selection in relation to sex. John Murray, London
Davidson D (1970) Mental events. In: Davidson D (1980) Actions and events, Clarendon Press, Oxford
Deacon TW (1990) Rethinking mammalian brain evolution. Am Zoologist 30:629–705
Deacon TW (1997) The symbolic species: the co-evolution of language and the brain. W.W. Norton & Company, New York
Deacon TW (2011) Incomplete nature: how mind emerged from matter. W.W. Norton & Company, New York
De Marco RJ, Menzel R (2008) Learning and memory in communication and navigation in insects. In: Byrne JH (ed) Learning and memory: a comprehensive reference, vol 1: Menzel R (ed) Learning theory and behavior. Academic (Elsevier), Amsterdam, Oxford, pp 477–498
Deaner RO, Isler K, Burkhart J, Van Schaik C (2007) Overall brain size, and not encephalization quotient, best predicts cognitive abilities across non-human primates. Brain Behav Evol 70:115–124
Delius JM, Siemann J, Emmerton L, Xia L (2001) Cognition of birds as products of evolved brains. In: Roth G, Wullimann M (eds) Brain evolution and cognition. Wiley, New York, pp 451–490
Dennett DC (1991) Consciousness explained. Little, Brown & Co., Boston
Deppe AM, Wright PC, Szelistowski WA (2009) Object permanence in lemurs. Anim Cogn 12:381–388
Descartes R (1648) La description du corps humain. Paris
De Quervain DJF, Fischbacher U, Treyer V, Schellhammer M, Schnyder U, Buck A, Fehr E (2004) The neural basis of altruistic punishment. Science 305:1254–1258
Dicke U, Roth G (2007) Evolution of the amphibian nervous system. In: Kaas JH, Bullock TH (eds) Evolution of nervous systems. A comprehensive review, vol 2, Non-mammalian vertebrates. Academic (Elsevier), Amsterdam, Oxford, pp 61–124
Dicke U, Heidorn A, Roth G (2011) Aversive and non-reward learning in the fire-bellied toad using familiar and unfamiliar prey stimuli. Curr Zool 57(6):709–716
Dobzhansky Th (1970) Genetics of the evolutionary process. Columbia University Press, New York
Dronkers NF, Redfern BB, Knight RT (2000) The neural architecture of language disorders. In: Gazzaniga MS et al (eds) The new cognitive neurosciences, 2nd edn. MIT Press, Cambridge, pp 949–958
Druckmann S, Gidon A, Segev I (2013) Computational neuroscience—capturing the essence. In Galizia G, Lledo P-M (eds) Neurosciences. Springe, Berlin
Dudel J, Menzel R, Schmidt RF (eds) (1996/2000) Neurowissenschaft. Springer, Berlin
Dunbar RIM (1995) Neocortex size and group size in primates—a test of the hypothesis. J Hum Evol 28:287–296
Dunbar RIM (1998) The social brain hypothesis. Evol Anthropol 6:178–190
Dunbar RIM, Shultz S (2007) Evolution in the social brain. Science 317:1344–1347
Eccles JC (1989) Evolution of the brain: creation of the self. Routledge, London
Eccles JC (1992) Evolution of consciousness. Proc Nat Acad Sci USA 89:7320–7324
Eccles JC (1994) How the self controls its brain. Springer, Berlin
Edelman GM, Tononi G (2000) Consciousness. How matter becomes imagination. Penguin Books, London
Egger, Feldmeyer (2013) Electrical activity in neurons, In: Galizia G, Lledo P-M (eds) Neurosciences. Springer, Berlin

Ehret G, Göpfert MC (2013) Auditory systems. In Galizia G, Lledo P-M Neurosciences. (eds) Springer, Berlin
Eigen M, Schuster P (1979) The hypercycle—a principle of natural self-organization. Springer, Heidelberg
Eisthen HE (1997) Evolution of vertebrate olfactory system. Brain Behav Evol 50:222–233
Elston GN (2002) Cortical heterogeneity: implications for visual processing and polysensory integration. J Neurocytol 31:317–335
Elston GN, Benatives-Piccione R, Elston A, Zietsch B, Defelipe J, Manger P et al (2006) Specializations of the granular prefrontal cortex of primates: implication for cognitive processing. Anat Rec A Discoveries Mol Cell Evol Biol 288A:26–35
Emery N, Clayton NS (2004) The mentality of crows: convergent evolution of intelligence in corvids and apes. Science 306:1903–1907
Enard W, Przeworski M, Fisher SE, Lai CSL, Wiebe V, Kitano T, Monaco AP, Pääbo S (2002) Molecular evolution of FOXP2, a gene involved in speech and language. Nature 418:869–872
Engel AK, König P, Singer W (1991) Direct physiological evidence for scene segmentation by temporal coding. Proc Nat Acad Sci USA 88:9136–9140
Evans PD, Gilbert SL, Mekel-Bobrov N, Vallender E, Anderson JR, Vaez-Azizi LM, Tishkoff SA, Hudson RR, Lahn BT (2005) Microcephalin, a gene regulating brain size, continues to evolve adaptively in humans. Science 309:1717–1720
Evans TA, Beran MJ, Harris EH, Rice DF (2009) Quantity judgments of sequentially presented food items by capuchin monkeys (*Cebus apella*). Anim Cogn 12:97–105
Falk D (2007) Evolution of the primate brain. In: Henke W, Tattersall I (eds) Handbook of paleaanthropology. Primate evolution and human origins, vol 2. Springer, Berlin, pp 1133–1162
Farris SM (2008) Evolutionary convergence of higher brain centers spanning the protostome-deuterostome boundary. Brain Behav Evol 72:106–122
Fehr E, Gächter S (2002) Altruistic punishment in humans. Nature 415:137–140
Fernald RD (1997) The evolution of eyes. Brain Behav Evol 50:253–259
Fichtel C, Kappeler PM (2010) Human universals and primate symplesiomorphies: establishing the lemur baseline. In: Kappeler PM, Silk J (eds) Mind the gap. Tracing the origins of human universals. Springer, Berlin, pp 395–426
Finlay BL, Darlington RB (1995) Linked regularities in the development and evolution of mammalian brains. Science 268:1578–1584
Fiorito G, Chicheri R (1995) Lesions of the vertical lobe impair visual discrimination learning by observation in *Octopus vulgaris*. Neurosci Lett 192:117–120
Fiorito G, Scotto P (1992) Observational-learning in *Octopus vulgaris*. Science 256:545–547
Fiorito G, Biederman GB, Davey VA, Gherardi F (1998) The role of stimulus preexposure in problem solving by *Octopus vulgaris*. Anim Cogn 1:107–112
Fitch WT (2011a) The evolution of syntax: an exaptationist perspective. Frontiers in evolutionary neuroscience 3:1–12
Fitch WT (2011b) Unity and diversity in human language. Phil Trans R Soc B 366:376–388
Fitch WT, Hauser MD (2004) Computational constraints on syntactic processing in a nonhuman primate. Science 303:377–380
Foelix RF (2010) Biology of spiders, 3rd edn. Oxford University Press, Oxford
Friederici AD (2009) Pathways to language: fiber tracts in the human brain. Trends Cogn Sci 13:175–181
Friederici AD (2011) The brain basis of language processing: from structure to function. Physiol Rev 91:1357–1392
Fujita K (2009) Metamemory in tufted capuchin monkeys (*Cebus apella*). Anim Cogn 12:575–585
Fuster JM (2008) The prefrontal cortex, 4th edn. Academic Press, London
Futuyma DJ (2009) Evolution, 2nd edn. Sinauer, Sunderland
Galizia GC, Lledo PM (2013) Olfaction In Galizia G, Lledo P-M (eds) Neurosciences. Springer (in press)

Gallese V, Goldman A (1998) Mirror neurons and the simulation theory of mind-reading. Trends Cogn Sci 2:493–501

Gallup GG Jr (1970) Chimpanzees: self-recognition. Science 167:86–87

Gardner RA, Gardner TB, van Cantfort TE (1989) Teaching sign language to chimpanzees. State University New York Press, New York

Gazzaniga MS (2008) The science behind what makes us unique. Ecco Press, New York

Genty E, Roeder JJ (2011) Can lemurs (*Eulemur fulvus* and *E. macaco*) use abstract representations of quantities to master the reverse-reward contingency task? Primates 52:253–260

Ghazanfar AA, Hauser M (1999) The neuroethology of primate vocal communication: substrate for the evolution of speech. Trends Cogn Sci 3:377–384

Ghysen A (2003) The origin and evolution of the nervous system. Int J Dev Biol 47:555–562

Gibson KR, Rumbaugh D, Beran M (2001) Bigger is better: primate brain size in relationship to cognition. In: Falk D, Gibson KR (eds) Evolutionary anatomy of the primate cerebral cortex. Cambridge University Press, Cambridge, pp 79–97

Giurfa M (2003) Cognitive neuroethology: dissecting non-elemental learning in a honeybee brain. Curr Opin Neurobiol 13:726–735

Goldman-Rakic PS (1996) Regional and cellular fractionation of working memory. Proc Nat Acad Sci USA 93:13473–13480

Götz M (2013) Biology and function of glial cells. In: Galizia G, Lledo P-M (eds) Neurosciences. Springer, Berlin

Goossens BMA, Dekleva M, Reader SM, Sterck EHM, Bolhuis JJ (2008) Gaze following in monkeys is modulated by observed facial expressions. Anim Behav 75:1673–1681

Gould SJ (1977) Ontogeny and phylogeny. Belknap–Harvard University Press, Cambridge

Gould SJ, Vrba ES (1982) Exaptation—a missing term in the science of form. Paleobiology 8:4–15

Grasso FW, Basil JA (2009) The evolution of flexible behavioral repertoire in cephalopod mollusks. Brain Behav Evol 74:231–245

Green RE et al (2010) A draft sequence of the neandertal genome. Science 328:710–725

Grimmelikhuijzen CJP, Carstensen K, Darmer D, McFarlane I, Moosler A, Nothacker HP, Reinscheid RK, Rinehart KL, Schmutzler C, Vollert H (1992) Coelenterate neuropeptides: structure, action and biosynthesis. Am Zool 32:1–12

Güntürkün O (2005) The avian 'prefrontal cortex and cognition. Curr Opin Neurobiol 15:686–693

Güntürkün O (2008) Wann ist ein Gehirn intelligent? Spektrum der Wissenschaft Nov 2008:124–132

Güntürkün O, von Fersen L (1998) Of whales and myths. Numerics of cetacean cortex. In: Elsner N, Wehner R (eds) New neuroethology on the move. Proceedings of the 26th Göttingen Neurobiology Conference, vol 2. Thieme, Stuttgart, p 493

Guttenplan S (1994) A companion to the philosophy of mind. Blackwell, Cambridge

Hakeem AY, Sherwood CC, Bonar CJ, Butti C, Hof PR, Allman JM (2009) Von economo neurons in the elephant brain. Anat Rec 292:242–248

Hammer M (1993) An identified neuron mediates the unconditioned stimulus in associative olfactory learning in honeybees. Nature 366:59–63

Hanus D, Call J (2007) Discrete quantity judgments in the great apes (*Pan paniscus, Pan troglodytes, Gorilla gorilla, Pongo pygmaeus*): the effect of presenting whole sets versus item-by-item. J Comp Psychol 121:241–249

Harbaugh WT, Mayr U, Burghart DR (2007) Neural responses to taxation and voluntary giving reveal motives for charitable donations. Science 316:1622–1625

Hart BL, Hart LA (2007) Evolution of the elephant brain: a paradox between brain size and cognitive behavior. In: Kaas JH, Krubitzer LA (eds) The evolution of nervous systems. A Comprehensive Review. vol 3, Mammals. Academic (Elsevier), Amsterdam, Oxford, pp 491–497

Hart BL, Hart LA, Pinter-Wollman N (2008) Large brains and cognition: where do elephants fit in? Neurosci Biobehav Rev 32:86–98

Haug H (1987) Brain sizes, surfaces, and neuronal sizes of the cortex cerebri: a stereological investigation of man and his variability and a comparison with some mammals (primates, whales, marsupials, insectivores, and one elephant). Am J Anat 180:126–142

Hawkins T (1950) Opening of milk bottles by birds. Nature 165(4194):435–436

Hayes KJ, Hayes C (1954) The cultural capacity of chimpanzee. Hum Biol 26:288–303

Haynes JD, Rees G (2005) Predicting the orientation of invisible stimuli from activity in human primary visual cortex. Nat Neurosci 8:686–691

Haynes JD, Rees G (2006) Decoding mental states from brain activity in humans. Nat Rev Neurosci 7:523–534

Heffner HE, Heffner RS (1995) Role of auditory cortex in the perception of vocalization by *Japanese macaques*. In: Zimmermann E, Newman JD, Jürgens U (eds) Current topics in primate vocal communication. Plenum Press, New York, pp 207–219

Heiligenberg W (1977) Principles of electrolocation and jamming avoidance in electric fish. A neuroethological approach. Springer, Heidelberg

Hennig W (1950) Grundzüge einer Theorie der phylogenetischen Systematik. Deutscher Zentralverlag, Berlin. English edition: Henning W (1966) Phylogenetic systematics. University of Illinois Press, Urbana

Henrich J, McElreath R, Barr A, Ensminger J, Barrett C, Bolyanatz A, Cardenas JC, Gurven M, Gwako E, Henrich N, Lesorogol C, Marlowe F, Tracer D, Ziker J (2006) Costly punishment across human societies. Science 312:1767–1770

Herculano-Houzel S (2009) The human brain in numbers: a linearly scaled-up primate brain. Front Hum Neurosci 3:31

Herculano-Houzel S (2012) Neuronal scaling rules for primate brains: the primate advantage. PBR, pp 325–340

Herculano-Houzel S, Collins CE, Wong P, Kaas JH (2007) Cellular scaling rules for primate brains. Proc Nat Acad Sci USA 104:3562–3567

Hille B (1992) Ionic channels of excitable membranes. Sinauer Assoc, Sunderland

Hirth F, Reichert H (2007) Basic nervous system types: one or many. In: Striedter GF, Rubenstein JL (eds) Evolution of nervous systems, vol 1, Theories, development, invertebrates. Academic (Elsevier), Amsterdam, Oxford, pp 55–72

Hochner B, Shomrat T, Fiorito G (2006) The Octopus: a model for a comparative analysis of the evolution of learning and memory mechanisms. Biol Bull 210:308–317

Hof PR, Van der Gucht E (2007) Structure of the cerebral cortex of the humpback whale, Megaptera novaeangliae (Cetacea, Mysticeti, Balaenopteridae). Anat Rec 290:1–31

Hofman MA (2001) Brain evolution in hominids: are we at the end of the road? In: Falk D, Gibson KR (eds) Evolutionary anatomy of the primate cerebral cortex. Cambridge University Press, Cambridge, pp 113–127

Hofman MA (2003) Of brains and minds. A neurobiological treatise on the nature of intelligence. Evol Cogn 9:178–188

Hofman MA (2012) Design principles of the human brain: An evolutionary perspective. PBR, pp 373–390

Hofmann MH, Northcutt RG (2008) Organization for major telencephalic pathways in an elasmobranch, the thornback ray *Platyrhinoidis triseriata*. Brain Behav Evol 72:307–325

Holekamp KE (2006) Questioning the social intelligence hypothesis. Trends Cogn Sci 11:65–69

Holland LZ, Short S (2008) Gene duplication, co-option and recruitment during the origin of the vertebrate brain from the invertebrate chordate brain. Brain Behav Evol 72:91–105

Horner V, Whiten A (2005) Causal knowledge and imitation/emulation switching in chimpanzees (*Pan troglodytes*) and children (*Homo sapiens*). Anim Cogn 3:164–181

Hunt GR, Holzhaider JC, Russell D (2007) Gray: spontaneous metatool use by new Caledonian crows. Curr Biol 17:1504–1507

Huttenlocher PR, Dabholkar AS (1997) Regional differences in synaptogenesis in human cerebral cortex. JCN 387:167–178
Inoue S, Matsuzawa T (2007) Working memory of numerals in chimpanzees. Curr Biol 17:1004–1005
Iriki A, Sakura O (2008) The neuroscience of primate intellectual evolution: natural selection and passive and intentional niche construction. Philos Trans R Soc Lond B Biol Sci 363:2229–2241
Ivry RB, Fiez JA (2000) Cerebellar contributions to cognition and imagery. In: Gazzaniga MS et al (eds) The new cognitive neurosciences, 2nd edn. MIT Press, Cambridge, pp 999–1011
Iwaniuk AN, Hurd PL (2005) The evolution of cerebrotypes in birds. Brain Beh Evol 65:215–230
James W (1890/1950) Principles of psychology. Dover Publications, New York
Jenkin SEM, Laberge F (2010) Visual discrimination learning in the fire-bellied toad *Bombina orientalis*. Learn Behav 38:418–425
Jensen K, Call J, Tomasello M (2007) Chimpanzees are rational maximizers in an ultimatum game. Science 318:107–109
Jerison HJ (1973) Evolution of the brain and intelligence. Academic, Amsterdam
Johnson-Frey SH (2003) The neural bases of complex tool use in humans. Trends Cogn Sci 8:71–78
Jones EG (2001) The thalamic matrix and thalamocortical synchrony. Trends Neurosci 24:595–601
Kaas JH (2007) Reconstructing the organization of neocortex of the first mammals and subsequent modifications. In: Kaas JH, Krubitzer LA (eds) Evolution of nervous systems. A comprehensive review, vol 3, Mammals. Academic (Elsevier), Amsterdam, Oxford, pp 27–48
Kaminski J, Call J, Fischer J (2004) Word learning in a domestic dog: evidence for "fast mapping". Science 304:1682–1683
Kandel ER (1976) Cellular basis of behavior—an introduction to behavioral neurobiology. WH Freeman, New York
Kandel ER, Schwartz JH, Jessell TM (2000) Neurowissenschaften. Spektrum Akademischer Verlag, Heidelberg
Karten HJ (1969) The organization of the avian telencephalon and some speculations on the phylogeny of the amniote telencephalon. Ann NY Acad Sci 167:164–179
Karten HJ (1991) Homology and evolutionary origins of the "neocortex". Brain Behav Evol 38:264–272
Kastner S, Ungerleider LG (2000) Mechanisms of visual attention in the human cortex. Annu Rev Neurosci 23:341–515
Kendal RL, Custance DM, Kendal JR, Vale G, Stoinski TS, Rakotomalala NL et al (2010) Evidence for social learning in wild lemurs (*Lemur catta*). Learn Behav 38:220–234
Kim J (1993) Supervenience and mind. Cambridge University Press, Cambridge
Kirschner M, Gerhardt J (2005) The plausibility of life: resolving Darwin's dilemma. Yale University Press, London
Knight RT, Grabowecky M (2000) Prefrontal cortex, time, and consciousness. In: Gazzaniga MS et al (eds) The new cognitive neurosciences, 2nd edn. MIT Press, Cambridge, pp 1319–1339
Koch C (2004) The quest for consciousness: a neurobiological approach. Roberts and Company, Greenwood
Koch C, Tsuchiya N (2012) Attention and consciousness: related yet different. Trends Cogn Sci 16:103–105
Korte M (2013) Cellular correlates of learning and memory. In Galizia G, Lledo P-M (eds) Neurosciences. Springer, Berlin
Kortlandt A (1968) Handgebrauch bei freilebenden Schimpansen. In: Rensch B (ed) Handgebrauch und Verständigung bei Affen und Frühmenschen. Hans Huber, Bern, pp 59–102
Krebs JR, Sherry DF, Healy SD, Perry H, Vaccarino AL (1989) Hippocampal specialization in food storing birds. Proc Nat Acad Sci USA 86:1388–1392

Kreiter AK, Singer W (1996) Stimulus dependent synchronization of neuronal responses in the visual cortex of the awake macaque monkey. J Neurosci 16:2381–2396
Kretzberg J, Ernst U (2013) Vision. In Galizia G, Lledo P-M (eds) Neurosciences. Springer, Berlin
Krusche P, Uller C, Dicke U (2010) Quantity discrimination in salamanders. J Exp Biol 213:1822–1828
Kuroshima H, Kuwahata H, Fujita K (2008) Learning from others' mistakes in capuchin monkeys (*Cebus apella*). Anim Cogn 11:599–609
Lamme VAF (2000) Neural mechanisms of visual awareness: a linking proposition. Brain Mind 1:385–406
Lamme VAF, Roelfsema PR (2000) The two distinct modes of vision offered by feedforward and recurrent processing. Trends Neurosci 23:571–579
Lefebvre L (1995) The opening of milk bottles by birds: evidence for accelerating learning rates, but against the wave-of-advance model of cultural transmission. Behav Process 34(1):43–53
Lefebvre L (2012) Primate encephalization. In: Hofman MA, Falk D (eds) Progress in Brain Research, vol 195. pp 393–412
Lefebvre L, Reader SM, Sol D (2004) Brains, innovations and evolution in birds and primates. Brain Behav Evol 63:233–246
Lefebvre L, Sol D (2008) Brains, lifestyles and cognition: are there general trends? Brain Behav Evol 72:135–144
Leopold DA, Logothetis N (1996) Activity changes in early visual cortex reflect monkeys' percept during binocular rivalry. Nature 379:549–553
Libet B (1978) Neuronal vs. subjective timing for a conscious sensory experience. In: Buser PA, Rougeul-Buser A (eds) Cerebral correlates of conscious experience. Elsevier, Amsterdam, pp 69–82
Libet B (1990) Cerebral processes that distinguish conscious experience from unconscious mental functions. In: Eccles JC, Creutzfeldt OD (eds) The principles of design and operation of the brain. Pontificae Academiae Scientiarum Scripta Varia, vol 78. pp 185–202
Lichtneckert R, Reichert H (2007) Origin and evolution of the first nervous systems. In: Kaas J, Bullock TH (eds) Evolution of nervous systems. A comprehensive review, vol 1, Theories, development, invertebrates. Academic (Elsevier), Amsterdam, Oxford, pp 289–315
Lindenfors P, Nunn CL, Barton RA (2007) Primate brain architecture and selection in relation to sex. BMC Biol 5:20
Lisney TJ, Yopak KE, Montgomery JC, Collin SP (2008) Variation in brain organization and cerebellar foliation in chondrichthyans. Brain Behav Evol 72:262–282
Logothetis NK, Pauls J, Augath M, Trinath T, Oeltermann A (2001) Neurophysiological investigation of the basis of the fMRI signal. Nature 412:150–157
Lorenz K (1973) Behind the mirror. A search for a natural history of human knowledge. Mariner Books, Boston
Lotto AJ, Hickok GS, Holt LL (2009) Reflections on mirror neurons and speech perception. Trends Cogn Sci 13:110–114
Lovejoy AO (1936) The great chain of being: a study of the history of an idea. Harvard University Press, Cambridge
Lüscher C, Petersen C (2013) The synapse. In: Galizia G, Lledo P-M (eds) Neurosciences. Springer, Berlin
Luthardt G, Roth G (1979) The relationship between stimulus orientation and stimulus movement pattern in the prey catching behavior of *Salamandra salamandra*. Copeia 1979:442–447
MacPhail EM (1982) Brain and intelligence in vertebrates. Clarendon Press, Gloucestershire
Macphail EM, Bolhuis JJ (2001) The evolution of intelligence: adaptive specializastions versus general process. Biol Rev 76:341–364
Maguire EA, Gadian DG, Johnsrude IS, Good CD, Ashburner J, Frackowiak RSJ, Frith CD (2000) Navigation-related structural change in the hippocampi of taxi drivers. Proc Nat Acad Sci USA 97:4398–4403

Margulis L (1970) Origin of eukaryotic cells. Yale University Press, New Haven
Marois R, Ivanoff J (2005) Capacity limits of information processing in the brain. Trends Cogn Sci 9:296–305
Martin RD (1996) Scaling of the mammalian brain: the maternal energy hypothesis. News Physiol Sci 11:149–156
Martínez S, Puelles E, Echevarria D (2013) Ontogeny of the vertebrate nervous system. In: Galizia G, Lledo PM (eds) Neurosciences. Springer, Berlin
Maturana H, Varela F (1980) Autopoiesis and cognition: the realization of the living. Reidel, Boston
Mausfeld R (2013) The biological function of sensory systems. In: Galizia G, Lledo P-M (eds) Neurosciences. Springer, Berlin
Mayr E (1974) Teleological and teleonomic, a new analysis. Boston Stud Philos Sci 14:91–117
McGrew WC (2010) Evolution. Chimpanzee technology. Science 328:579–580
McLaughlin B (1997) Emergence. In: Keil F, Wilson R (eds) MIT encyclopedia of cognitive sciences. MIT Press, Cambridge, pp 266–268
McLaughlin B, Beckermann A, Walter S (eds) (2011) The oxford handbook of philosophy of mind. Oxford University Press, Oxford
McLaughlin B, Bennett K (2005) Supervenience. In: Stanford encyclopedia of Philosophy
Medina L (2007) Do birds and reptiles possess homologues of mammalian visual, somatosensory, and motor cortices. In: Kaas J, Bullock TH (eds) Evolution of nervous systems. A comprehensive review, vol 2, Non-mammalian vertebrates. Academic (Elsevier), Amsterdam, Oxford, pp 163–194
Meek J, Schellart NAM (1998) A Golgi study of goldfish optic tectum. J Comp Neurol 182:89–122
Mehlhorn J, Hunt GR, Gray RD, Rehkämper G, Güntürkün O (2010) Tool-making new Caledonian crows have large associative brain areas. Brain Behav Evol 75:63–70
Mekel-Bobrov N, Gilbert SL, Evans PD, Vallender EJ, Anderson JR, Hudson RR, Tishkoff SA, Lahn BT (2005) Ongoing adaptive evolution of *ASPM*, a brain size determinant in Homo sapiens. Science 309:1720–1722
Mellars P, French JC (2011) Tenfold population increase in Western Europe at the neandertal–to–modern human transition. Science 333:623–627
Mendes N, Hanus D, Call J (2007) Raising the level: orangutans use water as a tool. Biol Lett 3:453–455
Menzel R (2012) In search of the engram in the honeybee brain (in press)
Menzel R (2013) Learning, memory and cognition: animal perspectives. In Galizia G, Lledo P-M (eds) Neurosciences. Springer, Berlin
Menzel R, Giurfa M (2001) Cognitive architecture of a mini-brain: the honeybee. Trends Cogn Sci 5:62–71
Menzel R, Brembs B, Giurfa M (2007) Cognition in invertebrates. In: Kaas JH (ed) Evolution of nervous systems. A comprehensive review, vol 1, Theories, development, invertebrates. Academic (Elsevier), Amsterdam, Oxford, pp 403–422
Mery F, Kawecki TJ (2002) Experimental evolution of learning ability in fruit flies. Proc Nat Acad Sci USA 99:14274–14279
Metzger W (1975) Gesetze des Sehens. Kramer, Frankfurt
Metzinger T (ed) (1995) Conscious experience. Ferdinand Schöningh, Paderborn
Miklósi A, Kubinyi E, Topa J, Gacsi M, Varnyi Z, Csanyi V (2003) A simple reason for a big difference: wolves do not look back at humans, but dogs do. Curr Biol 13:763–766
Mobbs PG (1985) Brain structure. In: Kerkut GA, Gilbert LI (eds) Comparative insect physiology, biochemistry and pharmacology. Pergamon Press, Oxford
Mooney R (2009) Neural mechanisms for learned birdsong. Learn Mem 16:655–669
Moroz LL (2009) On the independent origins of complex brains and neurons. Brain Behav Evol 74:177–190

Müller WA, Frings S (2009) Tier- und Humanphysiologie- Eine Einführung, 4th edn. Springer, Berlin
Mueller GB, Newmann SA (2003) Origin of organismal form: beyond the gene in developmental and evolutionary biology. MIT Press, Cambridge
Mulcahy NJ, Call J (2006) Apes save tools for future use. Science 312:1038–1040
Nässel DR, Larhammar D (2013) Neuropeptides and peptide hormones. In Galizia G, Lledo (eds) P-M Neurosciences. Springer, Berlin
Neiworth JJ, Steinmark E, Basile BM, Wonders R, Steely F, DeHart C (2003) A test of object permanence in a new-world monkey species, cotton top tamarins (*Saguinus oedipus*). Anim Cogn 6:27–37
New JG (1997) The evolution of vertebrate electroensory systems. Brain Behav Evol 50:244–252
Nieuwenhuys R, Voogd J, Van Huijzen C (1988) The human central nervous system. Springer, New York
Nieuwenhuys R, Ten Donkelaar HJ, Nicholson C (1998) The central nervous system of vertebrates, 3rd edn. Springer, Heidelberg
Nimchinsky EA, Gilissen E, Allman JM, Perl DP, Erwin JM, Hof PR (1999) A neuronal morphologic type unique to humans and great apes. Proc Nat Acad Sci USA 96:5268–5273
Niven JE, Farris SM (2012) Miniaturization of nervous systems review and neurons. Curr Biol 22:323–329
Nixon M, Young JZ (2003) The brains and lives of cephalopods. Oxford Biology, Oxford
Noesselt T, Hillyard SA, Woldorff MG, Schoenfeld A, Hagner T, Jäncke L, Tempelmann C, Hinrichs H, Heinze H-J (2002) Delayed striate cortical activation during spatial attention. Neuron 35:575–587
Nonacs P, Dill LM (1993) Is satisficing an alternative to optimal foraging theory? Oikos 67:371–375
Northcutt RG, Gans C (1983) The genesis of neural crest and epidermal placodes: a reinterpretation of vertebrate origins. Rev Biol 58:1–28
O'Connell S, Dunbar RIM (2003) A test for comprehension of false belief in chimpanzees. Evol Cogn 9:131–140
O'Rahilly R, Müller F (1999) Summary of the initial development of the human nervous system. Teratology 60:39–41
Osvath M, Osvath H (2008) Chimpanzee (*Pan troglodytes*) and orangutan (*Pongo abelii*) forethought: Self-control and pre-experience in the face of future tool use. Anim Cogn 11:661–674
Ottoni EB, Izar P (2008) Capuchin monkey tool use: overview and implications. Evol Anthropol Issues News Rev 17:171–178
Pahl M, Tautz J, Zhang S (2010) Honeybee cognition. In: Kappeler P (ed) Animal behaviour: evolution and mechanisms. Springer, Heidelberg
Pakkenberg B, Gundersen HJG (1997) Neocortical neuron number in humans: effect of sex and age. J Comp Neurol 384:312–320
Pauen M (2006) Feeling causes. J Cons Stud 13(1–2):129–152
Pearce JM (1997) Animal learning and cognition. Psychol Press, Exeter
Penn DC, Povinelli DJ (2007) On the lack of evidence that non-human animals possess anything remotely resembling a 'theory of mind'. Philos Trans R Soc Lond B Biol Sci 362:731–744
Pepperberg IM (2000) The Alex studies. Cognitive and communicative abilities of grey parrots. Harvard University Press, Cambridge
Phillips W, Barnes JL, Mahajan N, Yamaguchi M, Santos LR (2009) 'Unwilling' versus 'unable': capuchin monkeys' (*Cebus apella*) understanding of human intentional action. Dev Sci 12:938–945
Piaget J (1954) The construction of reality in the child. Basic Books, New York
Pickering R, Dirks PHGM, Jinnah Z de Ruiter DJ Churchill SE Herries AIR, Woodhead JD, Hellstrom JC, Berger LR (2011) *Australopithecus sediba* at 1.977 ma and implications for the origins of the genus *Homo*. Science 333:1421–1423

Pilbeam D, Gould SJ (1974) Size and scaling in human evolution. Science 186:892–901
Pinker S (1995) The language instinct: how the mind creates language. Harper Perennial, New York
Pinker S, Jackendorf R (2005) The faculty of language: what's special about it? Cognition 95:201–236
Plotnik JM, de Waal FBM, Reiss D (2006) Self-recognition in an Asian elephant. Proc Nat Acad Sci USA 103:17053–17057
Plowright CMS, Reid S, Kilian T (1998) Finding hidden food: behavior on visible displacement tasks by mynahs (*Gracula religiosa*) and pigeons (*Columba livia*). J Comp Psychol 86:13–25
Pollen AA, Dobberfuhl AP, Scace J, Igulu MM, Renn SCP, Shumway CA, Hofmann HA (2007) Environmental complexity and social organization sculpt the brain in Lake Tanganyikan clichlid fish. Brain Behav Evol 70:21–39
Pombal M, Megías M, Bardet SM, Puelles L (2009) New and old thoughts on the segmental organization of the forebrain in lampreys. Brain Behav Evol 74:7–19
Popper K, Eccles J (1984) The self and its brain. Springer, Heidelberg
Povinelli DJ (2000) Folk physics for apes. Oxford University Press, Oxford
Povinelli DJ, Vonk J (2003) Chimpanzee minds: suspiciously human? Trends Cogn Sci 7:157–161
Povinelli DJ, Nelson KE, Boysen ST (1990) Inferences about guessing and knowing by chimpanzees (*Pan troglodytes*). J Comp Psychol 104:203–210
Povinelli DJ, Rulf AB, Landau KR, Bierschwale DT (1993) Self-recognition in chimpanzees (*Pan troglodytes*): distribution, ontogeny, and patterns of emergence. J Comp Psychol 107:347–372
Premack D, Premack A (1983) The mind of an ape. Norton, New York
Premack D, Woodruff G (1978) Does the chimpanzee have a theory of mind? Behav Brain Sci 4:515–526
Preuss TM (1995) Do rats have a prefrontal cortex? The Rose-Woolsey-Akert program reconsidered. J Cogn Neurosci 7:1–24
Price CJ (2012) A review and synthesis of the first 20 years of PET and fMRI studies of heard speech, spoken language and reading. NeuroImage 62:816–847
Prior H, Schwarz A, Güntürkün O (2008) Mirror-induced behavior in the magpie (*Pica pica*): evidence of self-recognition. PLoS Biol 6:1642–1650
Puelles L, Rubenstein JL (1993) Expression patterns of homeobox and other putative regulatory genes in the embryonic mouse forebrain suggest a neuromeric organization. Trends Neurosci 16:472–479
Puelles L, Rubenstein JL (2003) Forebrain gene expression domains and the evolving prosomeric model. Trends Neurosci 26:469–476
Raff MC, Barres BA, Burne JF, Coles HS, Ishizaki Y, Jacobson MD (1993) Programmed cell death and the control of cell survival: lessons from the nervous system. Science 262:695–700
Rakic P (2002) Evolving concepts of cortical radial and areal specification. PBR 136:265–280
Rakic P (2009) Evolution of the neocortex: a perspective from developmental biology. Nat Rev Neurosci 10:204–219
Rakic P, Kornack DR (2001) Neocortical expansion and elaboration during primate evolution: a view from neuroembryology. In: Falk D, Gibson KR (eds) Evolutionary anatomy of the primate cerebral cortex. Cambridge University Press, Cambridge, pp 30–56
Reader SM, Laland KN (2002) Social intelligence, innovation, and enhanced brain size in primates. Proc Natl Acad Sci USA 99:4436–4441
Reader SM, Hager Y, Laland KN (2011) The evolution of primate general and cultural intelligence. Philos Trans Roy Soc Lond B Biol Sci 366:1017–1027
Reiner A, Perkel DJ, Bruce LL et al (2004) Revised nomenclature for avian telencephalon and some related brainstem nuclei. J Comp Neurol 473:377–414
Reiner A, Yamamoto K, Karten HJ (2005) Organization and evolution of the avian forebrain. Anat Rec Part A 287A:1080–1102

Reiss D, Marino L (2001) Mirror self recognition in the bottlenose dolphin: a case of cognitive convergence. Proc Nat Acad Sci USA 98:5937–5942
Rensch B (1968a) Manipulierfähigkeit und Komplikation von Handlungsketten bei Menschenaffen. In: Rensch B (ed) Handgebrauch und Verständigung bei Affen und Frühmenschen. Hans Huber, Bern, pp 103–126
Rensch B (1968b) Biophilosophie auf erkenntnistheoretischer Grundlage (Panpsychistischer Identismus). Gustav Fischer, Stuttgart
Rensch B, Altevogt R (1955) Zähmung und Dressurleistungen indischer Arbeitselefanten. Z für Tierpsychol 11:497–510
Rensch B, Döhl J (1967) Spontane Aufgabenlösung durch einen Schimpansen. Z Tierpsychol 24:476–489
Rizzolatti G, Fadiga L, Gallese V et al (1996) Premotor cortex and the recognition of motor actions. Cogn Brain Res 3:131–141
Rizzolatti G, Arbib MA (1998) Language within our grasp. Trends Neurosci 21:188–194
Rizzolatti G, Craighero L (2004) The mirror-neuron system. Annu Rev Neurosci 27:169–192
Roberts AC (2006) Primate orbitofrontal cortex and adaptive behavior. Trends Cog Sci 10:83–90
Rockel AJ, Hiorns W, Powell TPS (1980) The basic uniformity in structure of the neocortex. Brain 103:221–244
Rockland KS (2002) Non-uniformity of extrinsic connections and columnar organization. J Neurocytol 31:247–253
Rokas A (2008) The origins of multicellularity and the early history of the genetic toolkit for animal development. Annu Rev Genet 42:235–251
Roth G (1987) Visual behavior in salamanders. Springer, Berlin, Heidelberg, New York
Roth G (1996) Das Gehirn und seine Wirklichkeit. Suhrkamp, Frankfurt
Roth G (2000) The evolution and ontogeny of consciousness. In: Metzinger T (ed) Neural correlates of consciousness. MIT Press, Cambridge, pp 77–97
Roth G (2003) Fühlen, Denken, Handeln. Wie das Gehirn unser Verhalten steuert. Suhrkamp, Frankfurt
Roth G, Dicke U (2005) Evolution of the brain and intelligence. Trends Cogn Sci 9:250–257
Roth G, Dicke U (2012) Evolution of brain and intelligence in primates. Prog Brain Res 195:413–430
Roth G, Schwegler H (1995) Das Geist-Gehirn-Problem aus der Sicht der Hirnforschung und eines nicht-reduktionistischen Physikalismus. Eth Sozialwiss 6(1):69–156
Roth G, Wake DB (1989) Conservatism and innovation in the evolution of feeding in vertebrates. In: Wake DB, Roth G (eds) Complex organismal functions: integration and evolution in vertebrates. Wiley, London, New York, pp 7–22
Roth G, Wullimann MF (1996/2000) Evolution der Nervensysteme und Sinnesorgane. In: Dudel J, Menzel R, Schmidt RF (eds) Neurowissenschaft. Vom Molekül zur Kognition. Springer, Heidelberg-Berlin, pp 1–31
Roth G, Nishikawa KC, Wake DB (1997) Genome size, secondary simplification, and the evolution of the salamander brain. Brain Behav Evol 50:50–59
Roth G, Grunwald W, Dicke U (2003) Morphology, axonal projection pattern, and responses to optic nerve stimulation of thalamic neurons in the fire-bellied toad *Bombina orientalis*. J Comp Neurol 461: 91–110
Roth G, Naujoks-Manteuffel C, Nishikawa K, Schmidt A, Wake DB (2003) The salamander nervous system as a secondarily simplified, paedomorphic system. Brain Behav Evol 42:137–170
Roth G, Grunwald W, Mühlenbrock-Lenter S, Laberge F (2004) Morphology and axonal projection pattern of neurons in the telencephalon of the fire-bellied toad *Bombina orientalis*. J Comp Neurol. 478:35–61
Ruhl T, Dicke U (2012) The role of the dorsal thalamus in visual processing and object selection: a case of an attentional system in amphibians. Eur J Neurosci doi:10.1111/j.1460-9568.2012.08271.x

Ruiz A, Gómez JC, Roeder JJ, Byrne RW (2009) Gaze following and gaze priming in lemurs. Anim Cogn 12:427–434

Rumbaugh SR (1986) Ape language: from conditioned response to symbol. Columbia University Press, New York

Rütsche B, Meyer M (2010) Der kleine Unterschied—wie der Mensch zur Sprache kam. Z Neuropsychol 21:1–17

Russell MJ, Hall AJ (1997) The emergence of life from iron monosulphide bubbles at a submarine hydrothermal redox and pH front. J Geol Soc Lond 154:377–402

Russell S (1979) Brain size and intelligence: a comparative perspective. In: Oakley DA, Plotkin HC (eds) Brain, behavior and evolution. Methuen, London, pp 126–153

Sanfey AG, Rilling JK, Aronson JA, Nystrom LE, Cohen JD (2003) The neural basis of economic decisionmaking in the ultimatum game. Science 300:1755–1758

Sanz CM, Morgan DB (2009) Flexible and persistent tool-using strategies in honey-gathering by wild chimpanzees. Intern J Primatol 30:411–427

Savage-Rumbaugh ES (1984) Acquisition of functional symbol usage in apes and children. In: Roitblat HL, Bever TG, Terrace HS (eds) Animal cognition. Earlbaum, Hillsdale, pp 291–310

Savage-Rumbaugh ES (1986) Ape language: from conditioned response to symbol. Columbia University Press, New York

Schacter DL (1996) Searching for memory. The brain, the mind, and the past. Basic Books, New York

Scharff C, Petry J (2011) Evo-devo, deep homology and FoxP2: implications for the evolution of speech and language. Phil Trans R Soc B 366:2124–2140

Schlosser G, Wagner GP (2004) Modularity in development and evolution. University of Chicago Press, Chicago

Schuck-Paim C, Alonso WJ, Ottoni EB (2008) Cognition in an ever-changing world: climatic variability is associated with brain size in Neotropical parrots. Brain Behav Evol 71:200–215

Schülert N, Dicke U (2002) The effect of stimulus features on the visual orienting behavior in *Plethodon jordani*. J Exp Biol 205:241–251

Schülert N, Dicke U (2005) Dynamic response properties of visual neurons and context-dependent surround effects on receptive fields in the tectum of the salamander *Plethodon shermani*. Neuroscience 134:617–632

Schüz A (2001) What can the cerebral cortex do better than other parts of the brain? In: Roth G, Wullimann M (eds) Brain evolution and cognition. Wiley and Sons, New York, pp 491–500

Schüz A (2002) Introduction: homogeneity and heterogeneity of cortical structure: a theme and its variations. In: Schütz A, Miller R (eds) Cortical areas: unity and diversity. Taylor and Francis, London, pp 1–11

Selverston A, Elson R, Rabinovich M, Huerta R, Abarbanel H (2006) Basic principles for generating motor output in the stomatogastric ganglion. Ann New York Acad Sci 860:35–50

Semendeferi K, Lu A, Schenker N, Damasio H (2002) Humans and great apes share a large frontal cortex. Nat Neurosci 5:272–276

Seth AK, Dienes Z, Cleeremans A, Overgaard M, Pessoa L (2008) Measuring consciousness: relating behavioural and neurophysiological approaches. Trends Cogn Sci 12:314–321

Seyfarth RM, Cheney DL (2008) Primate vocal communication. In: Platt M, Ghazanfar AA (eds) Primate neuroethology. Oxford University Press, Oxford

Seyfarth RM, Cheney DL, Marler P (1980) Monkey responses to three different alarm calls: evidence of predator classification and semantic communication. Science 210:801–803

Shannon CE, Weaver W (1949) The mathematical theory of communication. The University of Illinois Press, Urbana

Shepherd SV, Platt ML (2008) Spontaneous social orienting and gaze following in ringtailed lemurs (*Lemur catta*). Anim Cogn 11:13–20

Sherwood CC, Subiaul F, Zawidzki TW (2008) A natural history of the human mind: tracing evolutionary changes in brain and cognition. J Anat 212:426–454

Sherry DF (2011) The hippocampus of food-storing birds. Brain Behav Evol 78:133–135

Shomrat T, Zarrella I, Fiorito G, Hochner B (2008) The octopus vertical lobe modulates short-term learning rate and uses LTP to acquire long-term memory. Curr Biol 18:337–342

Shumway CA (2008) Habitat complexity, brain, and behavior. Brain Behav Evol 72:123–134

Simon H (1956) Rational choice and the structure of the environment. Psychol Rev 63(2):129–138

Singer W (1999) Neuronal synchrony: a versatile code for the definition of relations. Neuron 24:49–65

Singer W, Gray CM (1995) Visual feature integration and the temporal correlation hypothesis. Annu Rev Neurosci 18:555–586

Singer T, Seymour B, O'Doherty J, Kaube H, JDolan JD, Frith C (2004) Empathy for pain involves the affective but not sensory components of pain. Science 303:1157–1162

Slobodchikoff CN (2002) Cognition and communication in prairie dogs. In: Beckoff M, Allen C, Burghardt GM (eds) The cognitive animal. A Bradford Book, Cambridge, pp 257–264

Smid HM, Wang G, Bukovinszky T, Steidle JLM, Bleeker MAK, van Loon JJA, Vet LEM (2007) Species-specific acquisition and consolidation of long-term memory in parasitic wasps. Proc R Soc B 274:1539–1546

Smith JD (2009) The study of animal metacognition. Trends Cogn Sci 13:389–396

Sol D, Duncan RP, Blackburn TM, Cassey P, Lefebvre L (2005) Big brains, enhanced cognition, and response of birds to novel environments. Proc Nat Acad Sci USA 102:5460–5465

Soon CS, Brass M, Heinze H-J, Haynes J-D (2008) Unconscious determinants of free decisions in the human brain. Nat Neurosci 11:543–555

Sowell ER, Thompson PM, Leonard CM, Welcome SE, Kann E, Toga A (1999) Longitudinal mapping of cortical thickness and brain growth in normal children. J Neurosci 24:8223–8231

Sporns O (2010) Networks of the brain. MIT, Cambridge

Squire LR, Kandel ER (1998) Memory: from mind to molecules. H. Holt and Company, New York

Steinmetz PRH, Kraus JEM, Larroux C, Hammel JU, Amon-Hassenzahl A, Houliston E, Wörheide G, Nickel M, Degnan BM, Technau U (2012) Independent evolution of striated muscles in cnidarians and bilaterians. Nature 487:231–234

Strausfeld NJ, Mok Strausfeld C, Loesel R, Rowell D, Stowe S (2006) Arthropod phylogeny: onychophoran brain organization suggests an archaic relationship with a chelicerate stem lineage. Proc Biol Sci 7:1857–1866

Strausfeld NJ, Hirth F (2013) Deep homology of arthropod central complex and vertebrate basal ganglia. Science 340:157–161

Striedter GF (2005) Brain evolution. Sinauer, Sunderland

Strong MK, Chandy KG, Gutman GA (1993) Molecular evolution of voltage-sensitive ion channel. Mol Biol Evol 10:221–242

Subiaul F, Cantlon JF, Holloway RL, Terrace HS (2004) Cognitive imitation in rhesus macaques. Science 305:407–410

Taylor K, Mandon S, Freiwald WA, Kreiter AK (2005) Coherent oscillatory activity in monkey area v4 predicts successful allocation of attention. Cerebral Cortex 15:1424–1437

Taylor AH, Hunt GR, Media FS, Gray RD (2009) Do new Caledonian crows solve physical problems through causal reasoning? Proc R Soc B 276:247–254

Teffer K, Semendeferi K (2012) Human prefrontal cortex: evolution, development, and pathology. PBR, pp 191–218

Terrace H (1987) Chunking by a pigeon in a serial learning task. Nature 325:149–151

Terrace H, Petitto LA, Sanders RJ, Bever TG (1979) Can an ape create a sentence? Science 206:891–902

Terry WS (2006) Learning and memory: basic principles, processes, and procedures. Pearson Education, Boston

Thiel A, Hoffmeister TS (2009) Decision-making dynamics in parasitoids of Drosophila. In: Prévost G (ed) Advances in parasitology, vol 70. Academic, Amsterdam, pp 45–66

Timpson N, Heron J, Smith GD, Enard W (2005) Comment on papers by Evans et al. and Mekel-Bobrov et al. on evidence for positive selection of MCPH1 and ASPM. Science 317:1936

Tinbergen N (1953) The study of instinct. Oxford University Press, Oxford

Tomasello M, Warneken F (2009) Varieties of altruism in children and chimpanzees. Trends Cogn Sci 13:397–402

Tomasello M, Call J, Hare B (2003) Chimpanzees understand psychological states—the question is which ones and to what extend. Trends Cogn Neurosci 7:153–156

Tomasello M, Hare B, Lehmann H, Call J (2007) Reliance on head versus eyes in the gaze following of great apes and human infants: the cooperative eye hypothesis. J Hum Evol 52:314–320

Tömböl T, Maglóczky ZS, Stewart MG, Csillag A (1988) The structure of chicken ectostriatum. J Hirnforsch 29:525–546

Treue S, Maunsell JHR (1996) Attentional modulation of visual motion processing in cortical areas MT and MST. Nature 382:539–541

Udell MAR, Dorey NR, Wynne CDL (2011) Can your dog read your mind? understanding the causes of canine perspective taking. Learn Behav 39:289–302

Uller C, Jaeger R, Guidry G, Martin C (2003) Salamanders (*Plethodon cinereus*) go for more: rudiments of number in an amphibian. Anim Cogn 6:105–112

Van Dongen PAM (1998) Brain size in vertebrates. In: Niewenhuys R (ed) The central nervous system of vertebrates. Springer, Heidelberg, pp 2099–2134

Van Gulick R (2004) Consciousness. Stanford encyclopedia of philosophy, pp 1–52

Vigneau M, Beaucousin V, Hervé P-Y, Jobard G, Petit L, Crivello F, Mellet E, Zago L, Mazoyer B, Tzourio-Mazoyer N (2011) What is right-hemisphere contribution to phonological, lexico-semantic, and sentence processing? insights from a meta-analysis. NeuroImage 54:577–593

Visalberghi E, Limongelli L (1994) Lack of comprehension of cause-effect relationships in tool-using capuchin monkeys (*Cebus apella*). J Comp Psychol 108:15–22

Visalberghi E, Addessi E, Truppa V, Spagnoletti N, Ottoni E, Izar P et al (2009) Selection of effective stone tools by wild bearded capuchin monkeys. Curr Biol 19:213–217

Von Bonin G (1937) Brain weight and body weight in mammals. J Gen Psychol 16:379–389

Von der Emde G (2013) Electroreception. In Galizia G, Lledo P-M (eds) Neurosciences. Springer, Berlin

Von Frisch K (1923) Über die Sprache der Bienen. Eine tierpsychologische Untersuchung. In: Zoologische Jahrbücher (Physiologie) 40:1–186

Von Frisch K (1965) Tanzsprache und Orientierung der Bienen. Springer, Berlin

Wächtershäuser G (1988) Before enzymes and templates: theory of surface metabolism. Microbiol Rev 52:452–484

Wächtershäuser G (2000) Origin of life: life as we don't know it. Science 289:1307–1308

Waddington CH (1956) Principles of embryology. George Allen & Unwin, London

Wake D, Wake MH, Specht CD (2011) Homoplasy: from detecting pattern to determining process and mechanism of evolution. Science 331:1032–1035

Wake DB, Roth G (eds) (1989) Complex organismal functions: integration and evolution in vertebrates. Wiley VCH, Weinheim

Walker R, Burger O, Wagner J, von Rueden CR (2006) Evolution of brain size and juvenile periods in primates. J Hum Evol 51:480–489

Wehner R, Menzel R (1990) Do insects have cognitive maps? A Rev Neurosci 13:403–414

Weiller C, Bormann T, Saur D, Musso M, Rijntjes M (2011) How the ventral pathway got lost—and what its recovery might mean. Brain and Language 118:29–39

Weir AAS, Chappell J, Kacelnik A (2002) Shaping of hooks in new Caledonian crows. Science 297:981

Weiskrantz L (1986) Blindsight: a case study and implications. Oxford University Press, Oxford

Wise SP (2008b) Forward frontal fields: phylogeny and fundamental function. Trends Neurosci 31:599–608

Withington PM (2007) The evolution of arthropod nervous systems. Insight from neural development in the Onychophora and Myriapoda. In: Kaas J, Bullock TH (eds) Evolution of nervous systems. A comprehensive review, vol 1, Theories, development, invertebrates. Academic (Elsevier), Amsterdam, Oxford, pp 317–336

Woolley SMN, More JM (2011) Coevolution in communication senders and receivers: vocal behavior and auditory processing in multiple songbird species. Ann NY Acad Sci 122:155–165

Wong P, Kaas JH (2009) An architectonic study of the neocortex of the short-tailed opossum (*Monodelphis domestica*). Brain Behav Evol 73:206–228

Wullimann MF, Vernier P (2007) Evolution of the nervous system in fishes. In: Kaas J, Bullock TH (eds) Evolution of nervous systems. A comprehensive review, vol 2, Non-mammalian vertebrates. Academic (Elsevier), Amsterdam, Oxford, pp 39–60

Young JZ (1971) The anatomy of the nervous system of *Octopus vulgaris*. Clarendon, Oxford

Yu F, Hill RS, Schaffner SF, Sabeti PC, Wang ET, Mignault AA, Ferland RJ, Moyzis RK, Walsh CA, Reich R (2007) Comment on "Ongoing Adaptive Evolution of *ASPM*, a brain size determinant in *Homo sapiens*". Science 316:370

Zhang K, Sejnowski TJ (2000) A universal scaling law between gray matter and white matter of cerebral cortex. Proc Nat Acad Sci USA 97:5621–5626

Zipfel B, DeSilva JM, Kidd RS, Carlson KJ, Churchill SE, Berger LR (2011) The foot and ankle of *Australopithecus sediba*. Science 333:1417–1420

Zimmermann H (2013) Cellular and molecular basis of neural function. In Galizia G, Lledo P-M Neurosciences. Springer, Berlin

Index

A

Absolute brain size, 223, 235, 241, 270, 273, 274, 276
Abstraction, 105, 192, 201, 263, 290
Action potential, 54–58, 65, 67, 74, 76, 82, 168–170
Amphibians, 4, 32, 36, 120, 123, 126, 127, 137, 139, 141, 142, 144, 145, 148, 151, 162, 169, 173, 175, 177–179, 185, 193, 197, 218, 221, 267
Anatomy and physiology of mind, 283
Animal language, 243
Annelids, 4, 86–89, 96, 98, 105
Archaea, 4, 33, 46, 47, 54, 69, 72, 73, 75, 107, 182
Arthropods, 4, 37, 87, 96, 98
Australopithecines, 243

B

Bacteria, 4, 28, 33, 41, 46, 54, 69, 70, 73–76, 107, 265, 287
Basic organization vertebrate brain, 131, 132
Behavioral flexibility, 7, 14, 22, 108, 155, 269, 270, 272
Birds, 4, 14, 16, 34, 77, 120, 123, 127, 128, 134, 139, 141, 144, 146, 147, 150, 153–155, 160, 162, 225, 240, 290
Brain-body relationship, 223

C

Canalization, 30, 31, 37, 284
Cephalopods, 14, 89, 91–93, 105, 107, 113, 122, 266, 269

Cerebellum, 22, 32, 125, 131, 132, 137, 138–140, 145, 151, 155, 162, 163, 175, 177, 196, 225, 233, 253, 267, 275
Chelicerates, 98, 99, 105, 183
Chlamydomonas, 74
Chondrichthyans, 124
Coelenterates, 4, 47, 55, 79–81, 265
Cognition amphibians, 193
Cognition teleosts, 193
Conscious attention, 209, 215, 217
Consciousness, 2–5, 7, 12, 13, 15–17, 20, 22, 23, 39, 49, 137, 146, 160, 193, 194, 200, 207, 209, 210, 215, 218, 221, 222, 243, 269, 283, 284–288, 295, 296
Convergent evolution, 33, 37, 119, 131, 160, 182, 266, 268
Corrected relative brain size, 223, 233, 273–276, 280
Cortex information processing capacity, 223
Cortex modularity, 223
Cytoarchitecture cortex, 155, 159, 162, 189, 243, 294

D

Darwinism, 25, 28
Deep homologies, 32, 37, 38, 88, 167
Determining factors brain evolution, 265
Diencephalon, 95, 118, 122, 131–136, 142–145, 147, 148, 162, 181, 189, 192, 267, 275
Dolphin intelligence, 209
Drosophila learning, 111
Dualism, 1, 7, 17, 18–22, 283, 285, 287

E

Ecdysozoa, 4, 73, 84, 87, 96–98, 105, 119, 131, 266, 270
Ecological intelligence, 3, 194, 195, 271–274
Elephant intelligence, 209
Encephalization quotient, 223, 231, 241, 267
Enlargement of human brain, 243
Escherichia coli, 70, 71, 73, 75, 76
Evolution *Homo sapiens*, 243
Evolution of cognitive-mental functions, 265, 269
Extra neurons, 223, 232, 234, 273

F

Flagellar motor, 70–72
Functional anatomy, 131, 155

G

Gaze following, 193, 199, 204, 205
General intelligence, 3, 5, 244, 264, 265, 271, 274, 275–277, 281

H

Homo erectus, 236, 246, 252, 261
Homo habilis, 243, 252, 260, 263
Homo neanderthalensis, 230, 243, 252, 261
Homoplasies, 35, 37
Honeybee learning, 107
Human language, 215, 222, 242, 243, 254, 256, 257–261, 264, 270, 276, 286

I

Identism, 7, 20, 22, 23, 283
Imitation, 1, 3, 8, 10–12, 149, 193, 194, 199, 204, 205–207, 214, 215, 243, 256–259, 262, 277, 285
Insect intelligence, 107–113
Insects, 63, 79, 87, 88, 96, 98, 100–105, 107, 108–111, 112, 115, 128, 139, 165, 168–172, 177, 183, 185, 191, 249, 266
Intelligence, 3, 5, 7, 13, 14, 22, 94, 107, 113, 153, 193, 194, 198, 202, 204, 209, 219–221, 223, 228, 231–234, 240, 242, 244, 261, 264, 265, 268–285
Invertebrate-protostome nervous systems, 84
Ion channels, 52, 54, 56, 64
Isocortex, 35, 131, 139, 142, 151, 154, 155, 160, 161, 163, 233, 237, 242, 267, 274

L

Language/speech centers, 157, 160, 161, 219, 261, 243
Learning, 3, 7, 8, 10, 11, 13, 105, 107, 108, 111–113, 193–199, 201–203, 205, 207, 209, 220, 254, 263, 266, 275, 293
Learning by observation *Octopus*, 114, 199
Learning parasitoid wasps, 111, 112, 115
Living systems, 39
Lophotrochozoa, 84, 87, 96

M

Machiavellian intelligence, 193, 204, 274
Mammals, 4, 7, 12, 31, 77, 93, 95, 107, 109, 117, 120, 122, 123, 127, 128–130, 137, 139, 142, 148–155, 160, 161–163, 169, 170, 173, 179, 189, 192, 193, 199, 204, 218, 227–242, 250, 254, 262, 267, 269, 271–273, 275, 280
Mammals-birds: tool use, 193
Mass extinctions, 32, 37, 92, 128
Medulla oblongata, 131–133, 137, 140, 144, 150, 155, 162, 175, 180, 233, 253
Medulla spinalis, 131, 136
Membrane excitability, 53
Memory, 2, 4, 8, 11–14, 49, 55, 62, 67, 70, 72, 74, 77, 105, 107, 109, 112, 153, 159, 163, 192–195, 199, 203, 217, 234, 239, 258, 260, 264, 269, 272, 281, 288, 291, 294
Mesencephalon/midbrain, 63, 68, 121, 131, 133, 142, 143, 162, 163, 181, 196
Mesonidopallium birds, 131, 160, 223, 240
Metacognition, 194, 209, 211, 221, 222, 263, 269
Mind, 2, 7, 17–19, 21–23, 216, 237, 246, 267, 279, 287–290, 292, 299, 300
Mirror self-recognition, 23, 194, 210, 211, 219, 220, 243, 263, 269, 275
Mollusks, 4, 79, 86, 87, 89, 92, 93, 105, 115
Monism, 17, 19–21
Multiple realization of mind, 283
  artificial mind/intelligence, 283
  true nature of mind, 283, 296, 297
Mushroom bodies, 16, 79, 88, 100, 169, 292, 294
Myxinoids, 79, 117, 122, 149

Index 319

N
Naturalism, 1, 7, 19, 20, 283, 295
Natural selection, 1, 25–30, 37, 165
Nematods, 4, 79, 96
Neodarwinism, 25, 27, 28, 29, 277
Neuronal information processing, 4, 49, 64, 82, 91, 271, 281
Neural transmission, 54, 55, 57, 61, 63
Non-Darwinian evolution, 25
Number cortical neurons, 223, 235, 236, 242

O
Object permanence, 193, 199, 202
Octopus brain, 79, 90, 95, 291, 292
Octopus intelligence, 113–115
Origin deuterostomes, 117–120
Origin of life, 25, 39, 45, 46
Origin of multicellular organisms, 75, 76
Origin vertebrate brain, 131
Osteichthyans, 124–126

P
Pallium birds, 165
Paramecium, 54, 73, 74
Petromyzontids, 229
Philosophy of mind, 17, 23, 296
Physicalism, 21, 23, 292, 299
Protozoans, 47, 55, 72, 75, 269

Q
Quantity representation, 193, 199, 201

R
Reasoning, 193, 199, 202, 203, 222, 264, 269, 272
Reconstruction of phylogeny, 33–37
Reductionism, 20, 22, 283, 287
Relationship intelligence-brains, 265
Relative brain size, 194, 223, 226, 227, 229, 230–233, 241, 273–276, 280
Reptiles, 34, 127, 141, 146, 147, 154, 163, 172, 228, 229, 233

S
Self-maintenance, 40–42, 48
Self-organization, 41, 42, 45, 48, 295, 296
Self-production, 40, 41, 48
Sense organs
    auditory system, 128, 142, 150, 151, 165, 177, 179, 219, 238, 257
    electroreception, 124, 138, 140, 144, 165, 170, 175
    general function, 167–169
    insect compound eye, 193
    lateral line system, 122, 126, 145, 169, 172, 175, 177, 182
    mechanical senses, 172
    olfaction, 123, 152, 165, 170–172, 194, 242
    parallel processing visual system, 191–193
    retina, 73, 74, 103, 144, 145, 165, 184, 188, 189, 190, 191, 192
    vertebrate eye, 37, 95, 119, 185, 187, 188
    visual system, 74, 93, 103, 184–193
Social behavior of humans, 243
Social intelligence, 3, 193, 199, 204, 205–209, 271, 272, 273–276
Spatial navigation, 107
Specialties cortex, 223
Sponges, 4, 76, 79–81, 96, 265
Strong emergentism, 19, 283, 285, 286, 287
Structural basis of intelligence
    birds, 14, 197, 202–204, 206, 208, 210, 213, 215, 225, 227, 244, 245, 276, 295, 296, 298–300
    honeybee, 4, 14, 107–109, 115, 169, 183, 269, 276, 290–295
    Octopus, 4, 107, 113–115, 283, 290, 291, 294
Structure of nerve cell, 50, 51
Synapse, 51, 60, 61, 64, 65, 240, 241, 290, 291, 293

T
Telencephalon/endbrain, 122, 134, 136, 138, 142, 145–147, 149–152, 154, 156, 162, 165, 200, 257, 271, 290
Theory of mind, 14, 194, 209, 212, 213, 221, 222, 243, 258, 262, 269, 275, 276
Tool fabrication, 3, 11, 14, 193, 199, 200, 201, 221, 249, 261, 269, 275, 276, 278

U
Urbilaterian brain, 131

V
Vitalism, 39, 40

W
Waggle dance honey bees, 107, 115
Working memory, 11, 16, 109, 159, 192, 193, 199, 202, 203, 212, 217, 234, 239, 258, 260, 263, 264, 270, 276, 281, 286, 288, 289, 294

Printed by Publishers' Graphics LLC
LMO130731.15.16.272